SP Pvt. Ltd.

Dedicated to

my parents

Finite Element Method

Theory, Software and Practice

Surendra Kumar
Central Mechanical Engineering Research Institute
Mahatma Gandhi Avenue
Durgapur – 713 209, West Bengal

2012

Studium Press (India) Pvt. Ltd.

Finite Element Method

Theory, Software and Practice

ISBN: 978-93-80012-31-5

Published by:

Studium Press (India) Pvt. Ltd.
4735/22, 2nd Floor, Prakash Deep Building
(Near Delhi Medical Association),
Ansari Road, Darya Ganj, New Delhi-110 002
Tel.: 91-11-43240200-15 (15 lines), Fax: 91-11-43240215;
jngovil@gmail.com; jngovil@hotmail.com

Printed at:
Salasar Imaging Systems
Delhi-110035 (India)

ABOUT THE AUTHOR

Surendra Kumar

Surendra Kumar is currently a principal scientist at Central Mechanical Engineering Research Institute, Durgapur; which is a constituent laboratory of Council of Scientific and Industrial Research, New Delhi, India. He obtained his B.Sc. (Engg.) and M.Tech. degrees from National Institute of Technology, Jamshe-dpur and Ph.D. degree in mechanical engineering from Indian Institute of Technology, Kharagpur. He has about 15 years of job experience which include research experience (National Metallurgical Laboratory, Jamshedpur and CSIR Centre for Mathematical Modeling and Computer Simulation, Bangalore) and teaching experience (National Institute of Foundry and Technology, Ranchi). His fields of research interest include Computational Mechanics, Finite Element Method, Composite Materials, Metal Working Processes and Software Development. He has published 20 research papers in different journals and conference proceedings. He has developed two large-size computer codes, (a) SB-FEM for finite element analysis of structures and metal working processes and (b) SB-Word for the purpose of word processing in bilingual (English-Hindi) typescripts, along with several small-size computer programs. He has recently launched and maintains a website www.softbodh.org, the purpose of which is to disseminate the knowledge and experience in software development and use.

PREFACE

The finite element method (FEM) is one of the most powerful numerical techniques and has become an indispensable part of engineering analysis and design. The method, developed in the past few decades, are now widely used in many branches of engineering for the analysis of structures, solid and fluids, and also in modeling and simulation of engineering processes. Accordingly, a number of textbooks on FEM have been published during this period. Most of these books emphasize the basic theoretical aspects of the method and present applications primarily to demonstrate the practical nature of the method. There are also some books available which are concerned with the applications of this method for a particular discipline of engineering.

Because of availability of a number of books, writing a new book on this subject obviously seemed to be a difficult decision for an author like me. However, the versatile nature of this method having many facets (concepts, theory, methodology, techniques, computer programming, software and the practical applications) permitted me to write this book in a different way. The purpose of this book is not only to provide the necessary concepts, theories and techniques of the method in the areas of structural analysis, heat transfer and metal working, but is also to be equipped with a user-friendly comprehensive computer code that can be used by a reader to solve different practical problems in these fields while reading the related theories. Although some finite element books present different subroutines of a finite element computer program usually included as an appendix, but the purpose of which is mainly to exhibit the necessary programming aspects of the derivations of element matrices and solution of finite element equations. Further, FEM requires a great emphasis on preprocessing (building the model and the mesh) and postprocessing (presentation and visualization of results) in its input and output stages, and a complete code must incorporate these facilities.

The book is divided into 13 chapters and 8 appendixes. Chapter 1 gives an introduction and overview of the finite element method. Chapter 2 in association with Appendixes 1 to 5 provides the general theoretical descriptions and techniques of the method, with the main focus on structural mechanics. As stated earlier, an essential feature of this book is the presentation of a versatile computer code which has been developed by me during working on this subject for over fifteen years. The essential features, capabilities and user's guidelines of the code, designated as SB-FEM, are discussed in Chapter 3 and in Appendixes 6 and 7. The code is user-friendly and requires minimum data input either in an interactive way using GUI or using a command input file. The command input file has a specific format and uses syntaxes same as those common in C programming language. The author is of the opinion that student community is nowadays familiar with C language syntaxes. It is obviously not possible to provide the full documentation of the code in this chapter, for which the appropriate documentation files inside the software may be referred to.

Chapters 4 to 8 are devoted to the specific structural analyses which include analyses of bars and beams, plane stress/plane strain and axisymmetric solids, plate bending, and three-dimensional problems of isotropic and composite structures. Chapter 9 deals with the transient dynamic analysis of structures. Geometrical nonlinear analyses of three dimensional solids and shells are described in Chapters 10 and 11 respectively. Chapter 12 shows the application of the finite element method to heat transfer problems. The book concludes with Chapter 13 in which finite element method has been applied to metal working analysis using flow (rigid-plastic) formulation.

There are several example problems solved at the end of each chapter starting with Chapter 4. Some selected SB-FEM input files are presented in Appendix 8 for the sake of completeness, although almost all the input files along with few others are available in a suitable directory within the given code.

I am indebted to various research workers and authors of text books and papers on this subject, which have been referred to during the working on this subject, development of the code and preparation of this book. The selected references of these works are given at the end of the book in Bibliography. Finally, several interactions with graduate students and colleagues on this subject are gratefully acknowledged.

Surendra Kumar

CONTENTS

1

Introduction to Finite Element Method

1.1 INTRODUCTION

Most of the practical engineering problems are very complex. Considering structural analysis problems, the complexities can arise due to one or a combination of the following factors:

a. Irregular/complex geometry.

b. Real (complex) boundary conditions.

c. Stress, temperature and time-dependent linear/non-linear behavior of constituent materials. The constituent materials may be non-homogeneous and anisotropic.

d. Space-time dependent loads.

e. Different types of structural response to be studied (linear, non-linear, dynamic stability, etc.).

The classical methods developed for analyzing structural components and systems can be broadly classified into two categories. In the first category, differential equations of equilibrium of the field are formulated considering stress-strain relations for the material and compatibility conditions at every point in the continuum, and these equations are solved for the required boundary conditions. In the second category, energy principles, particularly the principle of minimum potential energy is used to determine the deformation of the structure and then stresses in the structure are evaluated.

For problems involving complex shapes, material properties and complicated boundary conditions, it is difficult and in many cases unfeasible to obtain analytical solution that satisfies the governing differential equations and other requirements. Therefore, in general, closed from solutions are available in literature only for simple problems involving simple geometry and loading with homogeneous and isotropic linear material behavior. A few examples for which analytical solutions can readily be obtained, are stresses and displacement of an Euler beam, buckling of a prismatic column under axial loading, rectangular plate subjected to in-plane axial loading or bending, and so on. In case of complex problems, numerical techniques have to be employed to arrive at approximate solutions. Some of these techniques are finite difference method, Rayleigh-Ritz method and more recently, finite element method.

Finite difference method is generally employed to solve the differential equations of equilibrium. In this method, the body or the system is discretized by a grid of nodal points. The field variable is represented by the discrete values of the variable at the nodes (grid points). The governing differential equations and the boundary conditions are converted to finite difference forms which are then applied to each of the nodes in turn. As a result, a set of linear or non-linear algebraic equations in the discrete field variables is obtained which is then solved for the nodal values of the variable. On the other hand, Rayleigh-Ritz method can be employed for minimization of potential energy of the structure. In this method, the variation of the field variable is represented by a trail function and the unknown parameters are found out through minimization of potential energy. By this method, a continuous system of infinite degrees of freedom is reduced to a system of finite degrees of freedom.

The finite element method (FEM) is a powerful numerical method for solving problems of engineering and mathematical physics, and can be regarded as an extension of the Rayleigh-Ritz method since piecewise continuous trial functions are used here also. The solution provided by the method is obviously approximate since a continuum with infinite number of degrees of freedom is replaced with a discrete system with finite number of degrees of freedom. In this method of analysis, a complex region defining a continuum is discretized into simple geometric shapes called finite elements. The material properties and the governing relationships are considered over these elements and expressed in terms of unknown values at element corners called nodes. An assembly process (known as globalization), duly considering the loading and constraints, results in a set of equations. Solution of these equations gives us the approximate behavior of the continuum.

The FEM has become a powerful tool for the numerical solution of a wide range of engineering problems particularly those with complicated geometries, loadings, and material properties. Applications range from deformation and stress analysis of automotive, aircraft, spacecraft, missile, building, and bridge structures to field analysis of heat flux, fluid flow, magnetic flux, seepage, and other flow problems. With the advances in computer technology and CAD systems, complex problems can be modeled with relative ease. Several alternative configurations can be tried out on a computer before the first prototype is built. All of this suggests that we need to keep pace with these developments by understanding the basic theory, modeling techniques, and computational aspects of finite element method. Today, the developments in mainframe computers and availability of powerful microcomputers have brought this method within the reach of students, engineers, and scientists in research organizations.

1.2 Purpose of Using FEM

A number of distinctions can be made here as to what is the purpose of using finite element analysis (FEA). These features can be broadly classified into two categories.

1.2.1 *Study of operational impact*

The use of FEA mainly aims to improve the design of a product or a component by analyzing the effect of the loading (mechanical, thermal, or other) the structure will be subjected to during use. The result from the FEA is used to improve the design.

1.2.2 *Virtual manufacturing*

Product response

The use of FEA aim to investigate the response of the structure due to a manufacturing operation affecting the structural or global behavior of the product. An example can be studying the increase in torsional stiffness of a car body due to spot welding.

Manufacturing operation improvement

The use of FEA aim to predict the effect of a manufacturing operation on a product or component in terms of stress, plastic strain, displacements, temperature, microstructure, etc., so as to improve or optimize the manufacturing process itself.

Some of the specific advantages of FEM are as follows:

a. It can readily handle complex geometry.

b. It can handle complex analysis types (Vibration, Transients, Nonlinear, Heat transfer, Fluids, etc.).

c. It can handle complex loading such as node-based loading (point loads), element-based loading (pressure, thermal, inertial forces), time or frequency dependent loading, etc.

d. It can handle complex constraints.

e. Indeterminate structures can be analyzed.

f. It can handle bodies comprised of non-homogeneous materials (Every element in the model could be assigned a different set of material properties).

g. It can handle bodies comprised of non-isotropic materials (orthotropic, anisotropic).

h. Special material effects are handled such as temperature dependent properties, plasticity, creep, swelling, etc.

i. Special geometric effects can be modeled such as large displacements, large rotations, contact (gap) condition, etc.

Some of the limitations of FEM can be stated as follows:

a. A specific numerical result is obtained for a specific problem. A general closed-form solution, which would permit one to examine system response to changes in various parameters, is generally not produced.

b. The FEM is applied to an approximation of the mathematical model of a system and therefore, can inherit errors.

c. Experience and judgment are needed in order to construct a good finite element model.

d. A powerful computer and reliable FEM software are essential.

e. Numerical problems including round-off error and error accumulation can happen.

f. FEM may be susceptible to user-introduced modeling errors such as poor choice of element types, distorted elements, geometry not adequately modeled and so on.

g. Certain effects are not automatically included in the analysis such as buckling, large deflections and rotations, material nonlinearities and other nonlinearities.

1.3. Applications of FEM

The range of possible applications of the finite element method extends to all engineering disciplines, but civil, mechanical, and aerospace engineers are the most frequent users of this method. In addition to structural analysis other areas of applications include heat transfer, fluid mechanics, electromagnetism, biomechanics, geomechanics, and acoustics. The method is also useful in multidisciplinary problems where there is a coupling between two or more of the disciplines. Examples include thermal structures where there is a natural coupling between heat transfer and displacements, as well as aero elasticity where there is a strong coupling between external flow and the distortion of the wing. Although every branch of engineering is a potential user of the finite element method, this does not mean that it is the most practical solution technique. Often several other attractive techniques are available to solve a given problem. Each technique has its relative merits, and no technique enjoys the superior distinction of being "the best" for all problems.

In the aerospace industry FEM is used to analyze Static analyses, Modal analyses, Aerodynamics, Transient dynamics, Heat transfer , Fracture mechanics, Creep and plasticity analyses, Composite materials, Aero elasticity, Metal forming, Crashworthiness, etc. In the automotive industry FEM is used to investigate Static analyses, Modal analyses, Transient dynamics, Heat transfer, Mechanisms, Fracture mechanics, Metal forming, Crashworthiness, etc. In the architectural fields also, FEM is used to analyze Soil mechanics, Rock mechanics, Hydraulics, Fracture mechanics, Hydro elasticity, etc.

Based on the nature of the problem to be solved, applications of the finite element method can be divided into three categories, i.e.

a. Equilibrium or steady-state or time-independent problems

b. Eigenvalue problems and

c. Propagation or time-dependent problems.

1.3.1 *Equilibrium or time-independent problems*

The majority of applications of the finite element method fall into this category. For the solution of equilibrium problems in the solid mechanics area, we need to find the displacement distribution and the stress distribution for a given mechanical or thermal loading. Similarly, for the solution of equilibrium problems in fluid mechanics, we need to find pressure, velocity, temperature, and density distributions under steady-state conditions.

1.3.2 *Eigenvalue problems*

In the second category are the so-called eigenvalue problems of solid and fluid mechanics. In eigenvalue problems also, time will not appear explicitly. They may be considered as extensions of equilibrium problems in which critical values of certain parameters are to be determined in addition to the corresponding steady-state configurations. In these problems, we need to find the natural frequencies or buckling loads and mode shapes if it is a solid mechanics or structures problem, stability of laminar flows if it is a fluid mechanics problem and resonance characteristics if it is an electrical circuit problem.

1.3.3 *Time-dependent problems*

In the third category is the large number of time-dependent problems or propagation problems of continuum mechanics. This category is composed of the problems that result when the time dimension is added to the problems of the first two categories. This type of problem includes, for example, finding the response of a body under time-varying force in the area of solid mechanics and study of the heat transfer in a body under sudden heating or cooling.

1.4. Brief History of FEM

It is difficult to document the exact origin of the FEM, because the basic concepts have evolved over many years. Further, the idea of representing a given domain as a collection of discrete elements is not unique with the finite element method. For example, it was recorded that ancient mathematicians found the circumference of a circle by approximating it by the perimeter of a polygon inscribed in it. In 1941 Hrenikoff introduced the so-called frame work method, in which a plane elastic medium was represented as a collection of bars and beams. An approach similar to the finite element method, involving the use of piecewise continuous functions defined over triangular regions, was first suggested by Courant in 1943 for the solution of St. Venant torsion problem.

Although certain key features of the FEM can be found in these works of Hrenikoff (1941) and Courant (1943), the formal presentations of the finite element method as known today were presented in the papers of Argyris and Kelsey (1960) and Turner, Clough, Martin, and Topp (1956). In a series of articles appearing in 1954 and 1955, Argyris and Kelsey (1960) presented a formal unification of force and displacement methods using dual energy theorems. Although practical application of the duality proved momentary, the work systematized the concept of assembly of structural system equations from elemental components. This idea of space-continuum energy forms as basis for multi-dimensional element derivation, was successfully applied to triangular geometries by Turner *et al.* (1956). The term *finite element* was first coined by Clough in 1960. In the early 1960s, engineers used the method for approximate solution of problems in stress analysis, fluid flow, heat transfer, and other areas. The first book on the FEM by Zienkiewicz and Chung was published in 1967. In the late 1960s and early 1970s, the FEM was applied to a wide variety of engineering problems. The 1970s marked advances in mathematical treatments, including the development of new elements, and convergence studies. Several commercial FEM software packages (ABAQUS, ADINA, ANSYS, MARK, PAFEC) originated in the 1970s and 1980s.

According to Noor (1991), the development of modern finite element technology can be divided into four stages, each with a duration of eight to ten years, starting in the 1950s.

a. The first period is characerized by the method being mainly applied to structural mechanics problems.

b. In the second stage, the approach was expanded to address multifield problems while mathematical foundations were also given a great deal of attention, specifically on new efficient solution algorithms as well as techniques for solving large problems. During the 70s, some general purpose finite element codes were originated and released for public use.

c. In the 80s, the third stage, much of the effort by researchers was directed towards the development of new elements capable of handling unbounded domains, *e.g.* those used in flow calculations.

d. According to Noor, the fourth stage is characterized by new application fields, efficient algorithms for new computing systems (parallel computing), and the availability of finite element software on personal computers and workstations, as well as techniques for adaptive refinement of the mesh.

1.5. Basic Steps in FEA

The finite element method can be considered as a definite set of basic ingredients or steps. Regardless of the approach used in finite element formulation, the solution of a continuum problem by the finite element method always follows an orderly step-by-step process. In order, these are as follows:

a. Discretization

b. Interpolation

c. Element description

d. Assembly

e. Constraints

f. Solution

g. Computation of derived variables

1.5.1. *Discretization*

It is the process of converting a continuous model into a discrete model, which is an approximation of corresponding continuous model. As a result of this process, a mathematical model involving an ordinary differential equation is generally converted into a mathematical model involving algebraic equations. The continuum would thus be replaced by a series of simple, interconnected elements whose force-displacement characteristics are relatively easy to compute. In reality, these elements are connected to each other along their boundaries but in order to perform a theoretical approximation, the assumption is made that the elements are connected only at their corners called nodes. Finite elements of a continuum are special types of elements as depicted in Figure 1.1, that are constrained to maintain overall continuity of the assemblage. The continuity requirement forces the variation of the field variable along an element interface to be the same for adjacent elements. A wide variety of elements types in one, two, and three dimensions are well-established and documented. Engineering judgment is essential to determine not only which types of elements are appropriate for the problem at hand, but also the density required to sufficiently approximate the solution.

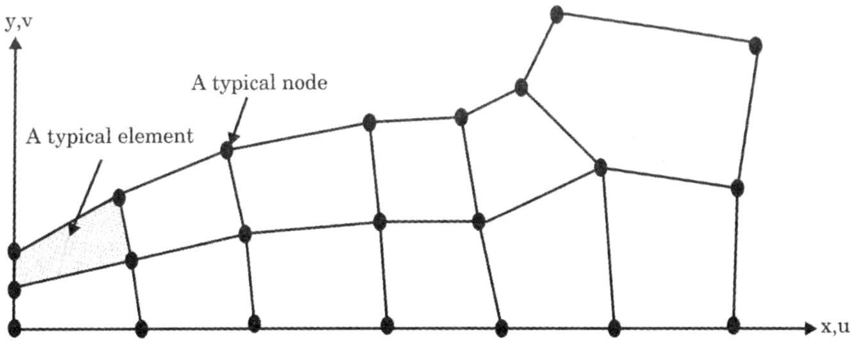

Fig. 1.1. Finite elements with nodal compatibility.

Discretization of irregular shaped regions has traditionally been performed manually, from sketches. Today, state-of-the-art software packages automate the meshing process. Caution is still advised, however, with any mesh-generation package, and the user's judgment and experience still remain important. Once a finite element mesh has been created, it must be checked to ensure that each element satisfies certain criteria for acceptability; for example distortion which may produce spurious results, must be avoided.

1.5.2. *Interpolation*

Once an element shape has been chosen, the analyst must determine how the variation of the field variable (*e.g.* displacement) across the element domain is to be represented or approximated. The unknown field variable for which the solution is sought is expressed in terms of a discrete number of unknowns at each node. Since a continuous solution is being represented, the solution must also be continuous throughout each element. The method of approximating the solution across each element is referred to as element interpolation. The functions used to represent the behavior of the solution within an element are called Interpolation functions. Interpolation functions are also referred to as shape functions or approximating functions. Writing mathematically, the approximation for a scalar field variable $u(x, y)$ within an element in a two-dimensional space is written as a combination of the (as yet unknown) field variables at the nodes belonging to that element, *i.e.*,

$$u(x,y) = \sum_{j=1}^{n} N_j(x,y)u_j \tag{1.1}$$

Here u_j is the field variable at jth node, $N_j(x,y)$ is jth interpolation function and n is the total number of nodes in the element.

In most cases, a polynomial interpolation function is used. The number of nodes assigned to an element dictates the order of the interpolation function which can be used. The degree to which the approximate solution is sufficient to accurately model the problem is affected by the type of interpolation, or the shape function used. The simplest method of approximation is to assume a linear distribution of the unknown function within the element domain.

1.5.3. *Element description (Computation of element properties)*

Once the interpolation functions have been chosen, the field variable in the domain of the element is approximated in terms of discrete values at the nodes. Consequently, a system of equations is formed which expresses the element properties in terms of quantities at the nodes and can be represented in matrix form. For example, in a structural analysis, the element equations relate the nodal forces to the nodal displacements through stiffness values. From the assumed field variable (displacement) model, the characteristics (stiffness) matrix $[K^e]$ and characteristics (load) vector $\{F^e\}$ of an element is derived by using either equilibrium condition or a suitable variational approach.

1.5.4. *Assembly*

Since the domain (structure) is composed of several finite elements, the assembly procedure combines each element approximation of the field variable (as defined in the previous step) to form a piecewise approximation of the behavior over the entire solution domain. Assembly is accomplished using the basic rule of compatibility that the value of the field variable at a node must be the same for each element that shares that node (Figure 1.2). This step is handled automatically by the finite element programs.

For example, in a structural analysis, the individual element stiffness matrices and load vectors are to be assembled in

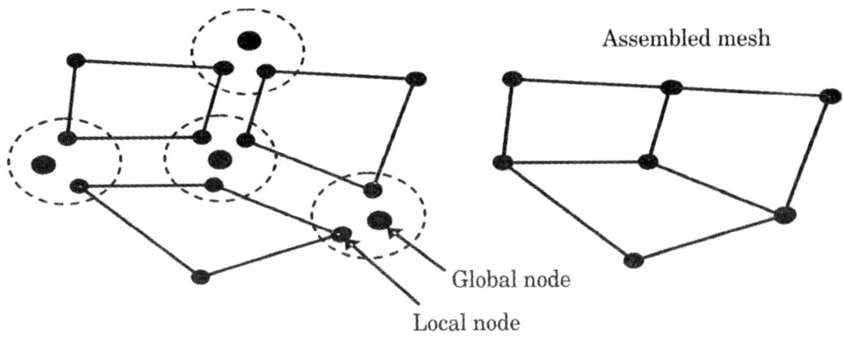

Fig. 1.2. Assemblage of elements.

suitable manners and the overall equilibrium equations have to be formulated as

$$[K] = \sum_e [K^e] \quad \text{and} \quad \{F\} = \sum_e \{F^e\}, \tag{1.2}$$

where $[K]$ is called assembled or global stiffness matrix, and $\{F\}$ is the vector of nodal forces for the complete domain (structure).

1.5.5. *Constraints*

The global system of equations created in the previous step cannot be solved without application of the boundary conditions. Mathematically, before applying the boundary conditions, the system of equations is indeterminate and does not have a unique solution.

In this step the required constraints (*e.g.* displacements and rotations in the required direction) are prescribed in the model at suitable locations. In finite element terminology, a node is prescribed if the value of the field variable is already known at this location. For example, a prescribed nodal displacement of zero is assigned to fixed supports of a structure. In a thermal analysis, a temperature is prescribed at one or more nodes. Usually these nodes lie on the boundaries of the finite element mesh.

1.5.6. *Solution*

Once the boundary conditions have been applied to the assembled matrix of equations, the linear simultaneous equations of equilibrium are solved for the unknown model field variable (*e.g.* displacements). The equilibrium equations can be expressed as

$$[K]\{U\}=\{F\}, \tag{1.3}$$

where {U} is the vector of field variables for the complete domain. Standard numerical techniques can be used to solve for the unknown field variable at each node. If the system of equations is linear, a Gaussian elimination or a Cholesky decomposition algorithm can be used to determine the vector {U}. The matrix equations generated by the finite element process are often sparsely populated and symmetrical. Many solution methods make use of these properties to provide fast and efficient computation algorithms which are now implemented in nearly all finite element packages. If the set of equations is non-linear, then the solution procedure is more difficult to obtain. For non-linear problems, the solution has to be obtained in a sequence of steps; each step involving the modification of the characteristics (stiffness) matrix[K] and/or the load vector {F}. The choice of solution method is very much dependent on the size of the problem as well as the type of analysis.

1.5.7. *Computation of derived variables*

From the known model field variables {U}, if required, the element resultants can be computed by using the necessary equations. For example in a structural analysis, additional calculations include the computation of element strains, stresses, principal stress, strain energy and dynamic responses.

1.6 Stages in Finite Element Programming

From the programming (computer implementation) point of view, a finite element analysis mainly consists of three stages of activity with logical interactions among them (Figure 1.3), namely preprocessing, processing and postprocessing.

1.6.1 *Preprocessing*

As the name indicates, preprocessing is something user does before processing the analysis. The preprocessing involves the preparations of data, such as nodal coordinates, connectivity, boundary conditions and loading and material information.

The preparation of data requires considerable efforts if all data are to be handled manually. If the model is small, the user can often just write a text file and feed it into the processor, but as the complexity of the model grows and the number of elements increases, writing the mesh information data manually can be very time consuming and error-prone. It is therefore necessary to have a preprocessor which helps in mesh generation, mesh plotting and boundary conditions plotting.

1.6.2 *Processing*

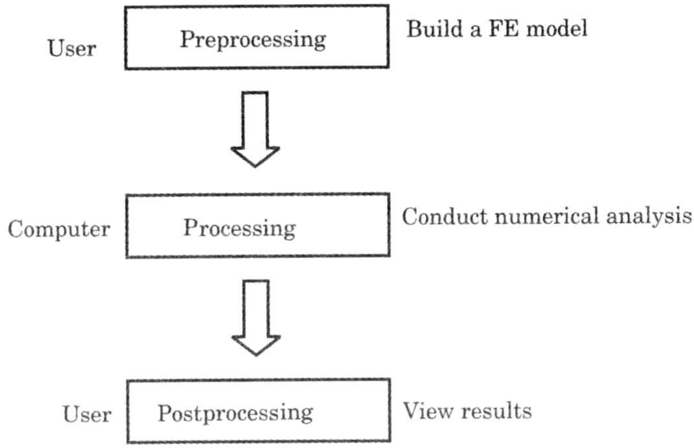

Fig. 1.3. Stages of FEM.

The processing stage involves generation of element matrices using numerical integration, assembly of element equations, imposition of the boundary conditions and solution of the global equations. This step results in the evaluation of primary variables at the nodal points. Other derived quantities, such as gradients or stresses, may be evaluated at this stage. This is a typical "black box" operation, as the user will see little of what is going on. We feed input data from the preprocessor, and we get output data after the processing operation by the computer.

1.6.3 *Postprocessing*

The postprocessing stage deals with the representation of output results in a desired format for printout and/or plotting. Typically, several secondary variables such as deformed configuration, mode shapes, temperature, and stress distribution are computed and displayed at this stage.

2

Ingredients of Finite Element Method

2.1 INTRODUCTION

This chapter describes the general steps included in a finite element formulation and solution to an engineering problem. We will use these steps as our guide in developing solutions for structural and nonstructural problems in subsequent chapters. For simplicity's sake, for the description of the steps to follow, we will consider mainly the structural problem. The nonstructural heat-transfer and metal working problems have analogies to the structural problem as far as steps/ ingredients of FEM are concerned.

Typically, for the structural stress-analysis problem, the engineer seeks to determine displacements and stresses throughout the structure, which is in equilibrium and is subjected to applied loads and suitable constraints. For many structures, it is difficult to determine the distribution of deformation using conventional methods, and thus the finite element method is necessarily used.

2.2 DISCRETIZATION OF THE DOMAIN

The first step of the finite element analysis involves the discretization of the irregular domain (problem geometry) into smaller and regular subdomains, known as finite elements. This is equivalent to replacing the domain having an infinite number of degrees of freedom by a system having finite number of degrees of freedom. The elements are assumed to be interconnected at a discrete number of nodal points situated on their boundaries. The shapes, sizes, number, and configurations of the elements have to be chosen carefully such that the original body or domain is simulated as closely as possible without increasing the computational effort needed for the solution. The choice of the type of element is mostly dictated by the problem geometry and the number of independent coordinates necessary to describe the system. The number of elements used in a problem depends mainly on the element type and accuracy desired. For the discretization of problems involving curved geometries, finite elements with curved sides are useful. The ability to model curved boundaries has been made possible by the addition of midside nodes. Finite elements with straight sides are known as linear elements, whereas those with curved sides are called higher order elements.

2.2.1. Basic element shapes

One-dimensional elements

If the geometry, material properties, and the field variable of the problem can be described in terms of only one spatial coordinate, we can use the one dimensional or line elements shown in Figure 2.1. The deformation of a bar under an axial load, the temperature distribution in a rod (or fin), the pressure distribution in a pipe flow, for example, can be determined using these elements. Although these elements have cross-sectional area (uniform or non-uniform), they are generally shown schematically as a line element (Figure 2.1).

For a simple analysis, one-dimensional elements are assumed to have two nodes, one at each end, with the corresponding value of the field variable chosen as the unknown degree of freedom. For the analysis of bars, the axial displacement is chosen as the unknowns (degrees of freedom) at each node as shown in Figure 2.1(a). However, for the analysis of beams, the values of the field variable (transverse displacement) and its derivative (slope) are chosen as the unknowns at each node as shown in Figure 2.1(b).

Two-dimensional elements

When two independent spatial coordinates must be required to describe the configuration and other details of the problem, two-dimensional elements are used as shown in Figure 2.2. These elements are generally used in the analysis of membrane, plate, shell, etc. The basic element useful for two-dimensional analysis is the triangular element. Although

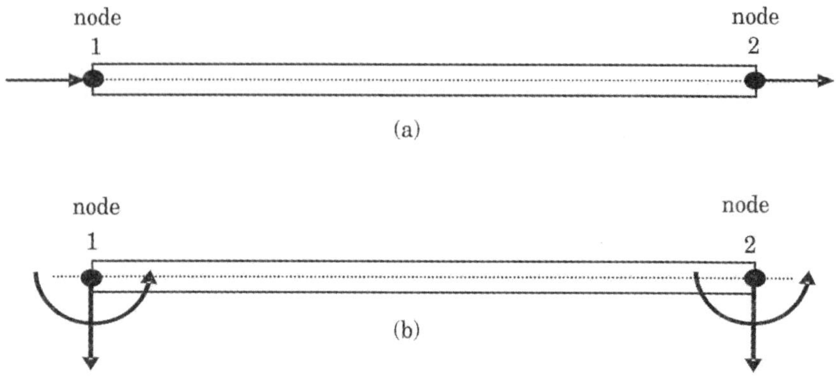

Fig. 2.1. One-dimensional elements.

a quadrilateral (or its special forms, rectangle and parallelogram) element can be obtained by assembling two or four triangular elements, in some cases the use of quadrilateral elements proves to be advantageous. For the bending analysis of plates and shells, multiple degrees of freedom such as transverse displacement and section rotations, are used at each node.

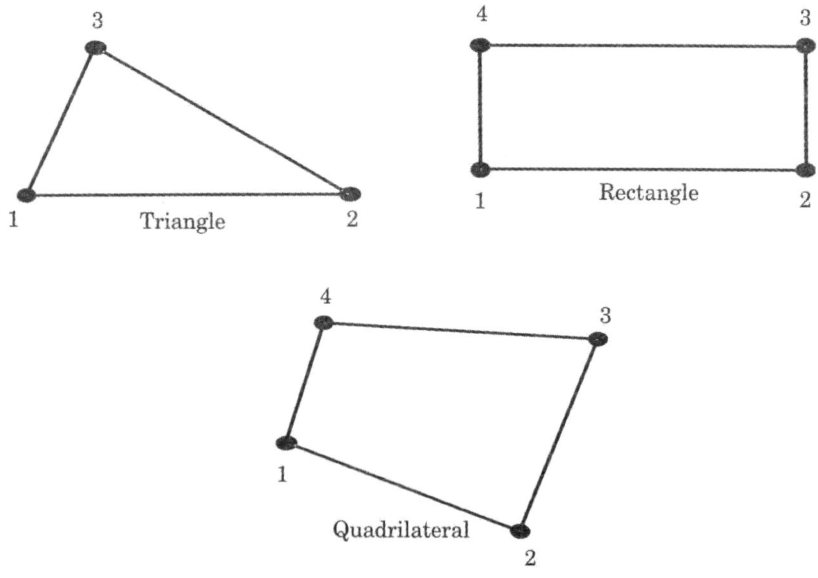

Fig. 2.2. Two-dimensional elements.

Three-dimensional elements

If the description of geometry, material properties, loading and other parameters of the body requires three independent spatial coordinates, three dimensional elements are used to idealize the body. Some three dimensional elements are depicted in Figure 2.3. The basic three-dimensional element, analogous to the triangular element in the case of two-dimensional problems, is the tetrahedron element. In several cases, the hexahedron element is used advantageously.

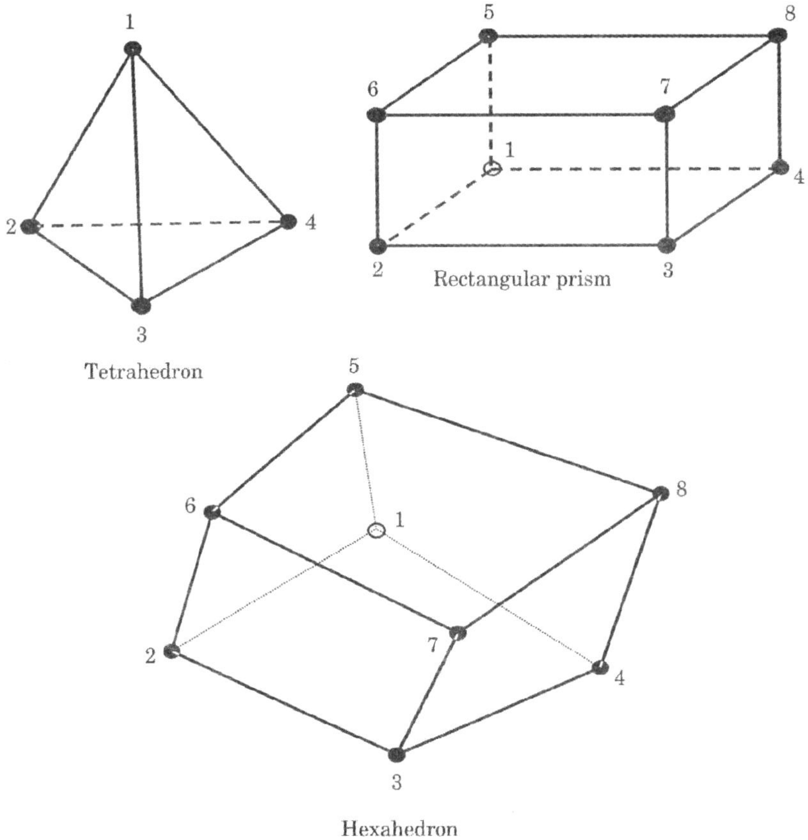

Fig. 2.3. Three-dimensional elements.

Axisymmetric elements

Several problems of bodies of revolution (axisymmetric solids) under axisymmetric loading are of considerable practical interest. Owing to symmetry about the circumferential (θ) direction, these problems can be described by only one or two independent coordinates instead of three coordinates. Thus with little modifications, any of the one-dimensional elements described earlier can be successfully employed in solving axisymmetric shell problems (Figure 2.4 (a)). Similarly any of the two-dimensional plane elements with slight adjustment can be used in solving axisymmetric solid problems (Figure 2.4 (b)). In case of stress analysis problems, the modifications required include the involvement of an additional non-zero circumferential (hoop) normal strain ε_θ due to the presence of non-zero radial displacement (See Section A2.2). The problems that possess axial symmetry include pistons, storage tanks, valves, rocket nozzles, and reentry vehicle heat shields.

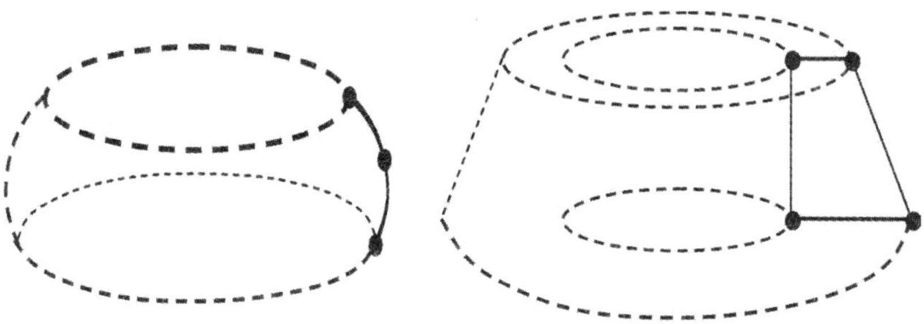

(a) (b)

Fig. 2.4. Axisymmetric elements used for axisymmetric problems: (a) A one-dimensional axisymmetric line element and (b) A two-dimensional axisymmetric quadrilateral element.

2.2.2 Size and number of elements

The number, size and density (*i.e.* mesh refinement) of elements used in a given problem depend on a number of considerations. The first consideration is to discretize the domain as close to exact as possible with elements that are admissible. In discretizing a domain, consideration must be given to elements interfaces where abrupt changes in geometry, material properties and loading occur. If the size of the elements is small, the final solution is expected to be more accurate. However, the use of small size elements will also lead to more computational time. The second consideration, which requires some engineering judgment, is to discretize the body or portion of the body using finer mesh (sufficiently small elements) where steep gradient of the solution is expected. For example, in the case of stress analysis of a plate with a hole, the size of elements has to be very small near the hole (where stress concentration is expected) compared to far away places.

Another characteristic related to the size of elements that affects the finite element solution is the aspect ratio of the elements. The aspect ratio describes the shape of the element in the assemblage of elements. For two-dimensional elements, the aspect ratio is taken as the ratio of the largest dimension of the element to the smallest dimension. Elements with an aspect ratio of nearly unity generally yield best results.

The number of elements to be chosen for idealization is related to the accuracy desired, size of elements, and the number of degrees of freedom involved. Although an increase in the number of elements generally provides more accurate results, for any given problem, there will be a certain number of elements beyond which the accuracy cannot be improved by any significant amount. Moreover, since the use of large number of elements involves a large number of degrees of freedom, computer memory requirement to store the resulting matrices will be significantly more.

2.2.3 Automatic mesh generation

Discretization of irregular shaped regions can be a tedious, time consuming and error-prone if done manually, particularly for three-dimensional geometry. In recognisation of this problem, a large number of methods have been devised to automate the mesh generation task. These automatic mesh generation algorithms are capable of discretizing general domains into an efficient finite element mesh without the need for user intervention. When the user supplies information on the surfaces and volumes of the material domains that make up the object, an automatic mesh generator generates the locations of the node points and elements, labels the nodes and elements, and provides the element-node connectivity relationships. The user can also specify the minimum permissible element sizes for different regions of the object. The automatic mesh generation techniques are usually tied to solid modeling and computer-aided design schemes. There are several papers which have attempted to provide an overall picture of the finite element mesh generation schemes by reviewing, classifying and comparing the mesh generation methods (*e.g.*, Shephard, 1988 and Ho-Le, 1988).

2.3. INTERPOLATION MODELS

The finite element formulation of a differential equation does not require a prior knowledge of the type of the element. The selection of an appropriate element can be made depending on the admissibility conditions implied by the finite element formulation, the geometry of the domain being modeled, and the degree of accuracy desired. On the other hand, the construction of interpolation functions does not depend on the specific differential equation being solved. The construction procedure depends on the type of element (geometry, number and position of the nodes and number of primary unknowns per node). Polynomial-type interpolation functions are most widely used due to the following reasons:

a. It is easier to formulate and computerize the finite element equations with polynomial-type interpolation functions. Specifically, it is easier to perform differentiation or integration with polynomials.

b. It is possible to improve the accuracy of the results by increasing the order of the polynomial.

One can develop a library of finite elements, which can be used wherever they are applicable. When the interpolation polynomial is of order one, the element is termed a linear element. A linear element is called a simplex element if the number of nodes in the element is 2, 3 and 4 in 1, 2, and 3 dimensions, respectively. If the interpolation polynomial is of order two or more, the element is known as a higher order element. In higher order elements, some secondary (midside and/or interior) nodes are introduced in addition to the primary (corner) nodes so that the number of nodes matches with the number of constants in the interpolation polynomial.

While choosing the order of the polynomial in a polynomial-type interpolation function, the following considerations have to be taken into account:

a. The interpolation polynomial should satisfy, as far as possible, the convergence requirements. That is, the finite element approximate solution must converge to the exact solution as the element size is reduced successively.

b. The pattern of variation of the field variable resulting from the polynomial model should not have any preferential direction. That is, the field variable representation within an element, and hence the polynomial, should not change with a change in the local coordinate system (when a linear transformation is made from one Cartesian coordinate system to another). This property is called geometric isotropy or geometric invariance or spatial isotropy. In order to ensure geometric isotropy, the interpolation polynomial should contain terms that do not violate symmetry. For two-dimensional elements, the number of terms required in the polynomial of different orders can be readily established from what is known as Pascal triangle. Pascal triangle contains polynomial terms of various degrees in the two variables x and y, as shown in Figure 2.5. For elements of three-dimensional type, the equivalence of the Pascal triangle is a Pascal tetrahedron which contains polynomial terms in three coordinates (Rao, 2004).

c. The number of generalized coordinates (total number of terms involved in the polynomial) should be equal to the number of nodes in the element. The satisfaction of this requirement enables us to express the polynomial coefficients in terms of the nodal unknowns of the element.

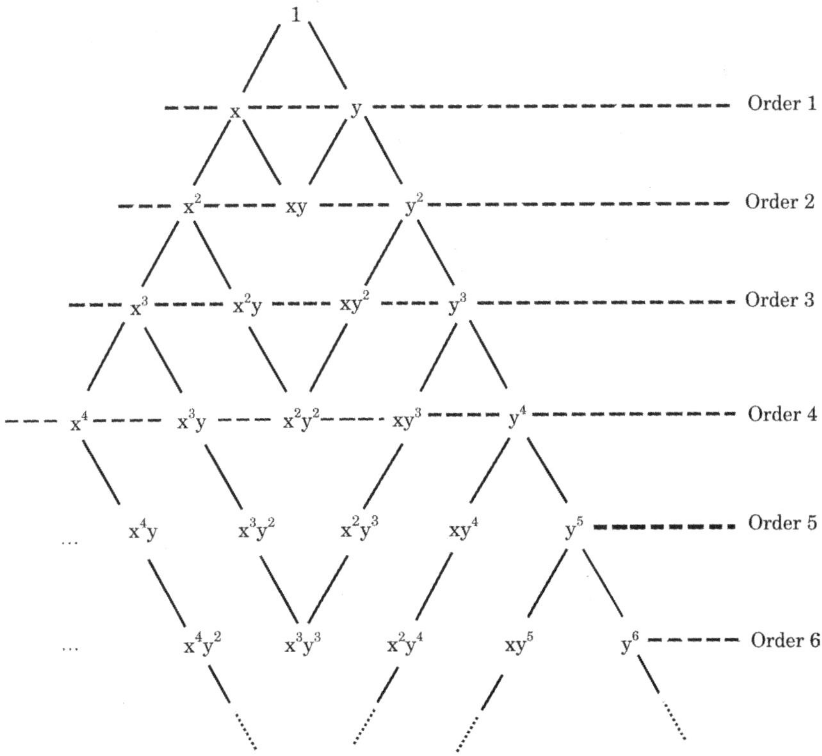

Fig. 2.5. The Pascal triangle.

2.3.1. Convergence and compatibility requirements

The finite element method provides a sequence of approximate numerical solutions to a complex problem as the element size is reduced successively. It may, therefore, be expected that this sequence must converge to the exact solution under certain circumstances. In structural analysis field, it can be shown that the displacement formulation of the method leads to an upper bound to the actual stiffness of the structure. Hence as the mesh is made finer the solution should converge to the correct result. The convergence would be achieved if the following three requirements are satisfied by the assumed interpolation function (model):

a. The field variable must be continuous within the element. This condition can easily be satisfied by choosing polynomials for the interpolation function.

b. The interpolation polynomial must be able to give a constant value of the field variable within the element, when the nodal values are numerically identical. In the case of structural problems, this requirement states that the assumed interpolation function must be capable of representing rigid body displacements of the element. That is when the nodes are given such displacements corresponding to a rigid body motion, the element should not experience any strain and hence must lead to zero nodal forces. The constant terms in the polynomials used for the interpolation models would usually ensure this condition.

c. Interpolation polynomial must be capable of representing uniform (constant) states of partial derivatives of the field variable up to a certain order (decided by the physics of the problem) within the element when, in the limit, the element size reduces to zero. The reason for this requirement in structural analysis can be understood if we imagine the condition when the body or structure is divided into smaller and smaller elements. As these elements approach infinitesimal size, the strains in each element also approach constant values. Hence the assumed displacement interpolation function should include terms for representing constant strain states. For one, two and three dimensional elasticity problems, the linear terms present in the polynomial satisfy this requirement.

Apart from convergence requirements, the interpolation function must assure compatibility condition between adjacent elements. That is when the elements deform there must not be any discontinuity between elements, i.e. elements must not overlap or separate. In the case of beam, plate and shell elements, this requirement would also include that there should be no sudden changes in slope across the inter-element boundaries.

Elements which satisfy all the three convergence requirements and compatibility condition are called 'compatible or conforming elements'. Further, elements which violate the compatibility requirement are known as 'incompatible elements'. There are many instances reported in the literature wherein incompatible elements have been successfully used. They have been found to give good results and are used to improve certain basic properties of the elements which are not possible otherwise.

2.3.2. Interpolation polynomial for two-dimensional elements

In order to demonstrate here, interpolation polynomials of two-dimensional rectangular elements (Figure 2.6) of the Lagrange family (*i.e.* elements with only one unknown, namely, the field variable itself, per node) can be developed from Pascal triangle.

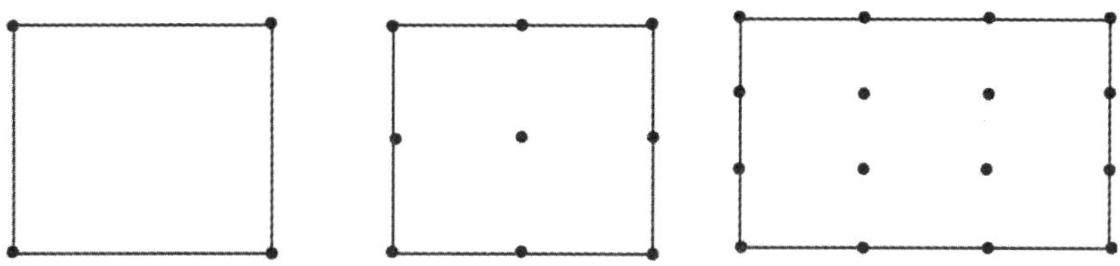

(a) Linear (b) Quadratic (c) Cubic

Fig. 2.6. Some members of rectangular elements.

Since a linear rectangular element has four corners (hence, four nodes), the polynomial should have the first four terms 1, x, y and xy (which form a parallelogram in the Pascal triangle). A quadratic rectangular element has nine nodes (four at the four corners, four at the midpoints of the sides, and one at the centre of the element), and the polynomial associated with the field variable $u(x, y)$ is given by

$$u(x,y) = a_1 + a_2x + a_3y + a_4xy + a_5x^2 + a_6y^2 + a_7x^2y + a_8xy^2 + a_9x^2y^2 \tag{2.1}$$

The polynomial contains the complete polynomial of the second degree plus the third-degree terms x^2y and xy^2 and also an x^2y^2 term.

In general, a pth-order Lagrange rectangular element has n nodes, with

$$n = (p+1)^2 \qquad p = 0, 1, \cdots.$$

Substituting coordinates of the various nodes of the element, a set of simultaneous equations can be obtained as

$$\begin{Bmatrix} u_1 \\ \vdots \\ u_9 \end{Bmatrix} = \begin{bmatrix} 1 & x_1 & y_1 & x_1y_1 & x_1^2 & y_1^2 & x_1^2y_1 & x_1y_1^2 & x_1^2y_1^2 \\ \vdots & \vdots & \vdots & \vdots & \vdots & \vdots & \vdots & \vdots & \vdots \\ 1 & x_9 & y_9 & x_9y_9 & x_9^2 & y_9^2 & x_9^2y_9 & x_9y_9^2 & x_9^2y_9^2 \end{bmatrix} \begin{Bmatrix} a_1 \\ \vdots \\ a_9 \end{Bmatrix} \tag{2.2}$$

or simply as

$$\{\mathrm{U^e}\} = [\mathrm{C}]\{\mathrm{a}\}. \tag{2.3}$$

Then $\{\mathrm{a}\} = [\mathrm{C}]^{-1}\{\mathrm{U^e}\}$.

We can rewrite Eq. (2.1) as

$$u = \begin{bmatrix} 1 & x & y & xy & x^2 & y^2 & x^2y & xy^2 & x^2y^2 \end{bmatrix}\{a\} = [P]\{a\}. \tag{2.4}$$

Then the shape function for the element defined by

$$u = [N_1 \quad N_2 \quad N_3 \quad N_4 \quad N_5 \quad N_6 \quad N_7 \quad N_8 \quad N_9]\{U^e\} = [N]\{U^e\} \tag{2.5}$$

can be found from $[N] = [P][C]^{-1}$. (2.6)

2.3.3. Lagrange interpolation functions

A systematic method of constructing shape functions of any order can be achieved by employing classical interpolation polynomials such as Lagrange and Hermite interpolation polynomials. When only the field variable function values are required to match at each point of interpolation (nodal point), the interpolation polynomial is called Lagrange interpolation formula. When the field variable function and its first derivative are to be assigned at each point of interpolation, the interpolation polynomial is called the Hermite interpolation formula. Lagrange polynomials are often used in the finite element formulation.

In the Lagrange interpolation, the assumed field variable function takes on the same values as the given field variable function at specified points. In the derivation of this interpolation function, it will be convenient to use local or natural coordinates defined for each element shape.

Interpolation functions in terms of natural coordinates

A natural coordinate system is a local coordinate system that permits the specification of any point inside the element by a set of non-dimensional numbers whose magnitude lies between 0 and 1 or between -1 and 1. Natural coordinate system (ξ, η) for a rectangle is shown in Figure 2.7 and is so chosen that on the opposite edges of the rectangle their values are ± 1. The interpolation (shape) functions are expressed in terms of the natural coordinates. The representation of geometry in terms of (nonlinear) shape functions can be considered as a mapping procedure that transforms a regular shape, such as a straight-sided rectangle (square) in the local coordinate system, into a distorted shape such as a curved-sided rectangle or quadrilateral in the global Cartesian coordinate system (Figure 2.7). The element defined in the natural coordinate system is sometimes called the master element or parent element. The geometrical transformation of a master element is based on the description of the coordinates of any point in that element in terms of the shape functions and the values of the nodal coordinates. When the shape functions used to interpolate the coordinates within

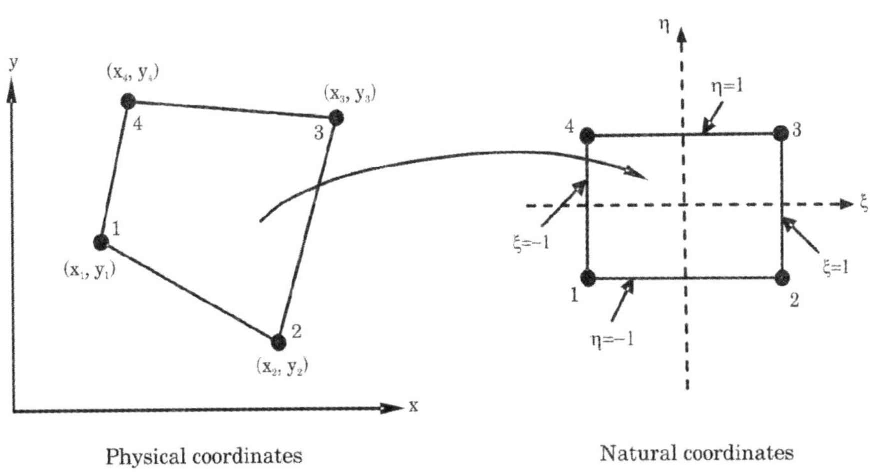

Fig. 2.7. Coordinate mapping between coordinate systems.

an element are the same as those used for interpolation of the field variable, such an element is called an isoparametric element. Isoparametric elements are widely used in practice due to the generality of approach and their ability to provide an accurate description of irregular domains (curved boundaries).

Lagrangian polynomial

Polynomials which have a value of unity at points of specified nodes and zero at all others satisfy all the inter-element continuity conditions. Polynomials in one coordinate system having this property are known as Lagrange polynomial and can be written down directly as

$$l_k^{n-1}(\xi) = \frac{(\xi-\xi_1)(\xi-\xi_2)\cdots(\xi-\xi_{k-1})(\xi-\xi_{k+1})\cdots(\xi-\xi_n)}{(\xi_k-\xi_1)(\xi_k-\xi_2)\cdots(\xi_k-\xi_{k-1})(\xi_k-\xi_{k+1})\cdots(\xi_k-\xi_n)}$$

$$= \prod_{\substack{i=1 \\ i\neq k}}^{n} \frac{(\xi-\xi_i)}{(\xi_k-\xi_i)}. \tag{2.7}$$

The superscript (n-1) in $l_k^{n-1}(\xi)$ denotes that the polynomial is of degree (n-1).

It is clear that the polynomial has a value of unity at point ξ_k and passes through zero at the remaining (n-1) points. The shape function for an n noded one-dimensional element can now be defined as

$$N_i = l_i^{n-1}(\xi). \tag{2.8}$$

Using Eq. (2.8), the shape function for a two-noded one-dimensional element shown in Figure 2.8 (a), can be derived as

$$\begin{aligned} N_1 &= \frac{1}{2}(1-\xi) \\ N_2 &= \frac{1}{2}(1+\xi). \end{aligned} \tag{2.9}$$

The quadratic shape function for the three-noded one-dimensional element shown in Figure 2.8 (b), can be derived as

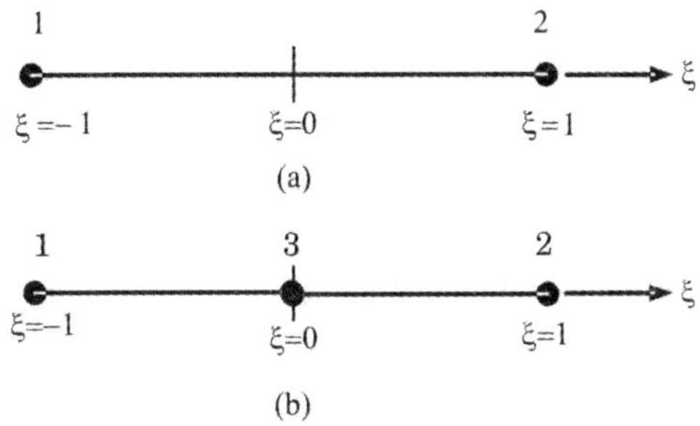

Fig. 2.8. Natural coordinates of one-dimensional elements (a) Two-noded and (b) Three-noded.

$$\begin{aligned} N_1 &= \frac{1}{2}\xi(\xi - 1) \\ N_2 &= \frac{1}{2}\xi(\xi + 1) \\ N_3 &= \left(1 - \xi^2\right). \end{aligned} \tag{2.10}$$

An easy and systematic method of generating Lagrange interpolation functions of any order in two or three dimensions can be achieved by taking the so-called tensor product of the one-dimensional interpolation functions in the two or three coordinates respectively. Thus, in two dimensions, if we label the node by its column and row numbers, I, J, we have

$$N_i \equiv N_{IJ} = l_I^{n-1}(\xi)\, l_J^{m-1}(\eta), \tag{2.11}$$

where n and m stand for the number of nodes in each coordinate direction of the element.

The elements shown in Figure 2.6 are a few members of this unlimited family of rectangular elements where m = n. For m = n = 2 (four noded rectangular element), we obtain the simple result, i.e.

$$N_i = \frac{1}{4}(1 + \xi_i\xi)(1 + \eta_i\eta), \quad i = 1 \text{ to } 4, \tag{2.12}$$

in which (ξ_i, η_i) are the natural (normalized) coordinates at node i.

Similar to the one-dimensional interpolation functions, the two-dimensional interpolation functions satisfy the following conditions:

$$N_i\left(\xi_j,\eta_j\right) = \delta_{ij} = \begin{cases} 1 & \text{if i = j} \\ 0 & \text{if i} \neq \text{j} \end{cases} \tag{2.13a}$$

and

$$\sum_{i=1}^{n} N_i\left(\xi,\eta\right) = 1. \tag{2.13b}$$

In the above equations: $i = j = 1,\cdots,n$, in which n is the number of nodal points in the element.

The Lagrange interpolation functions for three-dimensional hexahedral elements of different orders can similarly be derived. Here the master element is defined in natural coordinate (ξ,η,ζ) system, and is cube-shaped having 2 unit sides placed symmetrically with ξ, η, ζ coordinates as shown in Figure 2.9. As an example, shape functions for linear hexahedral element (having eight-nodes at each corner) (Figure 2.9) are determined to be

$$N_i = \frac{1}{8}\left(1+\xi_i\xi\right)\left(1+\eta_i\eta\right)\left(1+\zeta_i\zeta\right), \quad i = 1 \text{ to } 8, \tag{2.14}$$

where (ξ_i,η_i,ζ_i) represents the coordinates of node i of the element. Again the function N_i has the property that it is equal to unity at node i and is zero at all other nodes.

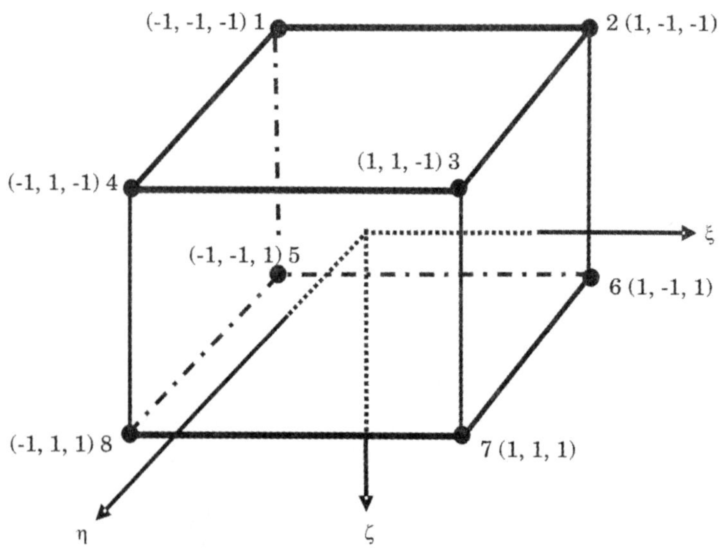

Fig. 2.9. Natural coordinate system for a three-dimensional hexahedral element.

2.3.4. The Serendipity elements

The higher order elements of the Lagrange family have internal nodes as may be seen for the nine-noded element shown in Figure 2.6 (b). Since these internal nodes do not contribute to the interelement connectivity, they can be condensed out at the element level so that the size of the element matrices is reduced. Alternatively, we can develop the so-called serendipity elements to avoid the internal nodes present in the Lagrange elements. The serendipity elements are those rectangular elements which have all the node points on the boundary of the element and have no internal nodes. The interpolation functions for serendipity elements cannot be obtained using tensor products of one-dimensional Lagrange interpolation functions. Instead, an alternative procedure that employs the conditions of the interpolation functions given by Eqs. (2.13a) and (2.13b), is used. As an example, the procedure to construct the interpolation functions for the eight-noded rectangular element is discussed later in Chapter 6.

2.3.5. Interpolation polynomials for vector quantities

In Eq. (2.5), the field variable u has been assumed to be a scalar quantity. In several problems the field variable may be a vector quantity having both magnitude and direction (*e.g.*, displacement $\{u\} = [u_1 \quad u_2 \quad u_3]^T$ in three-dimensional solid mechanics problems). In such cases, the usual procedure is to resolve the vector into components parallel to the coordinate axes and treat these components as the unknown quantities. Thus, there will be more than one unknown (degree of freedom) at a node in such problems. The number of degrees of freedom at a node will be one, two, or three depending on whether the problem is one-, two-, or three-dimensional. For example, in three dimensional solid mechanics the interpolation of displacement vector is represented by

$$\{u\} = [N]\{U^e\}, \tag{2.15}$$

where $[N]$ matrix in the three-dimensional case takes the form as

$$[N] = [[N_1] \cdots [N_n]]. \tag{2.16}$$

The element nodal displacement vector $\{U^e\}$ is given by

$$\{U^e\} = [u_{11}, u_{21}, u_{31}, \quad \cdots, \quad u_{1n}, u_{2n}, u_{3n}]^T . \tag{2.17}$$

In Eq. (2.16), a typical sub-matrix is given as $[N_r] = \begin{bmatrix} N_r & 0 & 0 \\ 0 & N_r & 0 \\ 0 & 0 & N_r \end{bmatrix}$

and n is the number of nodes in the element. In Eq. (2.17), u_{iI} is the ith displacement component of node I of the element.

2.4. DERIVATION OF ELEMENT MATRICES AND VECTORS

The characteristic matrices (stiffness matrices) and characteristic vectors (nodal load vectors) of finite elements can be derived by using any of the following approaches:

a. Direct approach
b. Variational approach
c. Weighted residual approach
d. Energy balance approach

2.4.1 Direct approach

In this method, direct physical reasoning is used to establish the element properties (characteristic matrices and vectors) in terms of pertinent variables. Considering structural problem, this method derives the stiffness matrix and element equations relating nodal forces to nodal displacements using force equilibrium conditions for a basic element, along with force/deformation relationships. Although the applicability of these methods is limited to simple types of elements such as spring and bar, a study of these methods enhances our understanding of the physical interpretation of the finite element method.

2.4.2 Variational approach

The variational approach establishes formulation in which a given differential equation is recast in an equivalent integral form by trading the differentiation between a test function and the dependent variable. The dependent variable of a given problem is approximated by a linear combination of appropriately chosen function. The variational solution is not differentiable enough to satisfy, a variational equation equivalent to the differential equation. Since most physical and engineering problems can be formulated in a variational form, the finite element method can be readily applied for finding their approximate solutions. The variational approach has been most widely used in the literature in formulating finite element equations. A major limitation of the method is that it requires the physical or engineering problem to be stated in the variational form, which may not be possible in all cases. Variational principle and its application to solid mechanics problem is discussed in Appendix 4. In the following discussion, the derivation of element stiffness matrix and load vector for stress analysis problems is done using this approach.

From Eq. (A4.16), we have

$$\delta I(\bar{u}) = \int_V \sigma_{ij}\delta\varepsilon_{ij}dV - \int_V b_i\delta u_i dV - \int_{S_F} t_i\delta u_i dS = 0\,. \tag{2.18}$$

Writing in matrix form by making use of the contracted (vector) form of stresses and strains, we get

$$\int_V \{\delta\varepsilon\}^T \{\sigma\}\, dV = \int_V \{\delta u\}^T \{b\}\, dV + \int_{S_F} \left\{\delta u^s\right\}^T \{t\}\, dS\,, \tag{2.19}$$

where

$$\begin{aligned} \{\varepsilon\} &= \begin{bmatrix}\varepsilon_1 & \varepsilon_2 & \varepsilon_3 & \gamma_{12} & \gamma_{23} & \gamma_{13}\end{bmatrix}^T, \\ \{\sigma\} &= \begin{bmatrix}\sigma_1 & \sigma_2 & \sigma_3 & \tau_{12} & \tau_{23} & \tau_{13}\end{bmatrix}^T, \\ \{b\} &= [b_1, b_2, b_3]^T \text{ and } \{t\} = [t_1, t_2, t_3]^T. \end{aligned} \tag{2.20}$$

In Eq. (2.19), The superscript s over u denotes that surface displacements are considered.

Now, the principle can be used effectively as a mechanism to generate finite element equations that govern the response of a structure or continuum.

Eq. (2.19) is rewritten as a sum of integration over the volume and areas of all finite elements, *i.e.*

$$\sum_e \int_{V^e} \{\delta\varepsilon\}^T [C]\{\varepsilon\}\, dV = \sum_e \int_{V^e} \{\delta u\}^T \{b\}\, dV + \sum_e \int_{S_F^e} \left\{\delta u^s\right\}^T \{t\}\, dS\,, \tag{2.21}$$

where e = 1,2, …, E and E = number of elements. In Eq. (2.21), constitutive relation $\{\sigma\} = [C]\{\varepsilon\}$ is used, where $[C]$ is the material property matrix relating the strains to the stresses in global coordinate system (See Appendix 2).

The displacements within each element *e* is related to its nodal displacements $\left\{U^e\right\}$ by its shape function matrix $\left[N\right]$ as given in Eq. (2.15). If we designate global displacements (set of all nodal displacements) as $\{U\}$, $\left\{U^e\right\}$ and $\{U\}$ can be related by an element connectivity matrix $\left[L^e\right]$ as

$$\left\{U^e\right\} = \left[L^e\right]\{U\}. \tag{2.22}$$

$\left[L^e\right]$ is of a size nedof × ngdof, nedof and ngdof being number of element degrees of freedom and number of global (total) degrees of freedom respectively. The entries in the connectivity matrix depend on the position of a particular element within the finite element mesh. In general, the connectivity matrix $\left[L^e\right]$ is just a Boolean matrix, with each component equal to 0 or 1. The latter is reserved for all the components that correspond to a particular node to which the element in question is attached.

The corresponding element strains can now be evaluated as

$$\{\varepsilon\} = \begin{Bmatrix} \dfrac{\partial u_1}{\partial x_1} \\[2mm] \dfrac{\partial u_2}{\partial x_2} \\[2mm] \dfrac{\partial u_3}{\partial x_3} \\[2mm] \dfrac{\partial u_1}{\partial x_2} + \dfrac{\partial u_2}{\partial x_1} \\[2mm] \dfrac{\partial u_2}{\partial x_3} + \dfrac{\partial u_3}{\partial x_2} \\[2mm] \dfrac{\partial u_3}{\partial x_1} + \dfrac{\partial u_1}{\partial x_3} \end{Bmatrix} = \begin{bmatrix} \dfrac{\partial}{\partial x_1} & 0 & 0 \\[2mm] 0 & \dfrac{\partial}{\partial x_2} & 0 \\[2mm] 0 & 0 & \dfrac{\partial}{\partial x_3} \\[2mm] \dfrac{\partial}{\partial x_2} & \dfrac{\partial}{\partial x_1} & 0 \\[2mm] 0 & \dfrac{\partial}{\partial x_3} & \dfrac{\partial}{\partial x_2} \\[2mm] \dfrac{\partial}{\partial x_3} & 0 & \dfrac{\partial}{\partial x_1} \end{bmatrix} \{u\} \tag{2.23}$$

Denoting the differential operator matrix as [Δ] and using the relation (Eq. (2.15)), we get

$$\{\varepsilon\} = [B]\{U^e\}, \tag{2.24}$$

where $[B]=[\Delta][N]$ is the strain displacement matrix. The rows of $[B]$ are obtained by appropriately differentiating and combining rows of the matrix $[N]$.

$[B]$ matrix can be written in a general form as

$$[B]=[[B_1] \dots [B_n]], \tag{2.25}$$

where $[B_r]$, r = 1 to n are strain-displacement submatrices of size 6 × 3 and are defined as

$$[B_r] = \begin{bmatrix} \frac{\partial N_r}{\partial x_1} & 0 & 0 & \frac{\partial N_r}{\partial x_2} & 0 & \frac{\partial N_r}{\partial x_3} \\ 0 & \frac{\partial N_r}{\partial x_2} & 0 & \frac{\partial N_r}{\partial x_1} & \frac{\partial N_r}{\partial x_3} & 0 \\ 0 & 0 & \frac{\partial N_r}{\partial x_3} & 0 & \frac{\partial N_r}{\partial x_2} & \frac{\partial N_r}{\partial x_1} \end{bmatrix}^T.$$

Substituting into Eq. (2.21) values for the element displacements and strains from Eqs. (2.15) and (2.24) respectively, we obtain

$$\begin{aligned} &\sum_e \{\delta U^e\}^T \left[\left(\int_{V^e} [B]^T [C][B]\, dV\right)\{U^e\}\right] = \\ &\sum_e \{\delta U^e\}^T \left[\left(\int_{V^e} [N]^T \{b\}\, dV\right) + \left(\int_{S_F^e} [N^s]^T \{t\}\, dS\right)\right]. \end{aligned} \tag{2.26}$$

In this equation, the surface displacement interpolation matrices $[N^S]$ are obtained from the volume displacement interpolation matrices $[N]$ by substituting element surface coordinates.

Now the terms in Eq. (2.26) can be defined as follows:

Element stiffness matrix

$$[K^e] = \int_{V^e} [B]^T [C][B]\, dV, \tag{2.27}$$

Element body force vector

$$\left\{F_b^e\right\} = \int_{V^e} [N]^T \{b\}\, dV \tag{2.28}$$

and Element surface force vector

$$\left\{F_s^e\right\} = \int_{S_F^e} \left[N^s\right]^T \{t\}\, dS. \tag{2.29}$$

Substituting Eq. (2.22) and (2.27-2.29) into Eq. (2.26) and rearranging, we get

$$\{\delta U\}^T \left[\left(\sum_e \left[L^e\right]^T \left[K^e\right]\left[L^e\right]\right)\{U\}\right] = \{\delta U\}^T \left[\sum_e \left[L^e\right]^T \left\{F_b^e\right\} + \sum_e \left[L^e\right]^T \left\{F_s^e\right\}\right]. \tag{2.30}$$

Since $\{\delta U\}$ is arbitrary, Eq. (2.30) can be written as

$$[K]\{U\} = \{F_b\} + \{F_s\}, \tag{2.31}$$

where global stiffness matrix is

$$[K] = \sum_e \left[L^e\right]^T \left[K^e\right]\left[L^e\right], \tag{2.32}$$

global body force vector is

$$\{F_b\} = \sum_e \left[L^e\right]^T \left\{F_b^e\right\} \tag{2.33}$$

and global surface force vector is

$$\{F_s\} = \sum_e \left[L^e\right]^T \left\{F_s^e\right\}. \tag{2.34}$$

Representing the sum of all load vectors in Eq. (2.31) as $\{F\}$, we have the following equations of equilibrium:

$$[K]\{U\} = \{F\}. \tag{2.35}$$

2.4.3. Weighted residual approach

In this method, the element matrices and vectors are derived directly from the governing differential equations of the problem without reliance on the variational statement of the problem. This method offers the most general procedure for deriving finite element equations and can be applied to almost all practical problems of science and engineering. Using this approach, it is possible to extend the use of FEM to solve problems where no functional is available.

Consider the differential equation

$$A(u) - f = 0, \tag{2.36}$$

where u is an unknown field variable and A is an operator (linear or non-linear), often a differential operator acting on u. f is a known function.

Assume that u is approximated by

$$u = \tilde{u} = \sum_{I=1}^{n} \phi_I u_I, \tag{2.37}$$

where ϕ_I is assumed approximating functions and u_I are the unknown parameters at n number of points.

If we put the above approximation in Eq. (2.36), we get

$$A(\tilde{u}) - f = R, \tag{2.38}$$

where R is the residual.

The method of weighted residuals seeks to determine the n unknown in such a way that the error R over the entire solution domain is small. This is achieved by forming a weighted average of the error and specifying that this weighted average vanishes over the solution domain, namely

$$\int_V R w_I dV = 0, \quad I = 1, 2, \cdots, n. \tag{2.39}$$

Here w_I are weighting functions and must be a linearly independent to make the set of equations in Eq. (2.39) linearly independent and hence solvable. The extension of this derivation for vector field variable is straightforward. In this case, each component of the vector field is assumed to be approximated by Eq. (2.37).

Within the weighted residual approach, different procedures, such as Galerkin method and least squares method, can be used in deriving the element equations. The Galerkin method is obtained easily from the method of weighted residuals by selecting the weighting functions same as basis functions. In least squares method, the integral of the weighted square of the residual over the domain is required to be a minimum. Galerkin method is relatively an easily adapted residual method and is discussed below for the derivation of element equations for mechanical problem.

The governing differential equation of stress equilibrium in a general three-dimensional form is given as (See Appendix 2)

$$\frac{\partial \sigma_{ij}}{\partial x_j} + b_i = 0 \cdot \tag{2.40}$$

Writing Galerkin theorem for this differential equation, we have

$$\int_V \left(\frac{\partial \tilde{\sigma}_{ij}}{\partial x_j} + b_i \right) \phi_I dV = 0, \qquad I = 1, 2, \cdots, n \,. \tag{2.41}$$

In the above equation, the summation is over index i and j.

Now, using integration by parts we have

$$\int_V \frac{\partial}{\partial x_j} \left(\tilde{\sigma}_{ij} \phi_I \right) dV - \int_V \tilde{\sigma}_{ij} \frac{\partial \phi_I}{\partial x_j} dV + \int_V b_i \phi_I dV = 0 \,. \tag{2.42}$$

Using divergence theorem (see Appendix 1) in the first term of Eq. (2.42) and rearranging, we have

$$\int_V \tilde{\sigma}_{ij} \frac{\partial \phi_I}{\partial x_j} dV - \int_V b_i \phi_I dV - \int_S \tilde{\sigma}_{ij} n_j \phi_I dS = 0 \,. \tag{2.43}$$

Using the boundary condition in terms of stress vector, i.e.

$$t_i = \sigma_{ij} n_j, \tag{2.44}$$

Eq. (2.43) can be rewritten as

$$\int_V \tilde{\sigma}_{ij} \frac{\partial \phi_I}{\partial x_j} dV - \int_V b_i \phi_I dV - \int_S t_i \phi_I dS = 0 \, . \tag{2.45}$$

If we use the one-dimensional vector notation for stresses, the foregoing equation can be written in matrix form as

$$\int_V ([\Delta][\phi])^T \{\tilde{\sigma}\} \, dV - \int_V [\phi]^T \{b\} \, dV - \int_S [\phi]^T \{t\} \, dS = 0 \, , \tag{2.46}$$

where $[\Delta]$ is as defined earlier. $[\phi]$ matrix is given as

$$[\phi] = \left[[\phi_1] \cdots [\phi_n] \right], \tag{2.47}$$

in which a typical sub-matrix is given by $[\phi_r] = \begin{bmatrix} \phi_r & 0 & 0 \\ 0 & \phi_r & 0 \\ 0 & 0 & \phi_r \end{bmatrix}$.

In Eq. (2.47), n is the number of points in the approximating function.

In matrix notation, the displacement vector can be written as

$$\{u\} = [\phi]\{U\} \, , \tag{2.48}$$

where $\{U\} = \left[u_{11}, u_{21}, u_{31}, \ \cdots, \ u_{1n}, u_{2n}, u_{3n} \right]^T$.

In this equation, u_{iI} is the ith displacement component of point I.

Now if we divide the domain into several sub-domains (elements), the displacement vector $\{u\}$ can be approximate within each element as given in Eq. (2.15).

From the equivalence of these two equations within the element, we find the relation

$$[\phi] \equiv [N]\left[L^e \right], \tag{2.49}$$

where $[L^e]$ is as defined earlier in Eq. (2.22).

Writing Eq. (2.46) as an assemblage of contributions from all the elements and using the relation above, we obtain

$$\sum_e \left[L^e\right]^T \int_{V^e} ([\Delta][N])^T \{\tilde{\sigma}\}\, dV = \sum_e \left[L^e\right]^T \left(\int_{V^e} [N]^T \{b\}\, dV + \int_{S^e} [N]^T \{t\}\, dS\right). \tag{2.50}$$

Using the relation in Eq. (2.24) and substituting the value $\{\tilde{\sigma}\}$ defined within the element as

$$\{\tilde{\sigma}\} = [C]\{\varepsilon\} = [C][B]\{U^e\} = [C][B]\left[L^e\right]\{U\},\quad \text{we get}$$

$$\left(\sum_e \left[L^e\right]^T \left(\int_{V^e} [B]^T [C][B]\, dV\right)\left[L^e\right]\right)\{U\} = \sum_e \left[L^e\right]^T \left(\int_{V^e} [N]^T \{b\}\, dV + \int_{S^e} [N]^T \{t\}\, dS\right). \tag{2.51}$$

It may be noted that Eq. (2.51) is identical to the Eq. (2.31) that was derived using variational principle.

2.4.4. Energy balance approach

The energy balance approach relies on the balance of the thermal and/or mechanical energy of a system. This approach like the weighted residual approach, requires no variational statements and hence broaden considerably range of possible application of finite element method.

2.4.5 Evaluation of element matrices and vectors

In the elasticity problems, element stiffness matrix $[K^e]$ given by Eq. (2.27) along with other matrices and vectors depend on shape function matrix $[N]$ and its derivatives with respect to global coordinates. This situation also arises in several other classes of problems. To evaluate such matrices, the two transformations are generally necessary. In the first place as N_i is defined in terms of local (natural) coordinates particularly for elements of arbitrary shape (isoparametric elements), it is necessary to devise some means of expressing the derivatives in global (x_1, x_2, x_3) coordinates in terms of derivatives in local (ξ, η, ζ) coordinates. In the second place the element of volume (or surface) over which the integration has to be carried out needs to be expressed in terms of local coordinates with an appropriate change of integration limits.

A function $f = f(x_1, x_2, x_3)$ can be considered to be an implicit function of ξ, η and ζ as $f = f[x_1(\xi, \eta, \zeta), x_2(\xi, \eta, \zeta), x_3(\xi, \eta, \zeta)]$. Using the chain rule of differentiation, we get

$$\begin{aligned}
\frac{\partial f}{\partial \xi} &= \frac{\partial f}{\partial x_1}\frac{\partial x_1}{\partial \xi} + \frac{\partial f}{\partial x_2}\frac{\partial x_2}{\partial \xi} + \frac{\partial f}{\partial x_3}\frac{\partial x_3}{\partial \xi} \\
\frac{\partial f}{\partial \eta} &= \frac{\partial f}{\partial x_1}\frac{\partial x_1}{\partial \eta} + \frac{\partial f}{\partial x_2}\frac{\partial x_2}{\partial \eta} + \frac{\partial f}{\partial x_3}\frac{\partial x_3}{\partial \eta} \\
\frac{\partial f}{\partial \zeta} &= \frac{\partial f}{\partial x_1}\frac{\partial x_1}{\partial \zeta} + \frac{\partial f}{\partial x_2}\frac{\partial x_2}{\partial \zeta} + \frac{\partial f}{\partial x_3}\frac{\partial x_3}{\partial \zeta}
\end{aligned} \tag{2.52}$$

or

$$\begin{Bmatrix} \dfrac{\partial f}{\partial \xi} \\ \dfrac{\partial f}{\partial \eta} \\ \dfrac{\partial f}{\partial \zeta} \end{Bmatrix} = [J] \begin{Bmatrix} \dfrac{\partial f}{\partial x_1} \\ \dfrac{\partial f}{\partial x_2} \\ \dfrac{\partial f}{\partial x_3} \end{Bmatrix}, \tag{2.53}$$

where $[J]$ is the Jacobian matrix of size 3 × 3 given as

$$[J] = \begin{bmatrix} \dfrac{\partial x_1}{\partial \xi} & \dfrac{\partial x_2}{\partial \xi} & \dfrac{\partial x_3}{\partial \xi} \\ \dfrac{\partial x_1}{\partial \eta} & \dfrac{\partial x_2}{\partial \eta} & \dfrac{\partial x_3}{\partial \eta} \\ \dfrac{\partial x_1}{\partial \zeta} & \dfrac{\partial x_2}{\partial \zeta} & \dfrac{\partial x_3}{\partial \zeta} \end{bmatrix} = \begin{bmatrix} J_{11} & J_{12} & J_{13} \\ J_{21} & J_{22} & J_{23} \\ J_{31} & J_{32} & J_{33} \end{bmatrix}. \tag{2.54}$$

To find the global derivatives, Eq. (2.53) can be inverted as

$$\begin{Bmatrix} \dfrac{\partial f}{\partial x_1} \\ \dfrac{\partial f}{\partial x_2} \\ \dfrac{\partial f}{\partial x_3} \end{Bmatrix} = [J]^{-1} \begin{Bmatrix} \dfrac{\partial f}{\partial \xi} \\ \dfrac{\partial f}{\partial \eta} \\ \dfrac{\partial f}{\partial \zeta} \end{Bmatrix}, \tag{2.55}$$

in which $[J]^{-1} = \dfrac{1}{\det[J]} \bullet \text{adj. of} [J] = \begin{bmatrix} J_{11}^* & J_{12}^* & J_{13}^* \\ J_{21}^* & J_{22}^* & J_{23}^* \\ J_{31}^* & J_{32}^* & J_{33}^* \end{bmatrix}$.

To transform the region (volume or surface) with respect to which the integration is made, a standard process is used which involves the determinant of the Jacobian matrix. This process, found in many text books, is described in Appendix 1.

Assuming that the inverse of [J] can be found, the element stiffness matrix $[K^e]$ is now evaluated by finding the following integral

$$\left[K^e\right] = \int_{-1}^{+1}\int_{-1}^{+1}\int_{-1}^{+1} [B]^T [C][B]\det[J]\, d\xi d\eta d\zeta, \tag{2.56}$$

in which Eq. (A1.21) is used.

In most of the cases, the above integral is calculated numerically using Gaussian quadrature (See Appendix - 5).

Evaluation of other element matrices are performed using a similar approach.

Nodal load vectors

In the finite element method all loading must be applied to the structure as equivalent nodal loads. Accordingly forms of loading such as gravity, pressure assigned to element surfaces must be reduced to equivalent nodal forces before solution can be processed. As can be seen in Eqs. (2.28) and (2.29), the calculation of equivalent nodal forces is not straightforward since area or volume integrations over arbitrarily shaped regions are generally involved. It is, therefore, essential in many cases to resort to the numerical integration technique. The integrals in these equations are first expressed in terms of local (natural) coordinates using the procedure similar to that used for the element stiffness matrix (Eq. 2.56) and are then evaluated using Gaussian quadrature.

2.5. ASSEMBLY OF ELEMENT MATRICES

In finite element analysis, the solid or structure is considered as an assemblage of 'finite elements' connected at the nodes. After computing the characteristics (stiffness) matrix and nodal load vector of the elements, the overall equation of equilibrium is formed by assembling these matrices and vectors into the overall structural stiffness matrix and load vector respectively. The connectivity information relating the degrees of freedom of an element and the corresponding global degrees of freedom is used to compute the contribution from that element to the global stiffness matrix and global load vector. The congruent transformation technique or direct stiffness method can be used for constructing the equation of equilibrium of the structure (Gallagher, 1975). However, the direct stiffness method is simple, elegant and is widely used in computer programs. From the computational point of view, this method is a process of transfer of the coefficients of element stiffness matrix and the nodal load vector into relevant positions in the global stiffness matrix and load vector by identifying the element degrees of freedom and the corresponding global degrees of freedom.

In order to introduce the element assembly process using direct stiffness method, a simple example of a rectangular plane divided into two elements is considered (Figure 2.10). Each element has two nodal degrees of freedom, *i.e.* u_x and u_y at each of the four nodes. The first element is shown in Figure 2.11 with local (element) node numbers and global node numbers. The element connectivity for the two elements is presented in Table 2.1 which specifies the global node numbers corresponding to the local node numbers for the elements.

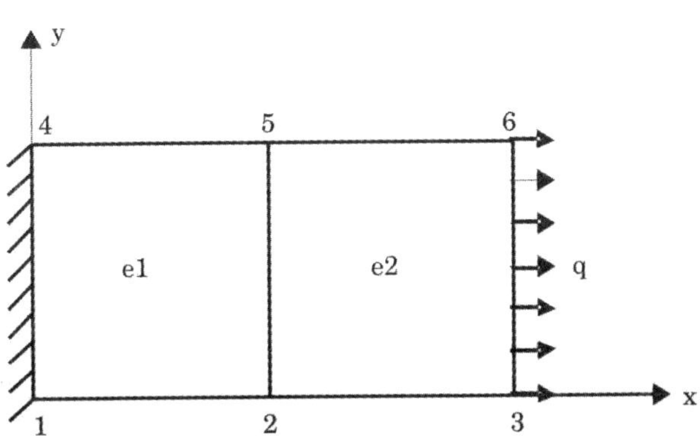

Fig. 2.10. Assemblage of rectangular elements.

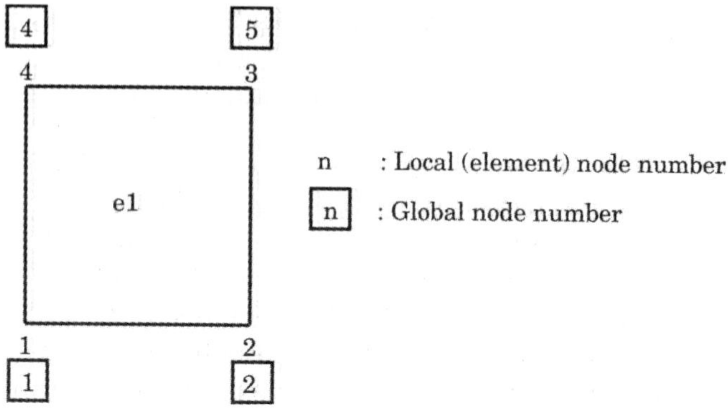

Fig. 2.11. Element number 1.

Table 2.1. Element connectivity table.

Element No.	*Node 1*	*Node 2*	*Node 3*	*Node 4*
1	1	2	5	4
2	2	3	6	5

The vector of unknown nodal degrees of freedom for the elements can be written as

$$\{U^{e1}\} = \begin{bmatrix} u_{x1} & u_{y1} & u_{x2} & u_{y2} & u_{x5} & u_{y5} & u_{x4} & u_{y4} \end{bmatrix}^T$$

and

$$\{U^{e2}\} = \begin{bmatrix} u_{x2} & u_{y2} & u_{x3} & u_{y3} & u_{x6} & u_{y6} & u_{x5} & u_{y5} \end{bmatrix}^T, \tag{2.57}$$

where $(u_{x\alpha}, u_{y\alpha})$ is the displacement at the α th node of the structure.

Further, the vector of global degrees of freedom is expressed as

$$\{U\} = \begin{bmatrix} u_{x1} & u_{y1} & u_{x2} & u_{y2} & u_{x3} & u_{y3} & u_{x4} & u_{y4} & u_{x5} & u_{y5} & u_{x6} & u_{y6} \end{bmatrix}^T. \tag{2.58}$$

We assume that the element stiffness matrices and element load vectors corresponding to element degrees of freedom may be written as follows:

$$\left[K^{e1}\right] = \begin{bmatrix} a_{11} & a_{12} & a_{13} & a_{14} & a_{15} & a_{16} & a_{17} & a_{18} \\ a_{21} & a_{22} & a_{23} & a_{24} & a_{25} & a_{26} & a_{27} & a_{28} \\ a_{31} & a_{32} & a_{33} & a_{34} & a_{35} & a_{36} & a_{37} & a_{38} \\ a_{41} & a_{42} & a_{43} & a_{44} & a_{45} & a_{46} & a_{47} & a_{48} \\ a_{51} & a_{52} & a_{53} & a_{54} & a_{55} & a_{56} & a_{57} & a_{58} \\ a_{61} & a_{62} & a_{63} & a_{64} & a_{65} & a_{66} & a_{67} & a_{68} \\ a_{71} & a_{72} & a_{73} & a_{74} & a_{75} & a_{76} & a_{77} & a_{78} \\ a_{81} & a_{82} & a_{83} & a_{84} & a_{85} & a_{86} & a_{87} & a_{88} \end{bmatrix},$$

$$\left[K^{e2}\right] = \begin{bmatrix} b_{11} & b_{12} & b_{13} & b_{14} & b_{15} & b_{16} & b_{17} & b_{18} \\ b_{21} & b_{22} & b_{23} & b_{24} & b_{25} & b_{26} & b_{27} & b_{28} \\ b_{31} & b_{32} & b_{33} & b_{34} & b_{35} & b_{36} & b_{37} & b_{38} \\ b_{41} & b_{42} & b_{43} & b_{44} & b_{45} & b_{46} & b_{47} & b_{48} \\ b_{51} & b_{52} & b_{53} & b_{54} & b_{55} & b_{56} & b_{57} & b_{58} \\ b_{61} & b_{62} & b_{63} & b_{64} & b_{65} & b_{66} & b_{67} & b_{68} \\ b_{71} & b_{72} & b_{73} & b_{74} & b_{75} & b_{76} & b_{77} & b_{78} \\ b_{81} & b_{82} & b_{83} & b_{84} & b_{85} & b_{86} & b_{87} & b_{88} \end{bmatrix},$$

$$\left\{F^{e1}\right\} = [A_1 \quad A_2 \quad A_3 \quad A_4 \quad A_5 \quad A_6 \quad A_7 \quad A_8]^T$$

and

$$\left\{F^{e2}\right\} = [B_1 \quad B_2 \quad B_3 \quad B_4 \quad B_5 \quad B_6 \quad B_7 \quad B_8]^T . \tag{2.59}$$

Having noted the contribution from all the elements, the global stiffness matrix and the equation of equilibrium can now be constructed. This is done by adding (superimposing) the individual element stiffnesses and load vectors directly to the overall stiffness matrix$[K]$ and the load vector $\{F\}$ in positions appropriate to the nodal degrees of freedom of the elements. From Eqs. (2.57) and (2.58), the positions of element degrees of freedom (e.d.o.f.) in the vector of global degrees of freedom (g.d.o.f.) are given in Table 2.2 for both the elements. A term in the position (i,j) of the element matrix will have a global position (I,J) in the overall matrix, where I and J are the global positions of ith and jth element degrees of freedom respectively.

Table 2.2. Global positions of element degrees of freedom.

Element No.	*e.d. o.f.1*	*e.d. o.f.2*	*e.d. o.f.3*	*e.d. o.f.4*	*e.d. o.f.5*	*e.d. o.f.6*	*e.d. o.f.7*	*e.d. o.f.8*
1	1	2	3	4	9	10	7	8
2	3	4	5	6	11	12	9	10

Thus, the resulting matrix and load vector can be determined as

$$[K]=\begin{bmatrix} a_{11} & a_{12} & a_{13} & a_{14} & 0 & 0 & a_{17} & a_{18} & a_{15} & a_{16} & 0 & 0 \\ a_{21} & a_{22} & a_{23} & a_{24} & 0 & 0 & a_{27} & a_{28} & a_{25} & a_{26} & 0 & 0 \\ a_{31} & a_{32} & a_{33}+b_{11} & a_{34}+b_{12} & b_{13} & b_{14} & a_{37} & a_{38} & a_{35}+b_{17} & a_{36}+b_{18} & b_{15} & b_{16} \\ a_{41} & a_{42} & a_{43}+b_{21} & a_{44}+b_{22} & b_{23} & b_{24} & a_{47} & a_{48} & a_{45}+b_{27} & a_{46}+b_{28} & b_{25} & b_{26} \\ 0 & 0 & b_{31} & b_{32} & b_{33} & b_{34} & 0 & 0 & b_{37} & b_{38} & b_{35} & b_{36} \\ 0 & 0 & b_{41} & b_{42} & b_{43} & b_{44} & 0 & 0 & b_{47} & b_{48} & b_{45} & b_{46} \\ a_{71} & a_{72} & a_{73} & a_{74} & 0 & 0 & a_{77} & a_{78} & a_{75} & a_{76} & 0 & 0 \\ a_{81} & a_{82} & a_{83} & a_{84} & 0 & 0 & a_{87} & a_{88} & a_{85} & a_{86} & 0 & 0 \\ a_{51} & a_{52} & a_{53}+b_{71} & a_{54}+b_{72} & b_{73} & b_{74} & a_{57} & a_{58} & a_{55}+b_{77} & a_{56}+b_{78} & b_{75} & b_{76} \\ a_{61} & a_{62} & a_{63}+b_{81} & a_{64}+b_{82} & b_{83} & b_{84} & a_{67} & a_{68} & a_{65}+b_{87} & a_{66}+b_{88} & b_{85} & b_{86} \\ 0 & 0 & b_{51} & b_{52} & b_{53} & b_{54} & 0 & 0 & b_{57} & b_{58} & b_{55} & b_{56} \\ 0 & 0 & b_{61} & b_{62} & b_{63} & b_{64} & 0 & 0 & b_{67} & b_{68} & b_{65} & b_{66} \end{bmatrix}$$

and

$$\{F\}=\begin{bmatrix} A_1 & A_2 & A_3+B_1 & A_4+B_2 & B_3 & B_4 & A_7 & A_8 & A_5+B_7 & A_6+B_8 & B_5 & B_6 \end{bmatrix}^T . \tag{2.60}$$

After assembling the element stiffness matrices and element nodal load vectors, the finite element system of equation can be written as

$$[K]\{U\}=\{F\}, \tag{2.61}$$

which is to be solved for $\{U\}$.

2.5.1 Equation numbering and storage

The global stiffness matrix may consist of both non-zero and zero coefficients. Since the structure nodes can be numbered in arbitrary sequence, different topology of global stiffness matrix can result with non-zero and zero coefficients placed at different locations in the matrix. Therefore, a topology is sought that favors compact storage and rapid solution of the structural equations. We can choose any of several solution algorithms, influenced by matrix topology, the number of equations, and whether they are ill-conditioned or not. Further, in most of the solution cases, the structural stiffness matrix is symmetric that must be exploited in both storage and solution.

One of the efficient schemes for reducing the main storage requirement is the skyline storage scheme. The skyline encloses the uppermost nonzero coefficients in each column of the matrix (Figure 2.12). Hence, all the coefficients above the skyline are zero. Because of symmetry, only the upper triangle of the matrix need be stored to retain all information in the matrix. Therefore, we can store all required coefficients in a one-dimensional array by taking information from the skyline down to and including the diagonal, from each column in order. All the zero coefficients, if any between the skyline and the diagonal are retained. This storage format is called active column storage and the number of terms stored in this format is sometimes termed as the profile of the matrix. Software must be told which terms in the one-dimensional array are diagonal terms of the global matrix of dimension, say $n \times n$. This information is stored in an auxiliary array of n numbers that contains the height of each column.

When the global stiffness matrix is unsymmetric, this storage scheme requires modification. In this case, we need to store all the data from the uppermost nonzero coefficient down to the lowermost nonzero coefficient, from each column of the matrix.

It is clear that skyline depends on the matrix topology and thus, on the node numbering scheme adopted while discretizing the structure. There are several investigations reported which deal with node renumbering algorithm that examines the node numbering provided by input data/generated during meshing, renumbers nodes for compact storage and efficient equation solving, and then converts back to the input numbering for presentation of results.

2.6. INCORPORATION OF BOUNDARY CONDITIONS

The system of equation given by Eq. (2.61) cannot be solved for $\{U\}$, since the matrix $[K]$ will be singular and its inverse does not exist. The physical significance of this in the case of solid or structural mechanics problems is that the loaded body or structure is free to undergo unlimited rigid body motion unless some support constraints are imposed to keep the body or structure in equilibrium under the loads. Hence, some

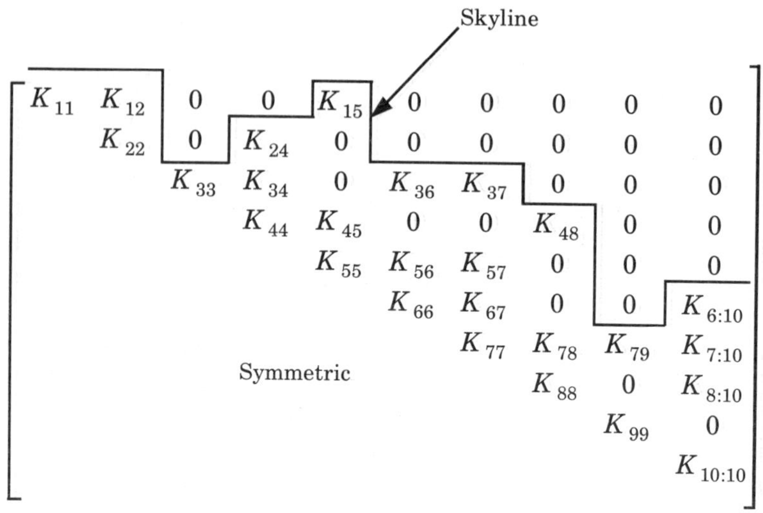

Fig. 2.12. Typical global stiffness matrix (symmetric) and its skyline.

kinematics boundary or support conditions have to be specified which are usually of the type

$$U_{d_1} = a_1, U_{d_2} = a_2, \cdots, U_{d_r} = a_r\,. \tag{2.62}$$

That is, the displacements along degrees of freedom $d_1, d_2, \cdots, d_r$ are prescribed to be equal to $a_1, a_2, \cdots, a_r$ respectively.

In addition, there are problems where multipoint constraints are to be imposed, which is of the type

$$lU_{d_1} + mU_{d_2} = n$$

or

$$lU_{d_1} + mU_{d_2} + nU_{d_3} = p\,, \tag{2.63}$$

where l, m, n and p are known constants.

2.6.1. Specified displacement boundary conditions

These boundary conditions can be incorporated by several methods. Two approaches (1) Condensation method and (2) Elimination method are discussed below. In addition, penalty approach can also be used for handling these boundary conditions.

Method 1 (Condensation method)

The stiffness matrix $[K]$ can be partitioned to correspond to unrestrained (free) displacements $\{U_F\}$ and to restrained displacements $\{U_R\}$. The equation of equilibrium can be written as

$$\begin{bmatrix} [K_{FF}] & [K_{FR}] \\ [K_{RF}] & [K_{RR}] \end{bmatrix} \begin{Bmatrix} \{U_F\} \\ \{U_R\} \end{Bmatrix} = \begin{Bmatrix} \{F_F\} \\ \{F_R\} \end{Bmatrix}. \tag{2.64}$$

The first set of equation can be written as

$$[K_{FF}]\{U_F\} + [K_{FR}]\{U_R\} = \{F_F\} \tag{2.65}$$

or

$$[K_{FF}]\{U_F\} = \{F_F\} - [K_{FR}]\{U_R\}. \tag{2.66}$$

Here $[K_{FF}]$ will not be singular if restrained displacements $\{U_R\}$ are sufficient in number and arrangement to prevent rigid body motion. Hence Eq. (2.66) can be solved to obtain unknown displacements $\{U_F\}$ as

$$\{U_F\} = [K_{FF}]^{-1} (\{F_F\} - [K_{FR}]\{U_R\}). \tag{2.67}$$

Once $\{U_F\}$ is known, the vector of unknown nodal reactions $\{F_R\}$ can be found using the second set of Eq. (2.64) as

$$[K_{RF}]\{U_F\} + [K_{RR}]\{U_R\} = \{F_R\}. \tag{2.68}$$

Alternative method (Elimination method)

The above theoretical procedure involves partitioning of the matrix and thus involves rearranging of stiffness matrix and also may affect the bandwidth/skyline of the matrix in actual computer implementation. A more practical way to impose boundary conditions must be adopted without destroying the symmetry of the matrix, changing its size or rearranging its terms. One of such techniques is the elimination method which is discussed here.

For simplicity, let the global equations be a 4 by 4 matrix as

$$\begin{bmatrix} K_{11} & K_{12} & K_{13} & K_{14} \\ K_{21} & K_{22} & K_{23} & K_{24} \\ K_{31} & K_{32} & K_{33} & K_{34} \\ K_{41} & K_{42} & K_{43} & K_{44} \end{bmatrix} \begin{Bmatrix} U_1 \\ U_2 \\ U_3 \\ U_4 \end{Bmatrix} = \begin{Bmatrix} F_1 \\ F_2 \\ F_3 \\ F_4 \end{Bmatrix}. \tag{2.69}$$

Assume the condition $U_2 = \Delta_2$ is to be imposed where Δ_2 is the prescribed degree of freedom.

Thus we have

$$\begin{bmatrix} K_{11} & 0 & K_{13} & K_{14} \\ K_{21} & 0 & K_{23} & K_{24} \\ K_{31} & 0 & K_{33} & K_{34} \\ K_{41} & 0 & K_{43} & K_{44} \end{bmatrix} \begin{Bmatrix} U_1 \\ U_2 = \Delta_2 \\ U_3 \\ U_4 \end{Bmatrix} = \begin{Bmatrix} F_1 - K_{12}\Delta_2 \\ F_2 - K_{22}\Delta_2 \\ F_3 - K_{32}\Delta_2 \\ F_4 - K_{42}\Delta_2 \end{Bmatrix}. \tag{2.70}$$

Since the matrix lacks symmetric, we replace the second equation with the trivial equation $U_2 = \Delta_2$ so that

$$\begin{bmatrix} K_{11} & 0 & K_{13} & K_{14} \\ 0 & 1 & 0 & 0 \\ K_{31} & 0 & K_{33} & K_{34} \\ K_{41} & 0 & K_{43} & K_{44} \end{bmatrix} \begin{Bmatrix} U_1 \\ U_2 \\ U_3 \\ U_4 \end{Bmatrix} = \begin{Bmatrix} F_1 - K_{12}\Delta_2 \\ \Delta_2 \\ F_3 - K_{32}\Delta_2 \\ F_4 - K_{42}\Delta_2 \end{Bmatrix}. \tag{2.71}$$

2.6.2. Multipoint constraints

Suppose, the constraint to be imposed in the global equation (Eq. (2.69)) is $lU_2 + mU_3 = n$, where l, m and n are known constants. The relation can also be written in the form as $U_2 + aU_3 = b$ or $U_2 = b - aU_3$.

Imposing this in the original matrix, we have

$$\begin{bmatrix} K_{11} & 0 & K_{13} - aK_{12} & K_{14} \\ K_{21} & 0 & K_{23} - aK_{22} & K_{24} \\ K_{31} & 0 & K_{33} - aK_{32} & K_{34} \\ K_{41} & 0 & K_{43} - aK_{42} & K_{44} \end{bmatrix} \begin{Bmatrix} U_1 \\ U_2 \\ U_3 \\ U_4 \end{Bmatrix} = \begin{Bmatrix} F_1 - bK_{12} \\ F_2 - bK_{22} \\ F_3 - bK_{32} \\ F_4 - bK_{42} \end{Bmatrix}. \tag{2.72}$$

The matrix lacks symmetry. Therefore, the following procedure is applied to make it symmetric.

First we modify third equation as

$$K'_{3j} = K'_{3j} - aK'_{2j} \text{ and } F'_3 = F'_3 - aF'_2,$$

where K'_{ij} and F'_i on the right-hand sides are the coefficients of the modified stiffness matrix and modified load vector as given in Eq. (2.72). Next, we replace the second equation with the trivial equation $U_2 + aU_3 = b$. The resulting system of equation, thus, becomes

$$\begin{bmatrix} K_{11} & 0 & K_{13} - aK_{12} & K_{14} \\ 0 & 1 & a & 0 \\ K_{31} - aK_{21} & 0 & \begin{matrix}(K_{33} - aK_{32} - \\ a(K_{23} - aK_{22}))\end{matrix} & K_{34} - aK_{24} \\ K_{41} & 0 & K_{43} - aK_{42} & K_{44} \end{bmatrix} \begin{Bmatrix} U_1 \\ U_2 \\ U_3 \\ U_4 \end{Bmatrix} = \begin{Bmatrix} F_1 - bK_{12} \\ b \\ \begin{matrix}(F_3 - bK_{32} - \\ a(F_2 - bK_{22}))\end{matrix} \\ F_4 - bK_{42} \end{Bmatrix}. \tag{2.73}$$

The matrix is still not symmetric. So, we add a trivial equation $aU_2 + a^2U_3 = ab$ in the third equation to give

$$\begin{bmatrix} K_{11} & 0 & K_{13} - aK_{12} & K_{14} \\ 0 & 1 & a & 0 \\ K_{31} - aK_{21} & a & \begin{matrix}(K_{33} - aK_{32} - \\ a(K_{23} - aK_{22}) + a^2)\end{matrix} & K_{34} - aK_{24} \\ K_{41} & 0 & K_{43} - aK_{42} & K_{44} \end{bmatrix} \begin{Bmatrix} U_1 \\ U_2 \\ U_3 \\ U_4 \end{Bmatrix} = \begin{Bmatrix} F_1 - bK_{12} \\ b \\ \begin{matrix}(F_3 - bK_{32} - \\ a(F_2 - bK_{22}) + ab)\end{matrix} \\ F_4 - bK_{42} \end{Bmatrix}. \tag{2.74}$$

The matrix is now symmetric.

In general, if we have to impose the condition: $U_j + aU_k + bU_l = c$, we have the following trivial equations which are to be added to the modified set of equations (obtained after modifying the corresponding rows and columns as discussed above) to make it symmetric:

$$\begin{bmatrix} 1 & a & b \\ a & a^2 & ab \\ b & ab & b^2 \end{bmatrix} \begin{Bmatrix} U_j \\ U_k \\ U_l \end{Bmatrix} = \begin{Bmatrix} c \\ ac \\ bc \end{Bmatrix}. \tag{2.75}$$

It is to be noted that if we know in advance that the global matrix is going to be unsymmetric, some or all of the steps used above to make the matrix symmetric can be avoided. The above procedure to impose the multipoint constraints can be implemented either at element level or at global level.

2.7. SOLUTION OF FINITE ELEMENT EQUATIONS

There are several standard numerical techniques which can be used to solve the finite element system of equations. The method adopted for the solution of equations is a major factor influencing the efficiency of a finite element program. Equation solvers can be classified as either direct or iterative. Direct methods include Gaussian elimination, LU decomposition and Choleski factorization that are closely related. In a direct solution such as Gaussian elimination, the number of operations required depends on matrix topology and the number of equations and can be calculated in advance. Finite element implementation of Gaussian elimination and LU decomposition methods is well described in several text books, for example, Krishnamoorthy (1994) and is not discussed here for the sake of brevity.

When using a direct solver, we need not assemble the entire structural matrix before starting to solve equations and steps of assembly can alternate with steps of solution. Such a solution procedure was proposed by Iron (1970) and is referred to as frontal solution method. In this method the coefficients of an equation are completely assembled from the contribution of all relevant elements and equations are reduced by Gaussian elimination at the same time. The finite elements that must be considered for processing equations corresponding to one specific node define the degrees of freedom currently active. This active set of degrees of freedom is referred to as a wavefront. The order in which equations are processed depend on element numbering and not node numbering. Hence, in the frontal solution method the numbering of elements is important as it determines the size of the wavefront. The frontal equation solver is well explained in detail in Hinton and Owen (1977).

In an iterative solution, the number of operations required is uncertain. Iterative cycle continues until a convergence norm is satisfied. Some of the iterative methods include Jacobi method, Gauss-Seidel method, Successive over-relaxation method and Conjugate gradient method.

2.8. COMPUTATION OF RESULTANT VARIABLES

From the known displacements $\{U\}$, several element resultants such as strains, stresses, principal stress and strain energy can be computed by using the necessary equations. In a general sense, the strains are calculated at the gauss points from strain-displacement matrix $[B]$. These strains are extrapolated to the nodes of the element using a suitable algorithm and a global smoothening of strain is performed for all nodes of the finite element model by averaging of contributions from the neighboring finite elements. Finally the stresses in each element are obtained using the smoothened element strains and stress-strain relation for the material belonging to that element. For homogeneous material domain, it is irrelevant whether strain smoothening is done first and then stresses are calculated at element nodal points, or stress is first calculated at Gauss points and stress smoothening is performed subsequently.

2.8.1 Stress smoothening

Although finite element methods based on the displacement approach give continuous displacements over the domain, the displacement derivatives, in many instances do not possess inter-element continuity, especially with lower-order elements. Therefore, the stresses near the boundaries of the domain and the inter-element joints are discontinuous, unlike the true stresses, which are continuous over the domain. This discontinuous behavior of stresses causes inaccuracies in calculating the nodal stresses, especially where there are high stress gradients. Therefore, to obtain a continuous stress distribution over the domain, it is necessary to use a stress smoothing technique with the finite element analysis.

Stress smoothing methods can be divided into two main categories, namely: global and local techniques. Local stress smoothing techniques are commonly used because they are computationally less intensive. Although global stress smoothing techniques are not complicated, they call for high computational effort because they need to solve a system of equations equal in number to the number of nodes in the finite element mesh. Thus the practical application of global techniques is limited in comparison with local procedures. There are several global and local stress smoothing methods available in the literature. A comparative study of the most important local stress smoothing methods has been described by Liyanapathirana *et. al.*, (2000).

In the displacement approach, polynomials are used to represent the displaced shape of the finite element model. These polynomials contain complete polynomial terms up to order n and some selected terms of higher order. If these polynomials are used to approximate a displacement field of order n, then an exact representation can be obtained for displacements and stresses. However, if a displacement field of order n+1 is approximated using this polynomial, only an approximation is obtained, as some higher order terms in the displacement field are not present in the polynomial. According to Barlow (1976), this results in erroneous stress distributions, which are discontinuous at inter-element and domain boundaries. However, some points in these discontinuous stress distributions are identified as giving stresses close to the true stress distribution. These points are known as superconvergent points. Barlow (1976) presented a rational approach to locate these points and found that they are the Gaussian points corresponding to the element displacement interpolation order. These points are usually termed as the optimal stress points because the order of accuracy of the stress on these points are higher than that of the stress solution itself. Barlow (1976) showed that for one-dimensional cubic beam elements, eight-node isoparametric plane elements and 20-node isoparametric solid elements, accurate stresses can be obtained at Gauss integration points if 2 × 2 integration is performed. For linear two-node elements, four-node quadrilateral elements and eight-node solid elements, accurate stresses can be obtained at the origin (centroid) and, for three-node flexural triangles, stresses are accurate at mid-sides.

Noticing the presence of the optimal stress points, many attempts have been made to recover or obtain a smooth stress solution which is more accurate than the stress derived directly from the displacement solution. Hinton and Campbell (1974) developed a local smoothing technique using a least squares method. In this technique, an error function is defined as the difference between the consistent (smoothed) stress field and stress field calculated by the finite element method and the value of this function is minimized by a least squares approach at an element level. This leads to a continuous stress field inside each element characterized by the set of nodal stresses. The authors (Hinton and Campbell, 1974) systematically chose the stress smoothing functions of degree one time lower than that of the displacement shape functions. Such choice has been found to be much reliable for elements of degree two, although it is not convenient for elements of degree one.

Specifically, Hinton and Campbell (1974) has shown for an 8-noded parallelogram element that if the distribution of each stress component within the element is taken as a bilinear surface given as

$$g\,(\xi,\eta) = a + b\xi + c\eta + d\xi\eta, \tag{2.76}$$

where a, b, c and d are a set of unknown coefficients; an error function in terms of $g\,(\xi,\eta)$ and $\sigma\,(\xi,\eta)$ can be defined as

$$\chi = \sum_{k=1}^{np}\left[\sigma(\xi_k,\eta_k) - g\,(\xi_k,\eta_k)\right]^2. \tag{2.77}$$

In the above equation, np is the number of Gauss integration points within the element and $\sigma(\xi_k,\eta_k)$ are the stresses calculated by the finite element method at the 2 x 2 Gauss integration points, which are also the superconvergent points for the stresses. Since stresses at the Gauss integration points are accurate, g is an exact least squares fit to the unsmoothed stresses σ. Using Eq. (2.77), unknown coefficients a to d can be evaluated and from which the smoothed corner node values of stresses can be obtained. In this context, this local smoothing method is simply a bilinear extrapolation of stresses sampled at the 2 x 2 reduced Gauss points.

Using this method, the authors (Hinton and Campbell, 1974 and Hinton *et al.*, 1975) gave the following smoothing matrix for the two-dimensional parabolic isoparametric element:

$$\begin{Bmatrix}\tilde{\sigma}_1\\ \tilde{\sigma}_2\\ \tilde{\sigma}_3\\ \tilde{\sigma}_4\end{Bmatrix} = \begin{bmatrix} 1+\frac{\sqrt{3}}{2} & -\frac{1}{2} & 1-\frac{\sqrt{3}}{2} & -\frac{1}{2} \\ -\frac{1}{2} & 1+\frac{\sqrt{3}}{2} & -\frac{1}{2} & 1-\frac{\sqrt{3}}{2} \\ 1-\frac{\sqrt{3}}{2} & -\frac{1}{2} & 1+\frac{\sqrt{3}}{2} & -\frac{1}{2} \\ -\frac{1}{2} & 1-\frac{\sqrt{3}}{2} & -\frac{1}{2} & 1+\frac{\sqrt{3}}{2} \end{bmatrix} \begin{Bmatrix}\sigma_{g1}\\ \sigma_{g2}\\ \sigma_{g3}\\ \sigma_{g4}\end{Bmatrix}, \tag{2.78}$$

where the $\tilde{\sigma}_i$ are the smoothed corner node stresses, σ_{gi} are the unsmoothed stresses at the Gauss integration points $g\,i$ as depicted in Figure 2.13. It should be noted that values at the midside nodes may be obtained by averaging the values at the nodes associated with the side in question, since the distribution of the smoothed stresses is linear along the sides of the element. At a given corner node, smoothed values from adjacent elements should subsequently be averaged to yield a single value. Thus a smoothed stress field at the global level is obtained.

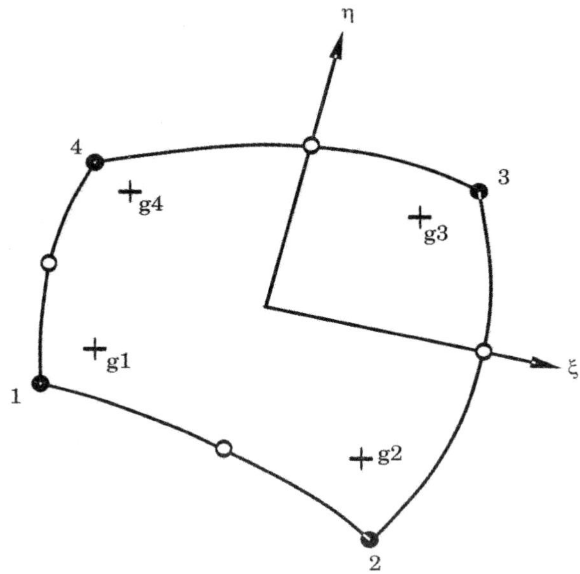

Fig. 2.13. A two-dimensional parabolic isoparametric element.

3

Software for Finite Element Analysis

3.1. INTRODUCTION

Finite element method is a computer-oriented numerical technique and application of the method to solve practical problems requires substantial computer programming. An essential feature of this book is the presentation of a computer code which was developed by the author for the finite element analysis of general engineering problems (Kumar, 2011a). The software designated as 'SB-FEM' has been developed in C^{++} programming language using object-oriented approach (Kumar, 2010c and Kumar, 2010d), the source code of which consisting of about 212 thousand lines of programming. The code can solve a good range of general kinds of structural analysis, heat transfer and metal working problems. Readers may find the code useful in solving several example problems given at the end of each subsequent chapter, and in visualizing the output results.

3.2. FEATURES OF THE COMPUTER CODE

The code 'SB-FEM' is capable of solving general structural analysis (static, multiple load steps, transient dynamics), heat transfer, metal working (two-dimensional and axisymmetric) and fluid flow (limited to incompressible viscous fluid only) problems. In structural stress analysis discipline, one-dimensional bars and beams, two-dimensional plane stress, plane strain and axisymmetric, plate bending, shell deformation and three-dimensional problems of isotropic materials and composites can be solved. Geometrical and material non-linearities can also be defined.

Like any general finite element program, the SB-FEM code has three distinctive capabilities, *i.e.*

a. Preprocessing capability in which the model is built and a finite element mesh is generated.

b. Processing capability in which constraints and loads are applied on the model or the mesh and the solution is obtained.

c. Post processing facility in which computational results are reviewed.

During preprocessing, the code takes minimum input data from the user, creates the finite element model and other data required for the analysis, and are capable of displaying the model for data-check and correction, if any, to be made by the user in an interactive mode. Other details about the finite element model such as boundary conditions and loadings can be specified using appropriate commands defined by the code in an interactive manner or as a batch input. The graphical post-processing facility available within the code helps in analyzing the finite element results and perceiving the physical consequences of the analysis.

The code has been provided with a sizeable Graphical User Interface (GUI) which consists of windows, menus, dialog boxes, tree controls and other components that allow an user to enter input data and execute various functions. The GUI mode is often referred to as an interactive mode. The communication with the code is done using commands which are either instructions that direct the program to perform the requested operation or input data for a particular function. Commands can be entered in an appropriate command input area in an interactive session or submitted via a batch input file. The code consists of several commands, each designed for a specific function. The commands have a specific format. A typical command contains a command name in the first field followed by several additional fields consisting of arguments. A comma separates each field of the command. The command syntax is as given below:

In the interactive session: COMMAND,ARG1,ARG2,ARG3,...
Inside the batch input file: COMMAND(ARG1,ARG2,ARG3,...);

Here COMMAND is the name of a SB-FEM command and ARGs are the arguments of the command. Command arguments may take a number (integer or real) or an alphanumeric character string, depending on their purpose. If any argument is to be assigned default value set by the code, that argument can be skipped from the command line by retaining the comma only. Although the command names are written in mixed case for clarity in the documentation of the code, all SB-FEM commands are case-insensitive.

The code also includes a text editing facility which can be used in preparing command input file inside the program and also in making reasonably well documentation related to an analysis.

3.3. PREPROCESSING (BUILDING THE MODEL)

SB-FEM has a reasonably good pre-processing capability that can be used to build a finite element model. In general, there are some definite steps which are often followed in sequence to develop the model. These steps include defining the element types, element attributes, material properties, and the model geometry.

3.3.1 Defining element type(s)

In any analysis, one or more element types are specified which are appropriate for the analysis. An element type determines the degree of freedom set, the type and dimension of the problem and some additional options. Each element type has a unique name that identifies the element category and attributes. The commands *Et, EtDele, EtList* and *SetCurEt* are associated with an element type among which *Et* defines the element type. For example, the command shown below defines in its simplest form an element type *stbeam3* which is a 3-noded beam element and assigns it a reference number 1.

```
Et(1,"stbeam3");
```

The SB-FEM program supports several element types whose characteristics are described briefly in Appendix 6.

3.3.2 Defining element attributes

There are several element attributes which depend upon the element type but may vary from element to element within the model. For example, element attributes for 2D bar (rectangular) elements are cross-sectional width and cross-sectional height. The command *EA* is used to specify element attributes while other commands associated with element attributes include *EADele*, *EAList* and *SetCurEA*. Each set of element attributes is identified by an identification number. In a model, different elements of the same type may have different sets of element attributes. The attributes within a particular element attribute set must be input in the order defined for the type (category) of the elements to which the set is to be assigned.

3.3.3 Defining material properties

An analysis requires material properties to be defined for the model. Within one analysis, it may be necessary to define multiple material property sets to correspond with multiple materials used in the model. The commands *MP, MPDele, MPList* and *SetCurMP* are associated with material properties among which the *MP* command defines the material properties. As with element types, each set of material properties has a material reference number which is used to identify the set. For example, the following command sets Young's modulus in X-direction for material reference number 1 to be equal to 1500:

```
mp(1,,"ex",1500);
```

Appropriate property labels are specified for each property. In the above example, "ex" label is used for Young's modulus in X-direction.

3.3.4 Constructing the model geometry

The next step in an analysis is to construct a finite element model consisting of nodes and elements that adequately describes the problem geometry. There are two methods to create the finite element model: direct generation and solid modeling. In a direct generation method, the location of each node and the connectivity of each element are manually defined. In solid modeling approach, the geometrical shape of the model is first described and then the code is instructed to automatically mesh the geometry with nodes and elements.

Direct generation method

Direct generation approach constructs the model strictly in terms of nodes and elements. Generally all the nodes are created first and then element generation follows. Direct generation of nodes may include defining nodes with given coordinates, generating additional nodes from existing nodes and reading a text file that contains nodal data. SB-FEM commands for direct generation of node include *N, NBN* and *NReadF*. Direct generation of an element requires that the minimum number of nodes necessary to define that element must have already been defined and proper element features (element type, material property, etc.) must have been set. An element can be directly defined by identifying its nodes using the *E* command. The number of nodes required for each element and the order in which they must be input are determined by the element type. The order in which nodes are input establishes the element normal direction. In addition, element data can be read from a text file using *EReadF* command.

Solid modeling approach

In this approach, the geometrical boundary of the model is first described. There are two approaches in defining the geometry : bottom-up approach and top-down approach. In a bottom-up approach, the keypoints (vertices of an object) are first created and then these keypoints are used to define the higher-order model entities, *i.e.* lines (edges of an object), areas (2D plane objects or faces of a 3D solid object) and volumes (3D solid objects). The SB-FEM commands associated with the model construction using bottom-up approach include *K, KBK, KL, LK, AK, AL, VK* and *VA*.

In the top-down approach, the model is assembled using geometric primitives, which are fully defined lines, areas and volumes. As a primitive is created, the program automatically creates all the lower entities associated with it. The primitives which can be created by the program include line primitives such as a straight line, arc, circle and spline; area primitives such as a rectangular area, elliptical area and polygonal area; and volume primitives such as block, cylinder and hollow cylinder. The commands associated with the top-down approach of the model construction include *line, line3D, arc, arc1, arc2, rect, triangle, ellipse, ellipse1, ellRing, block, blockHollow, cylinder, cylinder1* and *cylinderHollow*.

The code provides facility to construct additional areas from the existing area. An area (simple or complicated) that appears repetitively in the model need only be created once and can be moved and copied to a new location. It is also possible to scale an area up or down and copy to a new location. The command *ATrA* is used to perform this. The code also allows to perform Boolean operations (addition and subtraction) on 2D geometrical primitives or areas to create a complex area. The commands *aadd* and *asub* can be used to do Boolean operations on areas. Both types of areas created using the bottom up and top down approach may be used in transformations and Boolean operations.

Both bottom up and top down techniques can be freely combined, as appropriate, in a model construction. For example, a complicated area can be created using geometric primitives and necessary Boolean operations on these. This area can then be extruded along a given path to construct a volume. *vTwoAreas* and *vExtrude* are the two commands that can be used to construct a volume using one or more existing areas.

Mesh generation

In the solid modeling technique once the construction of the model is complete, the code can be instructed to mesh the model with nodes and elements. SB-FEM has the sizeable facility for automatic mesh generation using just the geometric description of the model. However, it is often desirable to set the mesh controls before meshing the solid model. Mesh controls allow the user to establish factors such as desirable element shape and size to be used in meshing the model. The mesh control commands available with the code include *Esize* and *LEsize*. There are two types of meshing available - mapped meshing and free meshing. A mapped mesh has usually a regular pattern and contains either all quadrilateral area elements, all triangular area elements or all hexahedral volume elements. In order to use mapped meshing, the geometry must be built as a series of fairly regular areas or volumes that can accept a mapped mesh. SB-FEM uses algorithms based on a blending functions technique to generate mapped mesh on an area or a volume. For an area to accept a mapped mesh, the following conditions must be satisfied:

a. The area must be bounded by either three or four lines.

b. The area must have equal number of element divisions specified on opposite edges (lines). Number of element divisions on the edges are specified using a *LEsize* command.

c. If the area is bounded by three lines, the number of element divisions specified on all the three edges must be equal and even.

d. If mapped meshing is to be used for an area bounded by more than four lines, some of the lines can be combined (concatenated) to reduce the total number of lines to four. The command *lineCcat* is used for concatenation of lines.

To use mapped meshing for a volume, the following conditions must be satisfied:

a. The volume must be enclosed by either five or six areas.

b. The volume must have equal number of element divisions specified on opposite faces (areas).

c. If the volume is bounded by five areas, the number of element divisions on triangular areas must be even.

In free meshing procedures, there is no special requirements which restrict the solid model and even irregular solid models can be meshed. For area meshing, a free mesh can contain only quadrilateral elements, only triangular elements or a combination of the two. SB-FEM uses an automatic triangulation technique and a quadtree method to generate free mesh on an area. Facility for an automatic free meshing on a volume is so far limited in the code. However, the volume created by extrusion of a complicated and irregular area or that enclosed between two complicated and irregular areas can be easily meshed.

The commands *Kmesh, Lmesh, Amesh* and *Vmesh* are available with the code to mesh keypoints, lines, areas and volumes respectively.

3.4. PROCESSING (APPLYING LOADS AND OBTAINING THE SOLUTION)

Activities in this step include defining the analysis type and analysis options, applying loads and obtaining the finite element solution.

3.4.1 Specifying analysis type and analysis options

The analysis type is generally based on the loading conditions. The code supports static, multiple load steps and transient dynamics analyses in the structural analysis discipline, and static and transient analyses in a heat transfer field. The command *AnType* is used to define the analysis type. Typical analysis options include the method of solution, options for geometrical non-linearity in a structural analysis and convergence norms. For example, the command *nlGeom* is used to specify geometrical non-linearity option, commands *kmatOpt* and *massOpt* specify the methods of calculation of stiffness matrix and mass matrix respectively, and the command *CnvTol* defines the convergence norms.

3.4.2 Applying loads

The next step after setting analysis type and analysis options, is to apply loads. The word 'loads' used here includes boundary conditions (degree of freedom constraints) as well as externally and internally applied loads (concentrated forces, surface loads and body loads). Most of these loads can be applied on either solid model entities (keypoints, lines and areas) or finite element model entities (nodes and elements).

Since all loads are required to be in terms of the finite element model during solution, the code automatically converts the loads applied on solid model entities into equivalent loads on nodes and elements, at the beginning of the solution. A degree of freedom (DOF) constraint fixes a degree of freedom to a specified value. Examples of constraints are specified displacements in a structural analysis, prescribed temperatures in a thermal analysis and specified velocities in a metal working analysis. The commands *DL, DA* and *DN* are used to set DOF constraints on lines, areas and nodes respectively.

A point force is a concentrated load applied at a node in the model. Examples are forces and moments in a structural analysis and heat flow rates in a thermal analysis. The command *FN* is used to prescribe the concentrated load at a node. A surface load is a distributed force over a surface (edges or faces). Examples are pressures in a structural analysis, and convections and heat fluxes in a thermal analysis. In SB-FEM, surface loads can be specified on the lines, areas, sets of selected nodes or faces of an element. The commands to specify surface loads include *SFN, SFL, SFA* and *SFE*.

A body force is a volumetric load distributed over a part of the body or the whole body. Examples include body weight and temperature in a structural analysis and heat generation rates in a thermal analysis. The command *BF* is used to specify body loads in the SB-FEM program.

3.4.3 Setting load step options

Load step options are the options which control how loads are applied during solution and can vary from load step to load step. The options include time at the end of a load step in a multiple step static analysis or transient analysis and number of substeps or time step size. *Time, noTimeSteps* and *DelTime* are used to establish the load step options.

3.4.4 Obtaining the solution

After applying loads and setting load step options the solution may be initiated. In the solution phase, the program generates a simultaneous set of finite element equations which are solved by the computer using a specified solver. The results of the solution are values for nodal degrees of freedom (displacements in a structural analysis, temperatures in a thermal analysis and so on) which form the primary solution, and values for derived quantities. Derived variables are results calculated from the primary variables, *e.g.* strains and stresses in a structural analysis and thermal gradients and fluxes in a thermal analysis. Derived data are mainly element results, and calculated first at all the integration points of the elements and then extrapolated to the nodes. In SB-FEM, there are two methods to solve the system of simultaneous equations: skyline solver and frontal solver. Both symmetric and unsymmetric sets of equations can be solved. The command *Solve* is used to commence the solution in SB-FEM.

During solution, the code writes and reads many files for data storage and retrieval. File names have the following pattern:

[Fname][Sufx].[Ext]

[Fname] is the file name specified using *OutFileName* command. [Sufx] is a suffix which may be appended to the file name and takes a character value 's'. [Ext] stands for the filename extension and is a unique character string identifier that recognizes the contents of the file. Table 3.1 lists the files created during the solution phase by SB-FEM.

Table 3.1. SB-FEM files created during the solution phase.

File contents	*File name*	*File format*
Element matrices	[Fname]s.EMT	Binary
Element strain-displacement matrices	[Fname]s.ESA	Binary
Composite layer information	[Fname]s.LAY	Binary
Finite element equations data	[Fname].EQR	
	[Fname].EQC	
	[Fname].RHS	Binary
Output results	[Fname]s.RSB	Binary

3.5. POSTPROCESSING (REVIEWING THE RESULTS)

Postprocessing is the step in which the results are evaluated and interpreted. The results can be reviewed either at a particulate time step over the entire model or the selected portion of the model, or at some specific points in the model over all time steps. The results can be reviewed through listings in tabular forms or through graphics displays.

3.5.1 Listings results in a tabular form

Tabular listing is an effective way of reviewing and documenting the analysis results. In SB-FEM, all the nodal and element solution data can be listed in a tabular form using appropriate commands and proper labels identifying the result items. The command *PrNSol* is used to list specified nodal solution data (primary as well as derived) while *PrESol* is used to list specified results for selected elements. In a similar way, *PrNSolT* command is used to print time-varying solutions for a particular node and *PrESolT* lists results over all time steps for a particular element.

3.5.2 Displaying results graphically

Graphics displays are perhaps the most effective way to review the results. The types of graphics displays available with the code include contour display, deformed shape display and vector display. Contour display shows the variation of a result item over the entire model as continuous contour lines. The commands *PlNSol* and *PlESol* are used to plot nodal and element solution data respectively. Derived solution results, which are usually discontinuous from element to element, are averaged at the nodes so that continuous contour lines can be displayed. In a structural analysis, a deformed shape can also be displayed along with the undeformed one to observe how the structure has deformed under the applied loads.

Vector plots use arrows to depict the variation of both magnitude and direction of a vector quantity in the model. Examples of vector quantities are displacements in a structural analysis, thermal flux in a thermal analysis, velocities in a metal working analysis, etc. SB-FEM provides *PlVNSol* command to produce a vector display.

3.6 SB-FEM COMMANDS AND COMMAND INTERPRETER

The code SB-FEM recognizes a number of commands which are used to supply data, or control to the program. There are several commands which have been grouped by functionality to become valid only within a particular processing operation (preprocessing, solution or postprocessing). However, there are several general commands which are not restricted to a specific processor and can be input anywhere. A list of SB-FEM commands along with their brief descriptions is mentioned in Appendix 7.

As stated earlier, the syntaxes for SB-FEM commands are little different in an interactive mode and in a batch input mode, although the command name and their input arguments are identical in the two formats. In the batch input mode, the arguments are enclosed in parentheses and a semi-colon is put at the end of the command. Furthermore the arguments can be both constants and variables (defined within the file) in this mode. It is thus necessary in this mode to enclose arguments in the form of constant character strings with double quotes. This is depicted below through *inpex* command which is used for executing a command file.

inpex, DIRNAME, FILENAME, EXT (interactive mode)

inpex ("DIRNAME", "FILENAME", "EXT"); (batch input mode)

In a batch input mode, the command may contain some arguments as variable addresses in order to retrieve data from the code. This is demonstrated below using the *ellipse* command that is used to create an ellipse or an elliptical segment.

int areaNo = ellipse(XCENT, YCENT, ZCENT, XRAD, YRAD, STANG, ENDANG, "NAME", WPNO, int* NoAreas, int* AreaNos);

In addition to creating the main area identified by *areaNo*, this command also generates different area segments whose identity numbers can be obtained by passing an array of integers of size >= 4. Here *NoAreas* is the pointer to an integer data which retrieves the number of areas generated, and *AreaNos* is the pointer to an array of integers of size >= 4 which retrieves identity numbers of the areas generated.

In a batch input mode, the command file can contain several statements in a syntax same as that of C programming language, along with SB-FEM commands expressed in a batch input format. There must be a *main*() function which is a entry point function and can drive other four functions *gen*(), *prep*(), *solu*() and *post*(), if defined. The function *gen*() consists of some general commands which define analysis title, input and output filenames and so on. In function *prep*(), all the commands required for building the finite element model need to be included. The commands within the function *solu*() describe analysis type, analysis options, boundary conditions and loadings, and initiate the solution process. Once the solution is over, the function *post*() is invoked which may consist of desirable postprocessor commands for viewing the finite element results either graphically or in a tabular form. All the three functions *prep*(), *solu*() and *post*() may also consist of some general commands including those listing/plotting the model entities. Further, several other general functions may be defined within the command file and called by these four program-defined functions as well as the *main*() function. However, the SB-FEM commands are only recognized inside these five functions. Global variables to be used in more than one function can be declared on the top of the command file.

3.6.1 SB-FEM command interpreter

Accordingly, the SB-FEM code has a built-in command interpreter for the interpretation of commands written in C language syntax. The following are salient features of the interpreter:

a. Integer, float and double data can be defined globally, locally within a function and locally within braces. The interpreter also recognizes one built-in data type as 'string' whose role is same as char*.

b. Almost all of the C language operators are supported including postfix/prefix incremental/decremental operators.

c. Several functions can be defined and called by other functions. Functions should return only the above data types. void is not supported. There must be a main function which acts as a driver to execute the subroutines.

d. *if...else, if...else if...else* and *for* statements are supported without braces and with braces. *return, continue* and *break* statements can also be used. Although *goto* statement is supported, it is not desirable to use this.

e. An array of above data types can be defined upto a maximum dimensions of five. Individual elements of the array can be assigned values in the same way as that in the C language. The array or its elements can be passed as function arguments.

f. Address of a data variable can be passed to the function as an argument using address-of operator (&) and its value can be modified within the function.

g. Declaration of pointer variables and their use is supported.

h. New and delete operators can be used for an individual data type or an one-dimensional array.

i. The following are not supported by the present implementation of the interpreter:

- Do and while statements.
- Structure type (struct), union type or enumerated type (enum).
- Bitwise operators.
- None of the C library functions are supported. Interpreter can only recognize library functions defined by SB-FEM program.

3.7. INSTALLATION OF SB-FEM

Some general guidelines for installation of the code are mentioned below. A detailed installation procedure can be found in an appropriate text file within the installation program.

3.7.1. Windows platform

SB-FEM installation program is available as <Program Name>.zip file. The steps required to install the program are as follows:

a. Unzip the file using winzip or other suitable programs. This will extract setup files into an appropriate directory.
b. Run setup.exe file inside the directory and follow the instructions.
c. Installation should also create proper shortcuts otherwise a shortcut can be manually created for the execution file <Program Name>.exe contained in the bin subdirectory of the target directory.

3.7.2. Linux platform

SB-FEM installation program is available in two different forms, (a) as a binary RPM package and (b) as an executable form, compiled using GNU C++ compiler on RedHat Linux. The steps required to install the program using the two forms of distribution are given below.

Using a binary RPM package

The binary RPM package will have a name that are of the form <Program Name>.rpm, in which <Program Name> specifies the name of the code suffixed with version number and others. The RPM file may be copied to a suitable working directory and the code can be installed using the *rpm* command, *i.e.* rpm -ivp --nodeps <Program Name>.rpm.

From an executable form

SB-FEM code is available in executable form as <Program Name>.tar.gz file ready to be installed. This tar file consists of executable, bitmap, input, help and other related files in appropriate subdirectories. The steps required to install the program are as follows:

a. Copy the tar file into a working directory in which the program is to be installed.

b. Uncompress the tar file using proper utility command. This will create a main directory as <Program Name> and the necessary subdirectories and will place all the files appropriately into these directories.

3.8. USING SB-FEM CODE

There are several example problems given at the end of each subsequent chapter. The SB-FEM command files for these example problems along with few others can be found in a suitable example subdirectory within the code. Some selected command files are also listed in Appendix - 8 for reference during reading of this book. To start with the code, these problems can be solved by using the following steps:

a. First go to menu: Command => Main.

b. There are two commands for running the input batch file : *input* and *inpex*.

   Syntaxes of these commands are:

   input, DIRNAME, FILENAME, FILEEXT

   inpex, DIRNAME, FILENAME, FILEEXT

   Although both are equivalent, they differ only when the directory name *DIRNAME* is not specified. Default directory name for *inpex* is the directory where all the example files are located.

c. Type in the command box: *inpex,,FILENAME,FILEEXT*, where *FILENAME* is the name of an example problem file and *FILEEXT* is the extension of the file. Then press OK (Do not use Enter or Esc keys).

d. Once program starts running, the progress of execution can be viewed in response window at the bottom of the main program window. Each file contains a few post-processor commands for viewing of results. Additional results can be viewed by typing appropriate commands in a command box that can be opened either using the path: Command=>Postprocessor or using the command tree.

e. For help on any command within the command file, an appropriate help file can be referred.

4

Analysis of Trusses/Bars

4.1 INTRODUCTION

Engineering structures generally consist of discrete parts like sheets, plates, beams, stiffeners etc., to form a complete system. In the process of constructing or building such a system, the components may be welded, riveted or bolted or connected by any other suitable means. A truss is an engineering structure composed of straight members, of a certain material, that are typically pin-joined at their ends. Such members are also called *two-force* members since they can only transmit or support force along their length or axis, whether in tension or compression. A plane truss is one where all the members and loads lie in one spatial plane opposite to a space or 3-D truss. Two-force members also deform along their length or axis and not transverse to it. The one-dimensional truss is also called a rod or an axial bar.

Finite element concepts described in previous chapter can be used to formulate the displacement method of analysis by treating each truss member as an element in the mesh.

4.2 BAR ELEMENT (LOCAL COORDINATES)

Consider a uniform prismatic bar element of length L, cross-sectional area A and elastic modulus E as shown in Figure 4.1. A node is located at each end of the bar. Here node is only allowed to displace only in the axial direction. Axial displacements at nodes are U_1 and U_2.

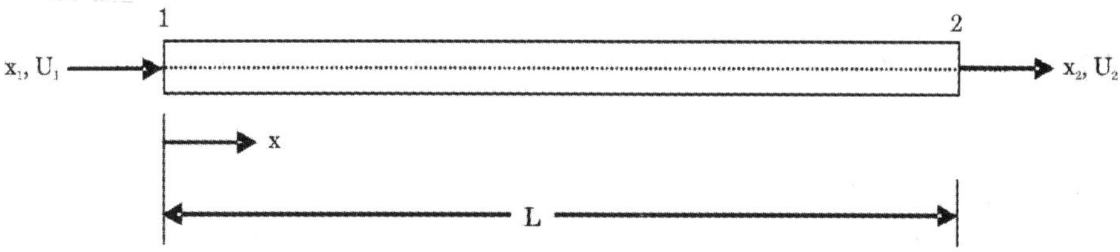

Fig. 4.1. Truss bar element lying along the local x-axis.

Assuming that the displacement u is varying linearly along the axis of the bar, we have

$$u = a_0 + a_1 x \ , \tag{4.1}$$

where a_0 and a_1 are unknown constants.

Displacements given by Eq. (4.1) should satisfy the following equations at each node:

$$U_1 = a_0 + a_1 x_1$$
$$U_2 = a_0 + a_1 x_2 \, .$$

Solving for the coefficients a_i from the above equation and putting the values in Eq. (4.1), we have

$$u = \left(1 - \frac{x}{L}\right) U_1 + \frac{x}{L} U_2 \, . \tag{4.2}$$

Strain in the bar can be expressed as

$$\varepsilon = \frac{du}{dx} = \frac{U_2 - U_1}{L} \tag{4.3}$$

and stress is calculated as

$$\sigma = E\,\varepsilon = E\,\frac{(U_2 - U_1)}{L} \, . \tag{4.4}$$

Force in the bar is given as

$$P = \sigma A \, \cdot \tag{4.5}$$

Considering the equilibrium of forces at nodes 1 and 2 as shown in the free body diagram (Figure 4.2), we have

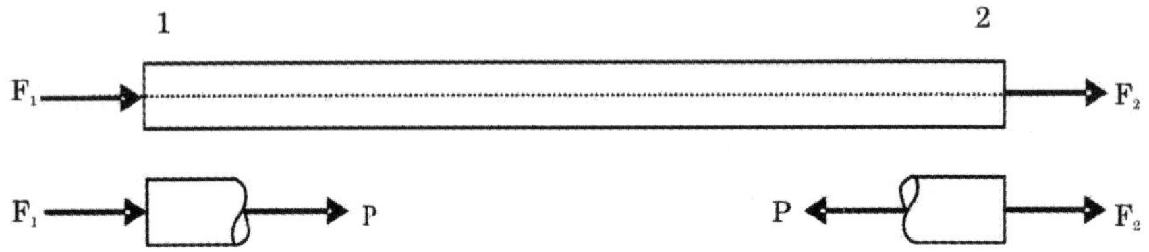

Fig. 4.2. Free body diagram.

$$\begin{Bmatrix} F_1 \\ F_2 \end{Bmatrix} = \begin{Bmatrix} -P \\ P \end{Bmatrix} = \frac{EA}{L}\begin{bmatrix} 1 & -1 \\ -1 & 1 \end{bmatrix}\begin{Bmatrix} U_1 \\ U_2 \end{Bmatrix}. \tag{4.6}$$

So, the element stiffness matrix is

$$\left[K^e\right] = \frac{EA}{L}\begin{bmatrix} 1 & -1 \\ -1 & 1 \end{bmatrix}. \tag{4.7}$$

Note that $\frac{EA}{L}$ can be regarded as k, the stiffness of a linear spring. A bar and a spring have the same behavior under axial load and are represented by the same stiffness matrix.

4.2.1 Formal approach

The above stiffness matrix can also be derived using a formal approach which can be applied to many other complicated situations.

If ξ *is* the natural coordinate so that x_1 is at $\xi = 0$ and x_2 is at $\xi = 1$, then the parametric linear shape functions are defined as

$$N_1(\xi) = 1 - \xi \text{ and } N_2(\xi) = \xi, \tag{4.8}$$

in which

$$\xi = \frac{x}{L}, \quad 0 \le \xi \le 1. \tag{4.9}$$

The value of shape function N_1 thus defines a straight line having value of 1 at node 1 and linearly reducing to 0 at node 2. N_2 has similar feature.

It is to be noticed that

$N_1 + N_2 = 1.$

We can write the displacements as

$u(x) = u(\xi) = N_1 U_1 + N_2 U_2$

$$\text{or } u = [N_1 \quad N_2]\begin{Bmatrix} U_1 \\ U_2 \end{Bmatrix} \text{ or } u = [N]\{U^e\}. \tag{4.10}$$

Strain is given as

$$\varepsilon = \left[\frac{d}{dx}[N]\right]\{U^e\} = [B]\{U^e\}, \tag{4.11}$$

where $[B]$ is element strain-displacement matrix given as

$$[B] = \frac{d}{dx}[N_1 \quad N_2] = [-1/L \quad 1/L]. \tag{4.12}$$

Stress can be written as

$$\sigma = E\varepsilon = E[B]\{U^e\}. \tag{4.13}$$

Now considering the strain energy stored in the bar element, we have

$$E_U = \frac{1}{2}\int_{V^e}\{\varepsilon\}^T\{\sigma\}\,dV = \frac{1}{2}\int_{V^e}\{U^e\}^T[B]^T E[B]\{U^e\}\,dV$$

$$= \frac{1}{2}\{U^e\}^T\left[\int_{V^e}[B]^T E[B]\,dV\right]\{U^e\}. \tag{4.14}$$

The work done by the two nodal forces is

$$W = \frac{1}{2}F_1U_1 + \frac{1}{2}F_2U_2 = \frac{1}{2}\{U^e\}^T\{F^e\}. \tag{4.15}$$

For conservative system, we have $E_U = W$; which gives

$$\left[\int_{V^e} [B]^T E [B] dV \right] \{U^e\} = \{F^e\}$$

or

$$\left[K^e\right]\{U^e\} = \{F^e\}, \tag{4.16}$$

where $\left[K^e\right] = \int_{V^e} [B]^T E [B] dV$.

Putting the value for matrix $[B]$, we get

$$\left[K^e\right] = \frac{EA}{L^2} \int_x [-1 \quad 1]^T [-1 \quad 1] dx = \frac{EA}{L} \begin{bmatrix} 1 & -1 \\ -1 & 1 \end{bmatrix}. \tag{4.17}$$

In Eq. (4.17), it was assumed that the cross-section area is constant throughout the length of the truss member. If the cross-section area of the element is not uniform, the variation can be expressed through a shape function and the stiffness matrix can be evaluated following the above procedure. Considering a truss member having area of cross-section at node 1 and 2 as A_1 and A_2 respectively, the variation in the cross-sectional area A can be expressed in terms of linear function given by

$$A(x) = A(\xi) = N_1 A_1 + N_2 A_2 .$$

The stiffness matrix $\left[K^e\right]$ is, therefore, given as

$$\left[K^e\right] = \frac{E}{L^2} \int_x [-1 \quad 1]^T [-1 \quad 1] A dx . \tag{4.18}$$

Substituting for variation of A and using Eq. (4.9) in the above equation, we obtain

$$\left[K^e\right] = \frac{E}{L} \begin{bmatrix} 1 & -1 \\ -1 & 1 \end{bmatrix} \int_0^1 (N_1 A_1 + N_2 A_2) d\xi = \frac{E(A_1 + A_2)}{2L} \begin{bmatrix} 1 & -1 \\ -1 & 1 \end{bmatrix}. \tag{4.19}$$

It is to be noted that there is a general formula for integration of quantities defined as polynomials of natural coordinate over the line length such as in Eq. (4.19), that is,

$$\int_x \xi^a dx = L \int_0^1 \xi^a d\xi = \frac{a!}{(a+1)!} L \,. \tag{4.20}$$

4.2.2 Isoparametric derivation of bar element stiffness matrix

If ξ is the natural coordinate so that x_1 is at $\xi = -1$ and x_2 is at $\xi = 1$, then the parametric linear shape functions are defined as

$$N_1(\xi) = \frac{1}{2}(1-\xi), \text{ and } N_2(\xi) = \frac{1}{2}(1+\xi) \,. \tag{4.21}$$

In isoparametric formulation, coordinates are also interpolated in the same manner, that is,

$$x = N_1 x_1 + N_2 x_2 \,. \tag{4.22}$$

As the formulation is done in terms of natural coordinates, it is necessary to establish the relationship between the derivative of a function in the natural coordinates and that in the Cartesian coordinates. This is derived by evaluating the Jacobian as

$$\frac{df}{d\xi} = \frac{dx}{d\xi}\frac{df}{dx} = J\,\frac{df}{dx} \,, \tag{4.23}$$

where f is a scalar function. Using Eq. (4.22), J can be evaluated as

$$J = \frac{x_2 - x_1}{2} = \frac{L}{2} \,.$$

The element strain-displacement matrix $[B]$ is given as

$$[B] = \frac{d}{dx}[N_1 \quad N_2] = \frac{1}{J}\frac{d}{d\xi}[N_1 \quad N_2] = [-1/L \quad 1/L] \,. \tag{4.24}$$

Substituting the value of $[B]$, the stiffness matrix $[K^e]$ can be written as

$$\left[K^e\right] = \int_{V^e} [B]^T E\,[B]\,dV = \frac{EA}{L^2}\int_x [-1 \quad 1]^T [-1 \quad 1]dx$$
$$= \frac{EA}{L^2}\int_{-1}^{1} [-1 \quad 1]^T [-1 \quad 1] J\, d\xi$$

or

$$\left[K^e\right] = \frac{EA}{L}\begin{bmatrix} 1 & -1 \\ -1 & 1 \end{bmatrix}. \tag{4.25}$$

It may be noted that the expression (Eq. (4.25)) for $[K^e]$ matrix here is same as Eq. (4.17).

4.3 BAR (TRUSS) ELEMENT (ORIENTED ARBITRARILY IN TWO-DIMENSIONAL PLANE)

Having derived the stiffness equations for a horizontal truss bar element lying along a local coordinate axis above, we now consider a general case of a bar element oriented arbitrarily in a two dimensional plane (Figure 4.3). We use coordinate transformation approach here.

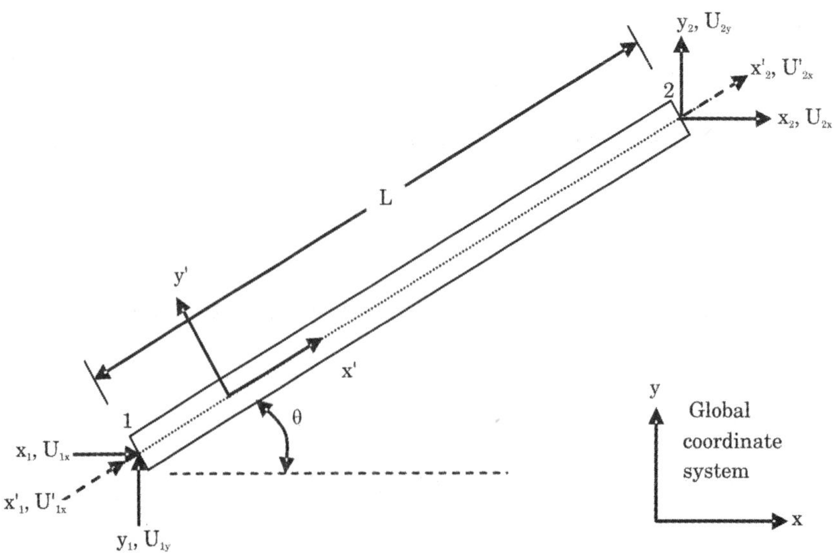

Fig. 4.3. Truss bar element lying in two-dimensional plane.

There are two sets of coordinates: the local coordinates (x', y') and the global (reference) coordinates (x, y).

The relationship between the local and global displacement components is given by

$$\begin{Bmatrix} U'_{1x} \\ U'_{1y} \\ U'_{2x} \\ U'_{2y} \end{Bmatrix} = \begin{bmatrix} l & m & 0 & 0 \\ -m & l & 0 & 0 \\ 0 & 0 & l & m \\ 0 & 0 & -m & l \end{bmatrix} \begin{Bmatrix} U_{1x} \\ U_{1y} \\ U_{2x} \\ U_{2y} \end{Bmatrix}, \tag{4.26}$$

where $l = \cos\theta$ and $m = \sin\theta$.

Writing symbolically, we have

$$\{U'^{(e)}\} = [T]\{U^e\}, \tag{4.27}$$

where $[T]$ is the coordinate transformation matrix. We can relate the four nodal forces in local coordinate directions to those in global coordinate directions using the same transformation matrix, that is,

$$\{F'^{(e)}\} = [T]\{F^e\}. \tag{4.28}$$

Now, stiffness equation in the local coordinate system is

$$\begin{Bmatrix} F'_{1x} \\ F'_{2x} \end{Bmatrix} = \frac{EA}{L}\begin{bmatrix} 1 & -1 \\ -1 & 1 \end{bmatrix}\begin{Bmatrix} U'_{1x} \\ U'_{2x} \end{Bmatrix}. \tag{4.29}$$

Because the transformation matrix $[T]$ is of an order 4×4, it is necessary to augment the above stiffness matrix from order 2×2 to 4×4. This can be done as follows:

$$\begin{Bmatrix} F'_{1x} \\ F'_{1y} \\ F'_{2x} \\ F'_{2y} \end{Bmatrix} = \frac{EA}{L}\begin{bmatrix} 1 & 0 & -1 & 0 \\ 0 & 0 & 0 & 0 \\ -1 & 0 & 1 & 0 \\ 0 & 0 & 0 & 0 \end{bmatrix}\begin{Bmatrix} U'_{1x} \\ U'_{1y} \\ U'_{2x} \\ U'_{2y} \end{Bmatrix},$$

or

$$\{F'^{(e)}\} = [K'^{(e)}]\{U'^{(e)}\}. \tag{4.30}$$

Substitution of Eqs. (4.27) and (4.28) into Eq. (4.30) gives

$$[T]\{F^e\} = \left[K'^{(e)}\right][T]\{U^e\}$$

or

$$\{F^e\} = [T]^{-1}\left[K'^{(e)}\right][T]\{U^e\}. \tag{4.31}$$

Because the transformation matrix is an orthogonal matrix (*i.e.* its inverse is the same as its transpose), we have

$$\{F^e\} = \left[K^e\right]\{U^e\} \tag{4.32}$$

with

$$\left[K^e\right] = [T]^T\left[K'^{(e)}\right][T]. \tag{4.33}$$

This stiffness matrix is in a congruent transformation form.

Substituting from Eqs. (4.26) and (4.30) for the terms in Eq. (4.33), we can write

$$\left[K^e\right] = \frac{EA}{L}\begin{bmatrix} [k_0] & -[k_0] \\ -[k_0] & [k_0] \end{bmatrix}, \tag{4.34}$$

where

$$[k_0] = \begin{bmatrix} l^2 & lm \\ lm & m^2 \end{bmatrix}. \tag{4.35}$$

It is to be noted that $[K^e]$ matrix is a singular matrix.

4.3.1 Stress calculation

We have the following stress-strain relation:

$$\sigma = E\,\varepsilon$$

or

$$\sigma = E\left(\frac{U'_{2x} - U'_{1x}}{L}\right) = \frac{E}{L}[-1 \quad 1]\begin{Bmatrix} U'_{1x} \\ U'_{2x} \end{Bmatrix} = \frac{E}{L}[-1 \quad 0 \quad 1 \quad 0]\begin{Bmatrix} U'_{1x} \\ U'_{1y} \\ U'_{2x} \\ U'_{2y} \end{Bmatrix}. \tag{4.36}$$

Using transformation relation (Eq. (4.26)), we can write

$$\sigma = \frac{E}{L}\begin{bmatrix} -l & -m & l & m \end{bmatrix}\{U^e\}. \tag{4.37}$$

4.4. BAR (TRUSS) ELEMENT (ORIENTED ARBITRARILY IN THREE-DIMENSIONAL SPACE)

In three dimensional space, the relationship between the local and global displacement components is given by

$$\begin{Bmatrix} U'_{1x} \\ U'_{1y} \\ U'_{1z} \\ U'_{2x} \\ U'_{2y} \\ U'_{2z} \end{Bmatrix} = \begin{bmatrix} l_1 & m_1 & n_1 & 0 & 0 & 0 \\ l_2 & m_2 & n_2 & 0 & 0 & 0 \\ l_3 & m_3 & n_3 & 0 & 0 & 0 \\ 0 & 0 & 0 & l_1 & m_1 & n_1 \\ 0 & 0 & 0 & l_2 & m_2 & n_2 \\ 0 & 0 & 0 & l_3 & m_3 & n_3 \end{bmatrix} \begin{Bmatrix} U_{1x} \\ U_{1y} \\ U_{1z} \\ U_{2x} \\ U_{2y} \\ U_{2z} \end{Bmatrix}, \tag{4.38}$$

where (l_1, m_1, n_1), (l_2, m_2, n_2) and (l_3, m_3, n_3) are the direction cosines between *x'*, *y'*, and *z'* axes and (x, y, z) axes respectively.

Writing symbolically, we have

$$\{U'^{(e)}\} = [T]\{U^e\}, \tag{4.39}$$

where $[T]$ is the coordinate transformation matrix. The six nodal forces in local coordinate directions can be related to those in global coordinate directions using the same transformation matrix, that is,

$$\{F'^{(e)}\} = [T]\{F^e\}. \tag{4.40}$$

Using the same procedure as in two-dimensional case, stiffness equation for a truss bar element oriented arbitrarily in three-dimensional space can be written as

$$\{F'^{(e)}\} = \left[K'^{(e)}\right]\{U'^{(e)}\}, \tag{4.41}$$

where

$$\left[K'^{(e)}\right] = \frac{EA}{L}\begin{bmatrix} 1 & 0 & 0 & -1 & 0 & 0 \\ 0 & 0 & 0 & 0 & 0 & 0 \\ 0 & 0 & 0 & 0 & 0 & 0 \\ -1 & 0 & 0 & 1 & 0 & 0 \\ 0 & 0 & 0 & 0 & 0 & 0 \\ 0 & 0 & 0 & 0 & 0 & 0 \end{bmatrix}. \tag{4.42}$$

Substituting Eqs. (4.39) and (4.40) into Eq. (4.41) and using the orthogonality of the transformation matrix, we get

$$\left[K^e\right] = [T]^T \left[K'^{(e)}\right][T]. \tag{4.43}$$

Substituting the values for different terms in the right hand side of the above equation, $\left[K^e\right]$ can be evaluated as

$$\left[K^e\right] = \frac{EA}{L}\begin{bmatrix} [k_0] & -[k_0] \\ -[k_0] & [k_0] \end{bmatrix}, \tag{4.44}$$

in which

$$[k_0] = \begin{bmatrix} l_1^2 & l_1 m_1 & l_1 n_1 \\ l_1 m_1 & m_1^2 & m_1 n_1 \\ l_1 n_1 & m_1 n_1 & n_1^2 \end{bmatrix}. \tag{4.45}$$

For simplicity, we can drop the subscript 1 from the direction cosine terms in the matrix $[k_0]$, which then becomes

$$[k_0] = \begin{bmatrix} l^2 & lm & ln \\ lm & m^2 & mn \\ ln & mn & n^2 \end{bmatrix}, \tag{4.46}$$

where l, m, n are the direction cosines between x' axes and the x,y,z axes respectively.

Here also, $[K^e]$ matrix is a singular matrix.

4.4.1 Stress calculation

In three-dimensional case, we have

$$\sigma = \frac{E}{L}[-1 \quad 1]\begin{Bmatrix} U'_{1x} \\ U'_{2x} \end{Bmatrix} = \frac{E}{L}[-1 \quad 0 \quad 0 \quad 1 \quad 0 \quad 0]\left\{U'^{(e)}\right\} \tag{4.47}$$

or

$$\sigma = \frac{E}{L}\begin{bmatrix}-l & -m & -n & l & m & n\end{bmatrix}\{U^e\}. \tag{4.48}$$

4.5. ASSEMBLY OF MULTIPLE ELEMENT CONTRIBUTIONS

The next step is to consider an assemblage of many truss elements connected by pin joints (Figure 4.4). Each element meeting at a joint, or node, will contribute a force there as dictated by the displacements of that element's nodes. To maintain static equilibrium, all element force contributions at a given node must sum to the force that is externally applied at that node.

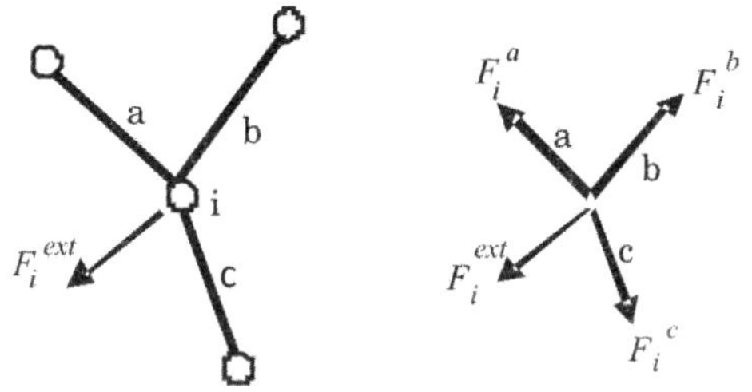

Fig. 4.4. Element contributions to total nodal force.

In the assembly process, each element stiffness matrix is added to the appropriate location of the overall (global) stiffness matrix that relates all of the truss displacements and forces.

4.6. EXAMPLE PROBLEMS

4.6.1. Example 1: A tapered aluminum alloy bar

A tapered aluminum alloy bar of a variable square cross section and length, L = 10 units is suspended from a ceiling (Figure 4.5). The cross section is a square of size w_1 =2 units at the top and w_2 =1 units at the bottom. An axial load of 10000 units is applied to the free end of the bar. Material properties are: E = 10.4×10^6 unit and $\nu = 0.3$. It is desired to calculate axial deflection and stress in the bar.

In order to solve the problem using one dimensional bar elements, the geometry of the bar can be approximated by several shorter brick-like elements or blocks each with a constant cross-section A_i and length l_i (Figure 4.5). The area of any of these elements is simply the average of the areas along its length. Since area is linearly changing, this would then simply be the average of the top cross-section area and the bottom cross-section area of the element.

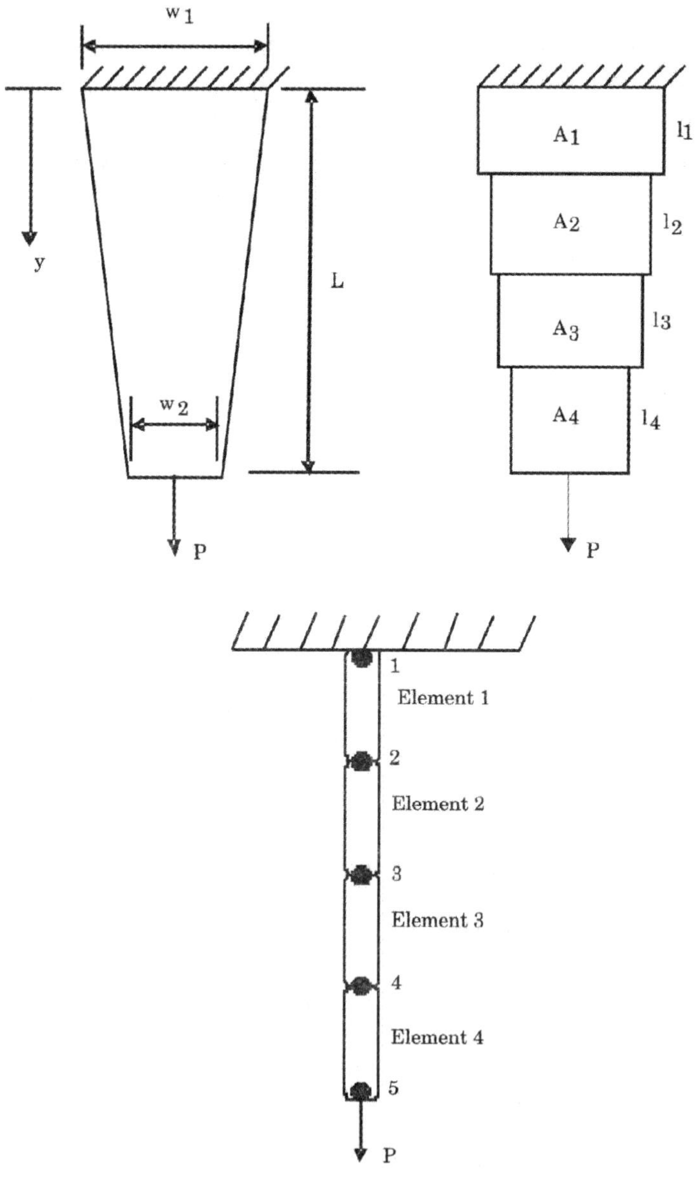

Fig. 4.5. A tapered aluminum alloy bar and its one-dimensional FEM idealization.

The SB-FEM command file is listed in Appendix 8. The tapered bar is modeled with 16 elements having varying cross-section areas. The present FEM results for maximum axial deflection (*i.e.* deflection at the bottom) in the bar and axial stress at mid-length (L/2) are given in Table 4.1 along with the theoretical results (Harris, 1959). It can be observed that an excellent agreement exists between the two results.

Table 4.1 Comparison of results for a tapered aluminum alloy bar.

	SB-FEM results	*Theoretical results (Harris, 1959)*
Maximum axial deflection	0.004805 units	0.0048077 units
Axial stress at mid-length	4450 units	4444 units

4.6.2. Example 2: A cantilever bar under axial load

In this example, a cantilever bar under uniformly distributed axial load is considered (Figure 4.6). The length, area of cross-section and Young's modulus of the material are taken as 10, 1 and 50 respectively. Consistent units are assumed. The intensity of axial pull is taken as 10 per unit length. The problem is analyzed using one dimensional bar elements and result for axial deflection in the bar is compared with theoretical value.

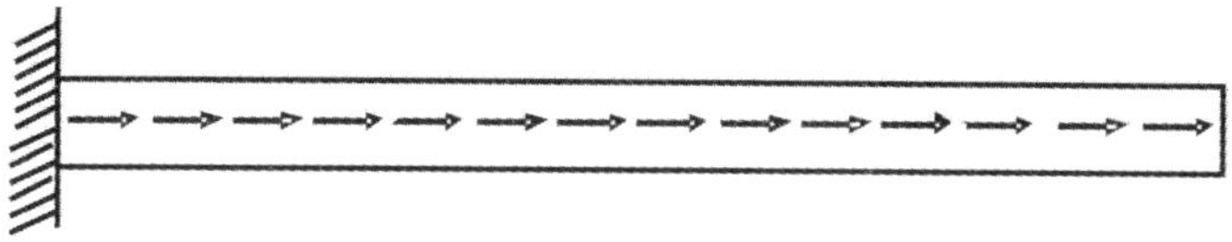

Fig. 4.6. A cantilever bar under uniformly distributed axial load.

The exact axial displacement at any point P of the cantilever bar under the given uniformly distributed tensile load is expressed as $u = 2x - \frac{x^2}{10}$, in which x is the distance of the point P from the fixed end.

The bar is modeled with 16 elements. The present FEM result for maximum axial deflection (*i.e.* deflection at the free end) in the bar is found to be 10 which is the exact theoretical result.

5

Analysis of Beams

5.1 INTRODUCTION

A beam is a bar element that is primarily subjected to transverse loads. The deformed shape of a beam is mainly described by transverse displacement (w) and slope (rotation) (θ) of the beam (Figure 5.1). The requirement for beam elements of good performance, which may be used to model either system of stand-alone beams or plane and space frames, need not be emphasized.

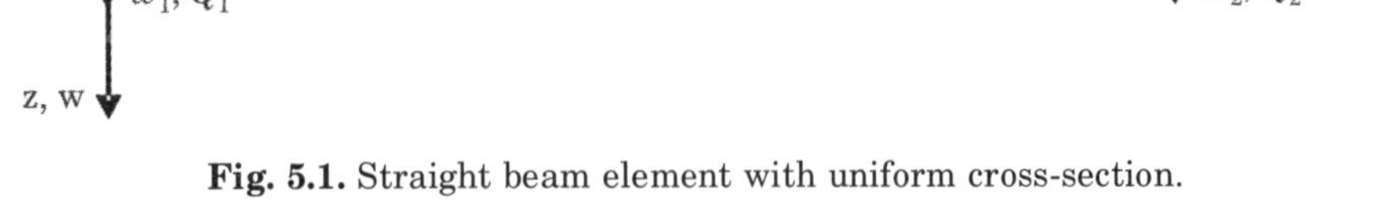

Fig. 5.1. Straight beam element with uniform cross-section.

5.2 EULER-BERNOULLI BEAM

Earlier formulations of the beam models employed the assumption of Euler-Bernoulli's theory. The Euler-Bernoulli Beam equation is based on the assumption that the plane sections normal to the neutral axis before deformation remain plane as well as normal to the neutral axis after deformation (Figure 5.2). This assumption denotes that the cross-section rotation, θ about the principal central y axis during deflection of the beam, is given by $-\frac{dw}{dx}$ (*i.e.* negative of the slope of deflection in terms of x) (Figure 5.2).

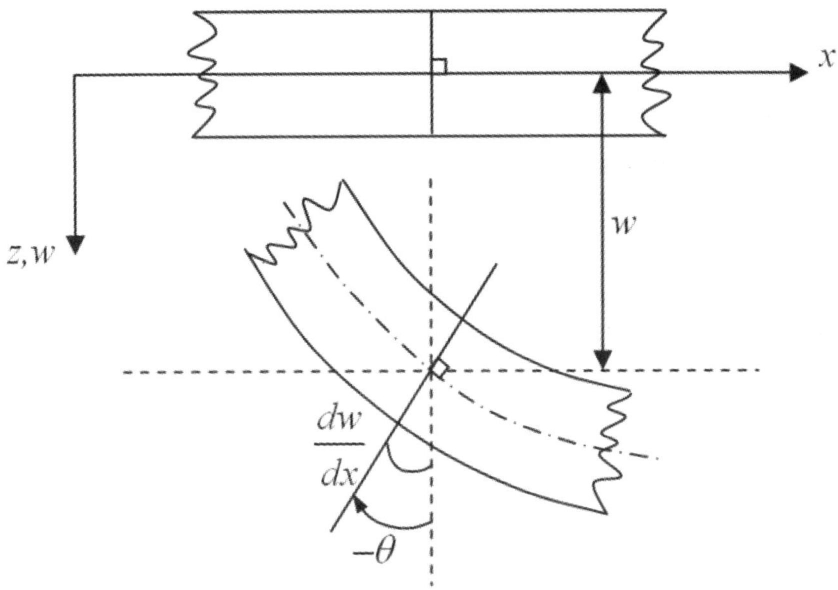

Fig. 5.2. Euler-Bernoulli beam

The axial displacement of any point in the cross-section at a distance z from the midplane is expressed as

$$u = z\theta = -z\frac{dw}{dx}. \tag{5.1}$$

The axial strain is given as

$$\varepsilon = \frac{du}{dx} = -z\frac{d^2w}{dx^2} \tag{5.2}$$

and the axial stress can be written as

$$\sigma = E\varepsilon = -Ez\frac{d^2w}{dx^2}. \tag{5.3}$$

The moment at the cross-section is given as

$$M = \int_{-h/2}^{h/2} \sigma z\, b\, dz = -EI\frac{d^2w}{dx^2}, \tag{5.4}$$

where b and h are width and height of the beam respectively, and I is the second moment of area (or moment of inertia) of the cross-section of the beam with respect to the y axis and is given as

$$I = \int_{-h/2}^{h/2} z^2 b \, dz \, .$$

I is assumed to be uniform for a prismatic beam member.

5.2.1 Governing equation

It is now worthwhile to derive governing equation in terms of transverse displacement, w. Equilibrium equations are obtained by free body diagram shown in Figure 5.3. In the figure, Q denotes shear force while q(x) denotes distributed load per unit length in the beam. Moment equilibrium about y axis and force equilibrium along z-axis yield after neglecting the higher order terms

$$\frac{dM}{dx} - Q = 0 \qquad (5.5)$$

and

$$\frac{dQ}{dx} + q(x) = 0 \, . \qquad (5.6)$$

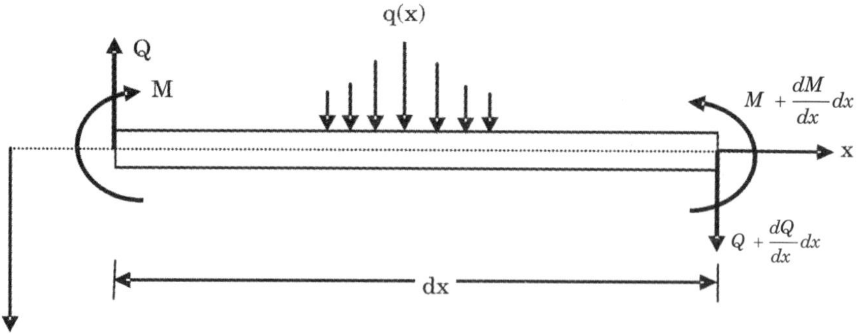

Fig. 5.3. Free body diagram of a differential element in a beam.

Eliminating the shear force from Eqs. (5.5) and (5.6) and using Eq. (5.4), we obtain the following governing equation:

$$\frac{d^2}{dx^2}\left(EI \, \frac{d^2 w}{dx^2} \right) = q \, . \qquad (5.7)$$

5.3 SHEAR DEFORMATION IN BEAMS

In the simple theory of bending of beams discussed in the earlier section, it is assumed that plane sections before bending remain plane even after bending. In other words, any deformation due to shear across the section is not accounted for (no shear deformation). When shear forces are present shear stresses are introduced which in turn cause angular distortion of the beam and as a consequence the plane sections no longer remain plane and neither remain perpendicular to the neutral axis after deformation. Also, warping of the adjacent sections will be different in the problems of varying shear force and this invariably introduces longitudinal stresses and strains. Thus, our assumption that the longitudinal stresses vary linearly across the section is no longer applicable.

5.3.1 Timoshenko beam

Timoshenko improved upon the Euler-Bernoulli theory of beams by adding the effect of transverse shear deformation into the beam equation. The kinematic assumptions of the Timoshenko theory are (a) normals to the axis of the beam remain straight after deformation and (b) there is no change in beam thickness after deformation. However, planes normal to the beam axis before deformation are not required to remain perpendicular to the axis after deformation. Figure 5.4 shows the deformation.

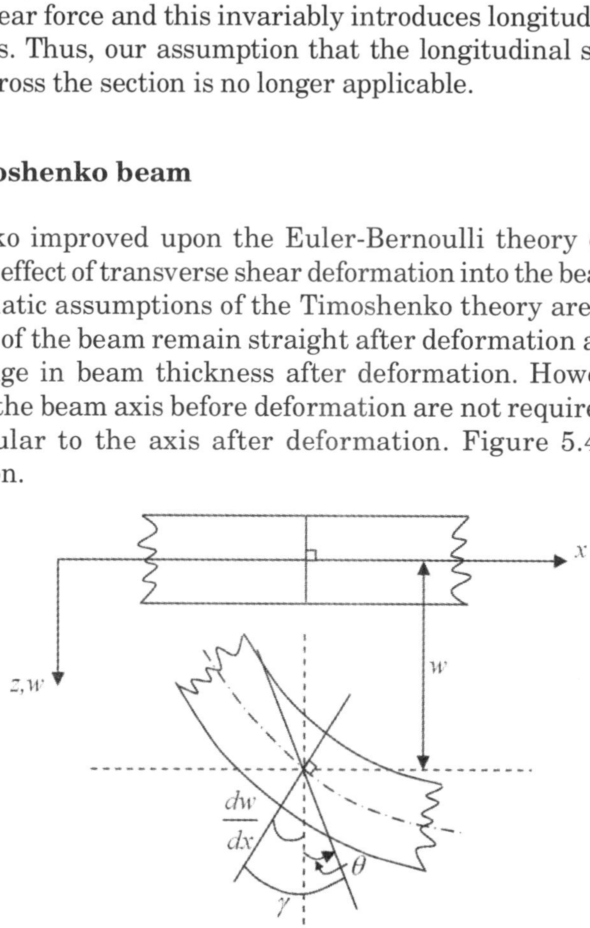

Fig. 5.4. Timoshenko beam.

Let u and w be the axial and transverse displacements of a point P in the cross-section of the beam at a distance x from the origin, respectively. Because of the transverse shear deformation, the rotation of the cross-section at a distance x after deformation will be different from $\frac{dw}{dx}$ and can be expressed as

$$\theta = -\frac{dw}{dx} + \gamma \,. \tag{5.8}$$

In Eq. (5.8), γ is the transverse shear strain. As a result, the displacement field in the Timoshenko beam can be written as

$$u(x, z) = z\,\theta(x),$$

$$w(x) = w \,. \tag{5.9}$$

From Eqs. (5.8) and (5.9), the axial and shear strains are given as

$$\varepsilon_x = \frac{du}{dx} = z\,\frac{d\theta}{dx} = z\,\kappa_x \,,$$

$$\gamma = \frac{dw}{dx} + \theta \,. \tag{5.10}$$

For Timoshenko beam, w and θ are independent variables and need to be interpolated within each element in order to derive element stiffness matrix.

5.4. THE 3-NODED TIMOSHENKO BEAM ELEMENT

The three-noded isoparametric quadratic shear flexible beam element based on Timoshenko theory is described here. Figure 5.5 shows the 3-noded beam element of length L^e with nodes at x_1, x_2 and x_3.

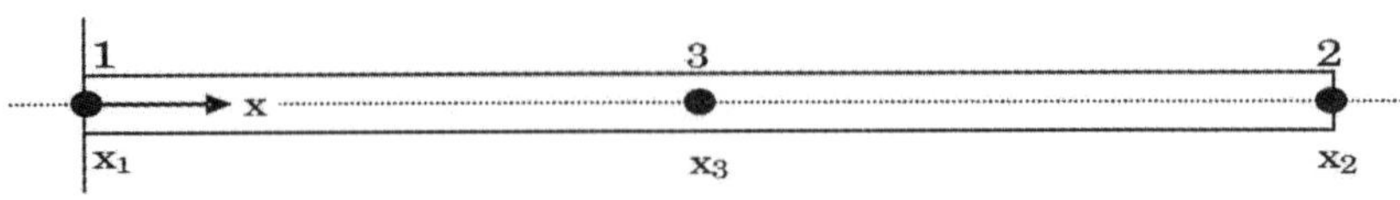

Figure 5.5. The 3-noded Timoshenko beam element.

If ξ *is* the natural coordinate so that x_1 is at $\xi = -1$, x_2 is at $\xi = 1$ and x_3 is at $\xi = 0$, then the parametric shape functions are defined as

$$N_1 = \frac{1}{2}\xi(\xi - 1),\; N_2 = \frac{1}{2}\xi(\xi + 1) \text{ and } N_3 = \left(1 - \xi^2\right). \tag{5.11}$$

Since w and θ are independent variables, we can interpolate them independently using the above shape functions, *i.e.*

$$w = \sum_{i=1}^{3} N_i w_i \tag{5.12}$$

$$\theta = \sum_{i=1}^{3} N_i \theta_i \;, \tag{5.13}$$

where w_i are the nodal vertical displacements and θ_i are the nodal rotation terms.

The parametric (better known as isoparametric) formulation requires the interpolation of the x coordinate as

$$x = \sum_{i=1}^{3} N_i x_i \;, \tag{5.14}$$

where x_i are the nodal coordinates.

Eqs. (5.12) and (5.13) can be represented in matrix form as

$$\{u\} = \begin{Bmatrix} w \\ \theta \end{Bmatrix} = [N]\{U^e\}, \tag{5.15}$$

where shape function matrix and the element degrees of freedom vector $\{U^e\}$ are given as

$$[N] = \begin{bmatrix} N_1 & 0 & N_2 & 0 & N_3 & 0 \\ 0 & N_1 & 0 & N_2 & 0 & N_3 \end{bmatrix},$$

$$\{U^e\} = [w_1 \;\; \theta_1 \;\; w_2 \;\; \theta_2 \;\; w_3 \;\; \theta_3]^T. \tag{5.16}$$

Using the degrees of freedom interpolation relation (Eq. (5.15)), the strain displacement relation can be written as

$$\{\varepsilon\} = \begin{Bmatrix} \varepsilon_x \\ \gamma \end{Bmatrix} = [B]\{U^e\}, \tag{5.17}$$

where $[B]$ is strain-displacement matrices of size 2 × 6 evaluated to be

$$[B] = \begin{bmatrix} 0 & z\dfrac{\partial N_1}{\partial x} & 0 & z\dfrac{\partial N_2}{\partial x} & 0 & z\dfrac{\partial N_3}{\partial x} \\ \dfrac{\partial N_1}{\partial x} & N_1 & \dfrac{\partial N_2}{\partial x} & N_2 & \dfrac{\partial N_3}{\partial x} & N_3 \end{bmatrix}. \qquad (5.18)$$

The constitutive relation is expressed as

$$\{\sigma\} = \begin{Bmatrix} \sigma_x \\ \tau \end{Bmatrix} = [C]\{\varepsilon\}, \qquad (5.19)$$

in which $[C]$ is the appropriate constitutive matrix.

5.4.1 Element stiffness matrix

The usual expression for the stiffness matrix for the element can be written as

$$\left[K^e\right] = \int_{V^e} [B]^T [C][B]dV\ . \qquad (5.20)$$

A typical submatrix of $[K^e]$ relating nodes i and j may be evaluated from the expression

$$\left[K^e_{ij}\right] = \int_{V^e} [B_i]^T [C]\left[B_j\right]dV$$

or

$$\left[K^e_{ij}\right] = \int_{V^e} \begin{bmatrix} 0 & z\dfrac{\partial N_i}{\partial x} \\ \dfrac{\partial N_i}{\partial x} & N_i \end{bmatrix}^T \begin{bmatrix} E & 0 \\ 0 & G \end{bmatrix} \begin{bmatrix} 0 & z\dfrac{\partial N_j}{\partial x} \\ \dfrac{\partial N_j}{\partial x} & N_j \end{bmatrix} dV\ . \qquad (5.21)$$

This may be written as

$$\left[K^e_{ij}\right] = \begin{bmatrix} \int_{V^e} G\dfrac{\partial N_i}{\partial x}\dfrac{\partial N_j}{\partial x}dV & \int_{V^e} GN_j\dfrac{\partial N_i}{\partial x}dV \\ \int_{V^e} GN_i\dfrac{\partial N_j}{\partial x}dV & \int_{V^e} z^2E\dfrac{\partial N_i}{\partial x}\dfrac{\partial N_j}{\partial x}dV + \int_{V^e} GN_iN_jdV \end{bmatrix}. \qquad (5.22)$$

Substituting $dV = dA\,dx$ into Eq. (5.22) and using the relation $dx = \det[J]d\xi$, we get

$$[K_{ij}^e] = \begin{bmatrix} \alpha GA\int_{-1}^{1}\frac{\partial N_i}{\partial x}\frac{\partial N_j}{\partial x}\det[J]d\xi & \alpha GA\int_{-1}^{1}N_j\frac{\partial N_i}{\partial x}\det[J]d\xi \\ \alpha GA\int_{-1}^{1}N_i\frac{\partial N_j}{\partial x}\det[J]d\xi & \begin{matrix} EI\int_{-1}^{1}\frac{\partial N_i}{\partial x}\frac{\partial N_j}{\partial x}\det[J]d\xi + \\ \alpha GA\int_{-1}^{1}N_iN_j\det[J]d\xi \end{matrix} \end{bmatrix}, \quad (5.23)$$

where $J = \frac{dx}{d\xi}$ represents the Jacobian, governing the transformation from x to ξ space. In foregoing equation, A is the area of cross-section of the beam, I is the second moment of area given as $I = \int_A z^2 dA$ and α has been introduced to represent correction factor for shear energy whose value is normally 5/6.

The integrals in Eq. (5.23) can be calculated explicitly, however it is more common to use a numerical integration technique such as Gaussian quadrature.

5.4.2 Element nodal force vector

A distributed lateral pressure load applied on the beam element can be represented in terms of nodal forces by making use of the general consistent nodal force formulae. The nodal force vector $\{F^e\}$ due to surface traction is given by the usual expression

$$\{F_s^e\} = \int_{S_F^e}[N^s]^T\{t\}dS. \quad (5.24)$$

If the beam is subjected to an uniformly distributed vertical load specified as intensity q per unit length, the above integral reduces to

$$\{F_s^e\} = \int_{L^e}[N^s]^T\begin{Bmatrix} q \\ 0 \end{Bmatrix}dx = \int_{-1}^{1}[N^s]^T\begin{Bmatrix} q \\ 0 \end{Bmatrix}\det[J]d\xi. \quad (5.25)$$

Substituting for $[N^s]$ which is same as $[N]$ (Eq. (5.16)), we can calculate element nodal force vector corresponding to six element degrees of freedom.

5.4.3 Alternative formulation

In case of beams, it is sometimes convenient to express the constitutive relation and finite element model in terms of stress-resultants.

If we set

$$\{\bar{\varepsilon}\} = \begin{Bmatrix} k_x \\ \gamma \end{Bmatrix}, \tag{5.26}$$

the constitutive relation can be written as

$$\{\bar{\sigma}\} = [\bar{D}]\{\bar{\varepsilon}\}, \tag{5.27}$$

where $\{\bar{\sigma}\}$ is the stress resultant vector and $[\bar{D}]$ is the stiffness matrix. These are given as

$$\{\bar{\sigma}\} = [M \quad Q]^T = \left[\int_A (\sigma z \quad \tau) dA\right]^T. \tag{5.28}$$

In the above equation, M is the bending moment and Q is the shear force. Matrix $[\bar{D}]$ is then given by

$$[\bar{D}] = \begin{bmatrix} D & 0 \\ 0 & S \end{bmatrix}, \tag{5.29}$$

where the stiffness coefficients are D=EI and S=αGA.

The strain-displacement relations can be represented as

$$\{\bar{\varepsilon}\} = [\bar{B}]\{U^e\}, \tag{5.30}$$

where strain-displacement matrix $[\bar{B}]$ is given as

$$[\bar{B}] = \begin{bmatrix} 0 & \frac{\partial N_1}{\partial x} & 0 & \frac{\partial N_2}{\partial x} & 0 & \frac{\partial N_3}{\partial x} \\ \frac{\partial N_1}{\partial x} & N_1 & \frac{\partial N_2}{\partial x} & N_2 & \frac{\partial N_3}{\partial x} & N_3 \end{bmatrix}. \tag{5.31}$$

In this case, the principle of virtual work for the beam can be written as

$$\delta I = \int_L \{\delta\bar{\varepsilon}\}^T \{\bar{\sigma}\} dx - \int_L \{\delta u^s\}^T \{t\} dx = 0 , \tag{5.32}$$

where traction vector $\{t\}$ is again given as $\{t\} = \begin{Bmatrix} q \\ 0 \end{Bmatrix}$.

Rewriting Eq. (5.32) as a sum of integration over the lengths of all finite elements and substituting for variations of the displacements and strains within each element, we obtain

$$\sum_e \{\delta U^e\}^T \left(\int_{L^e} [\bar{B}]^T \{\bar{\sigma}\} dx \right) = \sum_e \{\delta U^e\}^T \left(\int_{L^e} [N^s]^T \{t\} dx \right). \tag{5.33}$$

Since $\{\delta U\}$ is arbitrary, Eq. (5.33) can be written as the following equation of equilibrium:

$$\left[\sum_e \left(\int_{L^e} [\bar{B}]^T \{\bar{D}\} [\bar{B}] dx \right) \right] \{U\} = \sum_e \left(\int_{L^e} [N^s]^T \{t\} dx \right). \tag{5.34}$$

The stiffness matrix for an element is then

$$[K^e] = \int_{L^e} [\bar{B}]^T \{\bar{D}\} [\bar{B}] dx . \tag{5.35}$$

It may be noted in the above equation that the stiffness matrix is now reduced to a line integral as the integration across the area of cross-section has been done in Eq. (5.28) and accordingly the stress resultants have been considered.

Eq. (5.35) can again be evaluated explicitly or by using numerical integration.

5.4.4 Shear locking

The element stiffness submatrices $[K^e_{ij}]$ can be separated into two parts as

$$[K^e_{ij}] = \begin{bmatrix} 0 & 0 \\ 0 & EI \int_{-1}^{1} \frac{\partial N_i}{\partial x} \frac{\partial N_j}{\partial x} \det[J] d\xi \end{bmatrix} +$$
$$\begin{bmatrix} \alpha GA \int_{-1}^{1} \frac{\partial N_i}{\partial x} \frac{\partial N_j}{\partial x} \det[J] d\xi & \alpha GA \int_{-1}^{1} N_j \frac{\partial N_i}{\partial x} \det[J] d\xi \\ \alpha GA \int_{-1}^{1} N_i \frac{\partial N_j}{\partial x} \det[J] d\xi & \alpha GA \int_{-1}^{1} N_i N_j \det[J] d\xi \end{bmatrix}$$

or

$$\left[K_{ij}^{e}\right]=\left[K_{b(ij)}^{e}\right]+\left[K_{s(ij)}^{e}\right], \tag{5.36}$$

in which $\left[K_{b(ij)}^{e}\right]$ refers to bending stiffness and $\left[K_{s(ij)}^{e}\right]$ refers to shear stiffness.

The relation in Eq (5.36) shows the relative importance of the bending and shearing contributions to the stiffness matrix of an element. Here we can note that the factor $\alpha GA/EI$ can be very large when a thin element is considered and thus bending term becomes negligible compared to the shear term. This also means that as the beam becomes thin, the constraint of zero shear deformations ($\gamma=0$) will be approached. The imposition of this constraint leads to the deterioration of the stiffness matrix and overstiff results are obtained. The very stiff behavior exhibited by the thin elements is referred to as element shear locking. One of the simplest ways to avoid the shear locking is to relax the constraint imposed by the shear terms by under-integrating the coefficients of $\left[K_{s(ij)}^{e}\right]$ in the numerical integration. The remaining terms in the stiffness matrix can be exactly integrated in an attempt to retain the required rank of the overall stiffness matrix. This process is called selective integration.

5.5. EXAMPLE PROBLEMS

5.5.1 Example 1: A rectangular cantilever beam

A rectangular cantilever beam (dimension: a = 10, b = 2 and c = 2) is subjected to following different loading cases (Figure 5.6):

a. Concentrated bending moment of -4000 units at the free end (case 1),

b. Vertical tip force of 600 units at the free end (case 2) and

c. Uniformly distributed load per unit width of 80 units (case 3).

Material properties are taken as E = 1500 and ν = 0.00. Consistent units are assumed here. It is desired to calculate the deflection at the free end, bending stress at the outer fiber of the beam and shear stress at the fixed end using finite element method.

The SB-FEM command file for this example problem is listed in Appendix 8. The beam is modeled with 8 elements. Selective integration is used here. The FEM results are compared with analytical solutions (based on simple bending theory) as presented in Table 5.1 and are found to be in well agreement with analytical results.

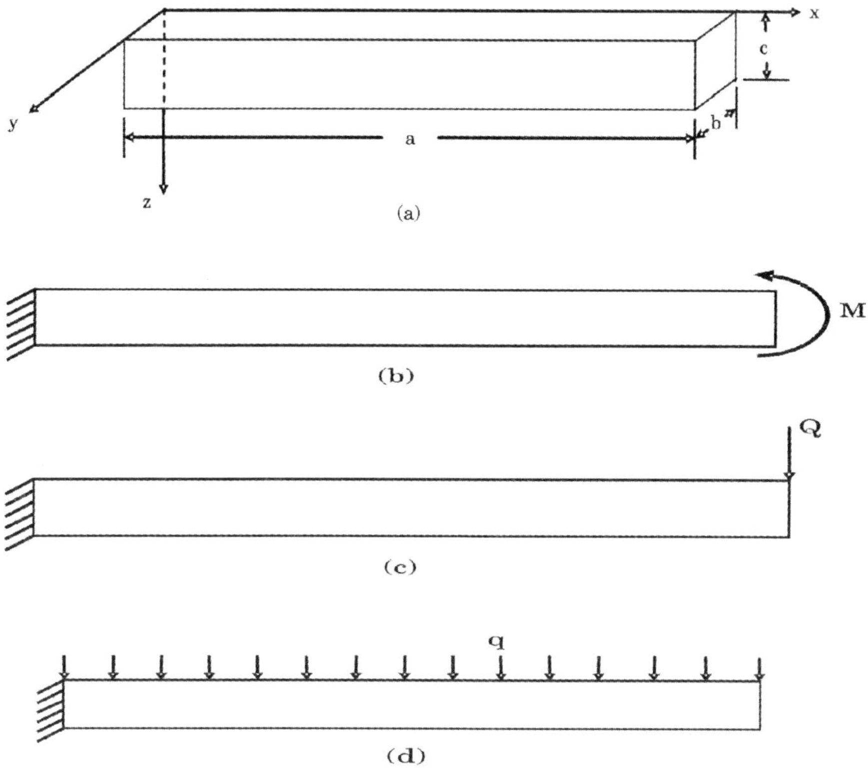

Fig. 5.6. Problem description of a rectangular cantilever beam: (a) geometry, (b) subjected to a bending moment (M), (c) subjected to a point load (Q) at the free end and (d) subjected to a uniformly distributed load (q).

Table 5.1 Comparison of results for a rectangular cantilever beam.

Loading case	*Deflection at the free end*		*Bending stress at the outer fibre*		*Shear stress at the fixed end*	
	Present	*Exact*	*Present*	*Exact*	*Present*	*Exact*
1	100	100	3000	3000	0	0
2	102	100	4500	4500	150	150
3	102.66	100	5937	6000	400	400

6

Two-dimensional Stress Analysis

6.1. INTRODUCTION

There are many problems in engineering which can be treated/idealized to 2D problems to obtain satisfactory results by carrying out analyses using 2D models, which can be much more efficient and cost-effective compared to conducting full 3D analyses. The governing system equation can also be drastically simplified. In stress analysis, the 2D problems comprise plane stress and plane strain cases. The formulations for plane stress and plane strain problems are very much the same, except for the difference in the material constant matrix. A 2D solid element, be it plane strain or plane stress, can be triangular, rectangular or quadrilateral in shape with straight or curved edges. The most often used elements in engineering practice are linear. Quadratic and higher order elements are also used for situations that require high accuracy in stress. The 2D formulation can easily be extended to the analysis of axisymmetric solids subjected to axisymmetric loading.

In this chapter, formulations of triangular elements (linear and quadratic) and quadrilateral elements (four noded, eight noded and nine noded) are described. Axisymmetric stress analysis using four noded quadrilateral element is also discussed.

6.2. TRIANGULAR ELEMENTS

Triangular elements are relatively simple in their formulation and prove to be quite versatile for the purpose of discretizing any two-dimensional continuum. Two types of triangular elements (linear and quadratic) are formulated in the following discussion.

6.2.1. Linear triangular element

This is the simplest two-dimensional element and is also called Constant-Strain Triangle (CST) element. For this element, we have three nodes at each vertex, which are numbered around the element in the counter-clockwise direction. Each node has two degrees of freedom (*e.g.* displacements) in the x- and y- directions (Figure 6.1). The element nodal displacements are represented by the vector

$$\{U^e\} = [u_1, v_1, \quad u_2, v_2, \quad u_3, v_3]^T, \tag{6.1}$$

where u_r and v_r are nodal displacements of node r in x and y directions respectively.

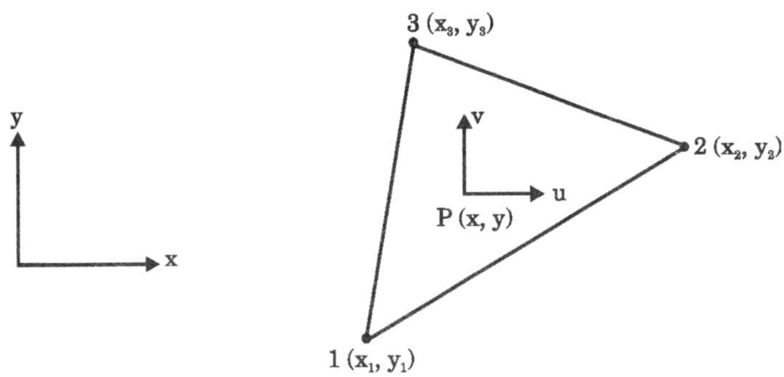

Fig. 6.1 Linear triangular element

The displacements u and v are assumed to be linear functions within the element such that at any point P we have

$$u = a_0 + a_1 x + a_2 y$$

$$v = b_0 + b_1 x + b_2 y, \tag{6.2}$$

where a_i and b_i (i = 0,1,2) are constants. From these, strains can be found as

$$\{\varepsilon\} = \begin{Bmatrix} \varepsilon_x \\ \varepsilon_y \\ \gamma_{xy} \end{Bmatrix} = \begin{Bmatrix} \dfrac{\partial u}{\partial x} \\ \dfrac{\partial v}{\partial y} \\ \dfrac{\partial u}{\partial y} + \dfrac{\partial v}{\partial x} \end{Bmatrix} = \begin{Bmatrix} a_1 \\ b_2 \\ a_2 + b_1 \end{Bmatrix}, \tag{6.3}$$

which are constants throughout the element.

Displacements given by Eq. (6.1) should satisfy the following equations at each node:

$$\left.\begin{aligned} u_r &= a_0 + a_1 x_r + a_2 y_r \\ v_r &= b_0 + b_1 x_r + b_2 y_r \end{aligned}\right\} \quad (r = 1,2,3). \tag{6.4}$$

Solving these equations, the coefficients a_i and b_i can be determined in terms of displacements and coordinates of nodes. Putting values of these coefficients into Eq. (6.2) and rearranging the terms we obtain

$$u(x,y) = \sum_{i=1}^{3} N_i(x,y) u_i$$

$$v(x,y) = \sum_{i=1}^{3} N_i(x,y) v_i, \tag{6.5}$$

where N_1, N_2 and N_3 are defined as shape functions and have the following values:

$$N_1 = \frac{1}{2A}\left\{(x_2 y_3 - x_3 y_2) + (y_2 - y_3)x + (x_3 - x_2)y\right\}$$

$$N_2 = \frac{1}{2A}\left\{(x_3 y_1 - x_1 y_3) + (y_3 - y_1)x + (x_1 - x_3)y\right\}$$

$$N_3 = \frac{1}{2A}\left\{(x_1 y_2 - x_2 y_1) + (y_1 - y_2)x + (x_2 - x_1)y\right\}. \tag{6.6}$$

Here $A = \frac{1}{2}\det\begin{bmatrix} 1 & x_1 & y_1 \\ 1 & x_2 & y_2 \\ 1 & x_3 & y_3 \end{bmatrix}$ is the area of the triangle.

In matrix notation, we can write

$$\{u\} = \begin{Bmatrix} u \\ v \end{Bmatrix} = [N]\{U^e\}, \tag{6.7}$$

where shape function matrix $[N]$ contains the appropriate N_i terms to satisfy Eq. (6.5) and given as

$$[N] = \begin{bmatrix} N_1 & 0 & N_2 & 0 & N_3 & 0 \\ 0 & N_1 & 0 & N_2 & 0 & N_3 \end{bmatrix}. \tag{6.8}$$

The Eq. (6.6) can also be written as

$$N_1 = \frac{A_1}{A},\ N_2 = \frac{A_2}{A},\ N_3 = \frac{A_3}{A}, \tag{6.9}$$

where $A_i (i = 1,2,3)$ is the area of triangle formed by nodes j and k and point P (Figure 6.2). The ratios of the areas of the subtriangles to that of the total triangle are also referred to as area coordinates.

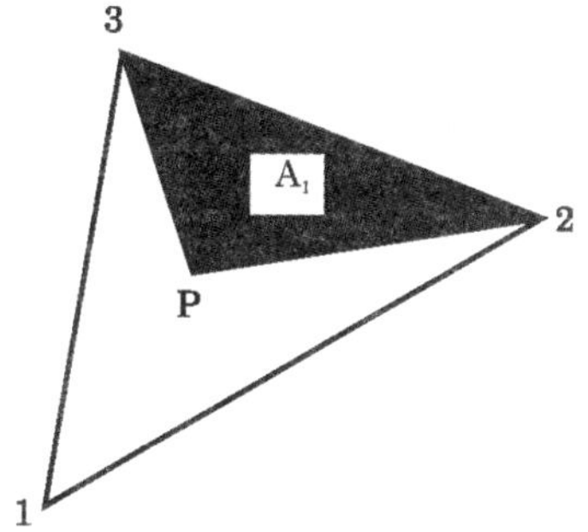

Fig. 6.2 Area coordinate system of the triangular element

The values of shape function N_1 thus define a plane surface shown shaded in Figure 6.3 and having a value of 1 at node 1 and linearly reducing to 0 at nodes 2 and 3. N_2 and N_3 have similar features.

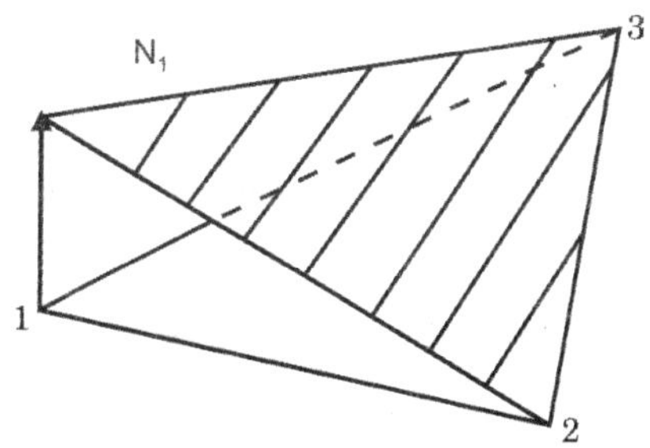

Fig 6.3 Shape function characteristics

It is to be noticed that

$$N_1 + N_2 + N_3 = 1. \tag{6.10}$$

Now, strain-displacement relation can be evaluated as

$$\{\varepsilon\} = \begin{bmatrix} \dfrac{\partial}{\partial x} & 0 \\ 0 & \dfrac{\partial}{\partial y} \\ \dfrac{\partial}{\partial y} & \dfrac{\partial}{\partial x} \end{bmatrix} \begin{Bmatrix} u \\ v \end{Bmatrix} = [\Delta]\{u\} \tag{6.11}$$

Substituting from Eq. (6.7), we get

$$\{\varepsilon\} = [B]\{U^e\}, \tag{6.12}$$

where

$$[B] = [\Delta][N] = \begin{bmatrix} \dfrac{\partial N_1}{\partial x} & 0 & \dfrac{\partial N_2}{\partial x} & 0 & \dfrac{\partial N_3}{\partial x} & 0 \\ 0 & \dfrac{\partial N_1}{\partial y} & 0 & \dfrac{\partial N_2}{\partial y} & 0 & \dfrac{\partial N_3}{\partial y} \\ \dfrac{\partial N_1}{\partial y} & \dfrac{\partial N_1}{\partial x} & \dfrac{\partial N_2}{\partial y} & \dfrac{\partial N_2}{\partial x} & \dfrac{\partial N_3}{\partial y} & \dfrac{\partial N_3}{\partial x} \end{bmatrix}. \tag{6.13}$$

By differentiating shape function terms above, we obtain

$$[B] = \frac{1}{2A} \begin{bmatrix} y_{23} & 0 & y_{31} & 0 & y_{12} & 0 \\ 0 & x_{32} & 0 & x_{13} & 0 & x_{21} \\ x_{32} & y_{23} & x_{13} & y_{31} & x_{21} & y_{12} \end{bmatrix}, \tag{6.14}$$

where $x_{ij} = x_i - x_j$ and $y_{ij} = y_i - y_j$ $(i, j = 1,2,3)$ as defined above.

Interpolation model in terms of local (natural) coordinates

We now introduce natural coordinates (ξ, η) on the triangle (Figure 6.4). We choose the unit right isosceles triangle as the master element. Then the shape functions for this linear element can be represented as

$$N_1 = \xi,\ N_2 = \eta,\ N_3 = 1 - \xi - \eta. \tag{6.15}$$

It can be noted again that the values of N_i in Eq. (6.16), we defines a plane surface whose value is one at node i and zero at other two nodes.

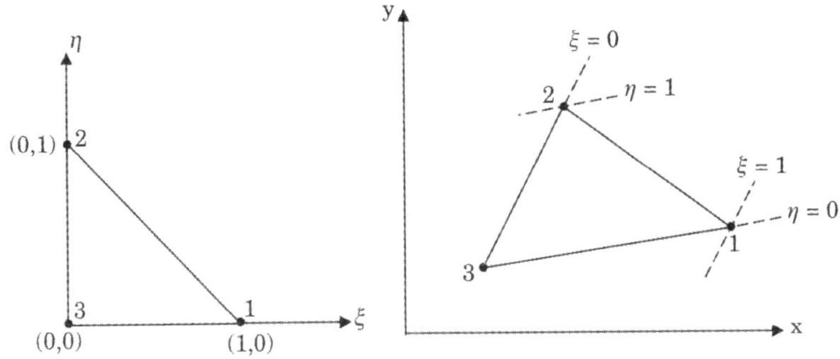

Fig. 6.4 Natural coordinates for a triangle

We have two coordinate systems for the element: the global coordinates (x,y) and the natural coordinates (ξ, η). The relation between the two can be given as

$$x(\xi,\eta) = \sum_{\alpha} N_{\alpha}(\xi,\eta)x_{\alpha}$$

$$y(\xi,\eta) = \sum_{\alpha} N_{\alpha}(\xi,\eta)y_{\alpha}, \tag{6.16}$$

where summation is over all three nodes.

Elements that use the same interpolation functions to define the element geometry as well as to describe the variation of the field variable within it are called isoparametric elements. The field variable within the element can be evaluated as

$$u = \sum_{\alpha} N_{\alpha}(\xi,\eta)u_{\alpha}$$

$$v = \sum_{\alpha} N_{\alpha}(\xi,\eta)v_{\alpha}. \tag{6.17}$$

Putting values of N_i in Eq. (6.16), we have

$$x = (x_1 - x_3)\xi + (x_2 - x_3)\eta + x_3 = x_{13}\xi + x_{23}\eta + x_3,$$

$$y = (y_1 - y_3)\xi + (y_2 - y_3)\eta + y_3 = y_{13}\xi + y_{23}\eta + y_3. \tag{6.18}$$

Displacement u or v on the element can be viewed as functions of (x,y) or (ξ,η). Using the chain rule of derivatives, we get

$$\begin{Bmatrix} \dfrac{\partial u}{\partial \xi} \\ \dfrac{\partial u}{\partial \eta} \end{Bmatrix} = \begin{bmatrix} \dfrac{\partial x}{\partial \xi} & \dfrac{\partial y}{\partial \xi} \\ \dfrac{\partial x}{\partial \eta} & \dfrac{\partial y}{\partial \eta} \end{bmatrix} \begin{Bmatrix} \dfrac{\partial u}{\partial x} \\ \dfrac{\partial u}{\partial y} \end{Bmatrix} = [J] \begin{Bmatrix} \dfrac{\partial u}{\partial x} \\ \dfrac{\partial u}{\partial y} \end{Bmatrix}, \tag{6.19}$$

where $[J]$ is called Jacobian matrix of coordinate transformation and is evaluated as

$$[J] = \begin{bmatrix} x_{13} & y_{13} \\ x_{23} & y_{23} \end{bmatrix}. \tag{6.20}$$

The equation can be inverted as

$$\begin{Bmatrix} \dfrac{\partial u}{\partial x} \\ \dfrac{\partial u}{\partial y} \end{Bmatrix} == [J]^{-1} \begin{Bmatrix} \dfrac{\partial u}{\partial \xi} \\ \dfrac{\partial u}{\partial \eta} \end{Bmatrix}, \tag{6.21}$$

in which the inverse of the Jacobian matrix can be evaluated as

$$[J]^{-1} = \frac{1}{\det[J]} \begin{bmatrix} y_{23} & -y_{13} \\ -x_{23} & x_{13} \end{bmatrix}, \tag{6.22}$$

where $\det[J] = x_{13}y_{23} - x_{23}y_{13} = 2A.$

Strain-displacement relation can be evaluated as

$$\{\varepsilon\} = \begin{Bmatrix} \dfrac{\partial u}{\partial x} \\ \dfrac{\partial v}{\partial y} \\ \dfrac{\partial u}{\partial y} + \dfrac{\partial v}{\partial x} \end{Bmatrix} = [A] \begin{bmatrix} \dfrac{\partial u}{\partial \xi} & \dfrac{\partial u}{\partial \eta} & \dfrac{\partial v}{\partial \xi} & \dfrac{\partial v}{\partial \eta} \end{bmatrix}^T, \tag{6.23}$$

where [A] is a 3×4 matrix whose co-efficients are found out by putting appropriate terms from the inverse of Jacobian matrix as

$$[A] = \begin{bmatrix} J^*_{11} & J^*_{12} & 0 & 0 \\ 0 & 0 & J^*_{21} & J^*_{22} \\ J^*_{21} & J^*_{22} & J^*_{11} & J^*_{12} \end{bmatrix}. \tag{6.24}$$

In Eq. (6.24), J^*_{ij} are the coefficients of $[J]^{-1}$ matrix.

Also we have

$$\begin{Bmatrix} \frac{\partial u}{\partial \xi} \\ \frac{\partial u}{\partial \eta} \\ \frac{\partial v}{\partial \xi} \\ \frac{\partial v}{\partial \eta} \end{Bmatrix} = [G]\{U^e\}, \tag{6.25}$$

where [G] is a 4 × 6 matrix whose coefficients are evaluated by differentiating shape function terms with respect to natural coordinates (ξ, η) and is given as

$$[G] = \begin{bmatrix} \frac{\partial N_1}{\partial \xi} & 0 & \frac{\partial N_2}{\partial \xi} & 0 & \frac{\partial N_3}{\partial \xi} & 0 \\ \frac{\partial N_1}{\partial \eta} & 0 & \frac{\partial N_2}{\partial \eta} & 0 & \frac{\partial N_3}{\partial \eta} & 0 \\ 0 & \frac{\partial N_1}{\partial \xi} & 0 & \frac{\partial N_2}{\partial \xi} & 0 & \frac{\partial N_3}{\partial \xi} \\ 0 & \frac{\partial N_1}{\partial \eta} & 0 & \frac{\partial N_2}{\partial \eta} & 0 & \frac{\partial N_3}{\partial \eta} \end{bmatrix}. \tag{6.26}$$

Hence, strain can be expressed as

$$\{\varepsilon\} = [A]\ [G]\ \{U^e\} = [B]\ \{U^e\}, \tag{6.27}$$

where

$$[B] = [A]\ [G]. \tag{6.28}$$

$[B]$ is strain displacement matrix of size 3 × 6 and can be calculated as

$$[B] = \frac{1}{\det[J]} \begin{bmatrix} y_{23} & 0 & y_{31} & 0 & y_{12} & 0 \\ 0 & x_{32} & 0 & x_{13} & 0 & x_{21} \\ x_{32} & y_{23} & x_{13} & y_{31} & x_{21} & y_{12} \end{bmatrix}. \tag{6.29}$$

The above equation for $[B]$ matrix resembles Eq. (6.14) derived earlier.

Element stiffness matrix

Using the general relationship, we obtain element stiffness matrix for this CST element as

$$\left[K^e\right] = \int_{V^e} [B]^T [C][B]\, dV = hA\left([B]^T [C][B]\right), \tag{6.30}$$

in which h is the thickness of the element. It is to be noted that the integrand is constant in Eq. (6.30) since the matrix $[B]$ as given by Eq. (6.29) is constant and also the elasticity matrix $[C]$ is assumed to be uniform over the element.

Element force vectors

Body force : A body force $\{b\} = \left[b_x \quad b_y\right]^T$ which is distributed force per unit volume, contributes to the global load vector $\{F\}$. The element body force vector is given by

$$\left\{F_b^e\right\} = \int_{V^e} [N]^T \{b\}\, dV = h\int_{A^e} [N]^T \{b\}\, dA$$

or

$$\left\{F_b^e\right\} = \left[hb_x \int_{A^e} N_1 dA, \quad hb_y \int_{A^e} N_1 dA, \quad \ldots\right]^T. \tag{6.31}$$

A typical integrand in the above vector can be evaluated as

$$\int_{A^e} N_1 dA = \int_0^1 \int_0^{1-\xi} N_1 \det[J]\, d\eta d\xi = 2A\int_0^1 \int_0^{1-\xi} \xi d\eta d\xi = \frac{1}{3}A. \tag{6.32}$$

It is useful to note that there is a general formula for integration of quantities defined as polynomials of natural coordinates over the triangular region, that is,

$$\int_{A^e} \xi^a \eta^b dA = \int_0^1 \int_0^{1-\xi} \xi^a \eta^b \det[J]\, d\eta d\xi = \frac{a!b!}{(a+b+2)!} 2A. \tag{6.33}$$

It is easy to show that all other integrands in Eq. (6.31) are evaluated to be same, *i.e.* $\frac{1}{3}A$. Finally, we have

$$\left\{F_b^e\right\} = \frac{hA}{3}\left[b_x \quad b_y \quad b_x \quad b_y \quad b_x \quad b_y\right]^T. \tag{6.34}$$

Traction force : Let us now consider uniformly distributed traction force applied on the boundary surface. The boundary surface of a triangular element is a line (edge). In order to be consistent with the order of listing of nodal connectivity numbers in the element topology definition, the two nodes forming the loaded edge is also listed in a counter-clockwise sequence with respect to the element. Thus with reference to Figure 6.5 the loaded element edge would be defined by the nodal sequence 1–2. Assuming that a traction force $\{t\} = \left[t_x \quad t_y\right]^T$, a force per unit area, is applied on the line 1–2 and l_{1-2} is the length of the line, the element traction force vector $\left\{F_s^e\right\}$ is given by in which

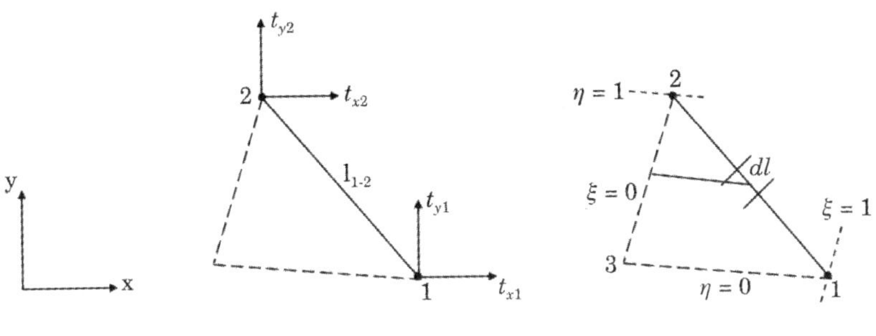

Fig. 6.5 Traction force on a triangle

$$\left\{F_s^e\right\} = h\int_{l_{1-2}} \left[N^s\right]^T \{t\}\, dl, \tag{6.35}$$

$[N^s]$ sis the surface interpolation matrix pertaining to line $1-2$ and can be derived from the shape function matrix $[N]$ by imposing $N_3 = 0$ (which is equivalent to $\eta = 1 - \xi$) along the line . Thus we have

$$\left[N^s\right] = \begin{bmatrix} N_1 & 0 & N_2 & 0 & 0 & 0 \\ 0 & N_1 & 0 & N_2 & 0 & 0 \end{bmatrix}. \tag{6.36}$$

With reference to the Figure 6.5, the elementary length dl is given as: $dl = -l_{1-2}d\xi$.

Assuming that the traction force $\{t\}$ is constant on the line $1-2$, Eq. (6.35) can be evaluated as

$$\left\{F_s^e\right\} = -hl_{1-2}\int_1^0 \left[N^s\right]^T \{t\}\, d\xi = \frac{hl_{1-2}}{2}\left[t_x \quad t_y \quad t_x \quad t_y \quad 0 \quad 0\right]^T. \tag{6.37}$$

It can be seen that the element traction forces $\left\{F_s^e\right\}$ are equally distributed between the two nodes 1 and 2, which define the loaded edge.

The traction forces $\{t\}$ need not be uniform but can vary (independently) along the element edge. Since the element is a linear one, at best a linear loading distribution of $\{t\}$ can be accommodated. The linear variation of $\{t\}$ on the line 1–2 can be represented by the following interpolation function (Figure 6.5):

$$t_x = N_1 t_{x1} + N_2 t_{x2}$$

$$t_y = N_1 t_{y1} + N_2 t_{y2}, \tag{6.38}$$

where $\left(t_{x\alpha}, t_{y\alpha}\right)$ is the traction force at the αth node.

Writing the above relations in matrix form, we have

$$\{t\} = \left[N^s\right]\left\{T^e\right\}, \tag{6.39}$$

in which

$$\left\{T^e\right\} = \left[t_{x1} \quad t_{y1} \quad t_{x2} \quad t_{y2} \quad 0 \quad 0\right]^T. \tag{6.40}$$

Substituting Eq. (6.39) into Eq. (6.35), we have

$$\left\{F_s^e\right\} = \left(h\int_{l_{1-2}} \left[N^s\right]^T \left[N^s\right] dl\right)\left\{T^e\right\}. \tag{6.41}$$

Eq. (6.41) contains following integrands which can be evaluated as:

$$\int_{l_{1-2}} N_1^2 dl = \int_{l_{1-2}} \xi^2 dl = -l_{1-2}\int_1^0 \xi^2 d\xi = \frac{l_{1-2}}{3}$$

$$\int_{l_{1-2}} N_2^2 dl = \int_{l_{1-2}} \eta^2 dl = -l_{1-2}\int_1^0 (1-\xi)^2 d\xi = \frac{l_{1-2}}{3}$$

$$\int_{l_{1-2}} N_1 N_2 dl = \int_{l_{1-2}} \xi\eta dl = -l_{1-2}\int_1^0 \xi(1-\xi) d\xi = \frac{l_{1-2}}{6}. \tag{6.42}$$

Substituting the values from Eq. (6.42) into Eq. (6.41), we get

$$\left\{F_s^e\right\} = \frac{hl_{1-2}}{6}\left[2t_{x1} + t_{x2} \quad 2t_{y1} + t_{y2} \quad t_{x1} + 2t_{x2} \quad t_{y1} + 2t_{y2} \quad 0 \quad 0\right]^T. \tag{6.43}$$

In Eq. (6.43), $l_{1-2} = \sqrt{(x_2 - x_1)^2 + (y_2 - y_1)^2}$.

If pressure p is acting on the boundary of the element, it has to be converted into suitable traction terms for the computation of element traction force vector. With reference to Figure 6.6, we have

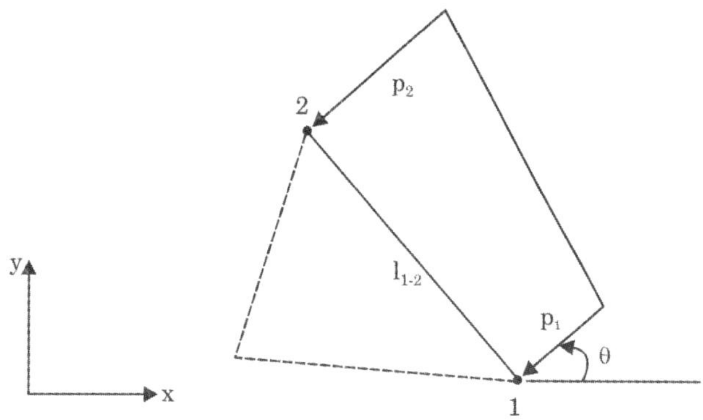

Fig. 6.6 Pressure on a triangle

$$\cos\theta = \frac{y_2 - y_1}{l_{1-2}}, \sin\theta = \frac{x_2 - x_1}{l_{1-2}}$$

$$t_{x1} = -p_1\cos\theta, \quad t_{x2} = -p_2\cos\theta$$

$$t_{y1} = -p_1\sin\theta, \quad t_{y2} = -p_2\sin\theta. \qquad (6.44)$$

Point load/Concentrated force : Let a point or concentrated force $\{P\} = \left[P_x, P_y\right]^T$ is acting on the body. Point loads are considered in the usual manner by having a structural node at that point and simply adding to global load vector.

6.2.2. Quadratic triangular element

A Quadratic triangular element has primary nodes at each vertex and secondary nodes at the mid-sides (Figure (6.7)) and the shape functions can be calculated as discussed below.

The quadratic variation of a displacement component, say u in a six noded triangular element can be expressed by

$$u(\xi,\eta) = \alpha_1 + \alpha_2\xi + \alpha_3\eta + \alpha_4\xi\eta + \alpha_5\xi^2 + \alpha_6\eta^2, \qquad (6.45)$$

where α_i $(i = 1 \text{ to } 6)$ are constants.

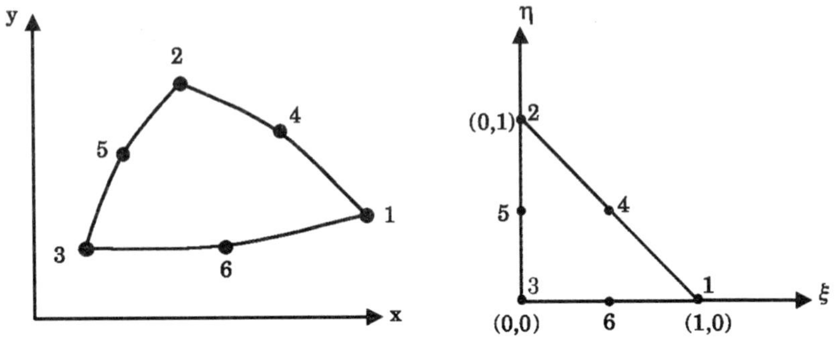

Fig. 6.7 Quadratic triangular element and its master element

Substituting the nodal coordinates for all six nodes in the above equation, we get

$$\begin{Bmatrix} u_1 \\ u_2 \\ u_3 \\ u_4 \\ u_5 \\ u_6 \end{Bmatrix} = \begin{bmatrix} 1 & 1 & 0 & 0 & 1 & 0 \\ 1 & 0 & 1 & 0 & 0 & 1 \\ 1 & 0 & 0 & 0 & 0 & 0 \\ 1 & 1/2 & 1/2 & 1/4 & 1/4 & 1/4 \\ 1 & 0 & 1/2 & 0 & 0 & 1/4 \\ 1 & 1/2 & 0 & 0 & 1/4 & 0 \end{bmatrix} \begin{Bmatrix} \alpha_1 \\ \alpha_2 \\ \alpha_3 \\ \alpha_4 \\ \alpha_5 \\ \alpha_6 \end{Bmatrix},$$

or simply as

$$\{U_u\} = [A]\{\alpha\}. \tag{6.46}$$

Then,

$$\{\alpha\} = [A]^{-1}\{U_u\}. \tag{6.47}$$

The inverse of the matrix $[A]$ can be shown to be

$$[A]^{-1} = \begin{bmatrix} 0 & 0 & 1 & 0 & 0 & 0 \\ -1 & 0 & -3 & 0 & 0 & 4 \\ 0 & -1 & -3 & 0 & 4 & 0 \\ 0 & 0 & 4 & 4 & -4 & -4 \\ 2 & 0 & 2 & 0 & 0 & -4 \\ 0 & 2 & 2 & 0 & -4 & 0 \end{bmatrix}. \tag{6.48}$$

We can write Eq. (6.45) as

$$u = \begin{bmatrix} 1 & \xi & \eta & \xi\eta & \xi^2 & \eta^2 \end{bmatrix} \{\alpha\} = [P]\{\alpha\}. \tag{6.49}$$

Then the shape function for the element defined by

$$u = \begin{bmatrix} N_1 & N_2 & N_3 & N_4 & N_5 & N_6 \end{bmatrix} \{U_u\} = [N]\{U_u\}, \tag{6.50}$$

can be found from $[N] = [P][A]^{-1}$. Thus we obtain

$$\begin{aligned} &N_1 = (2\xi - 1)\xi, \quad N_2 = (2\eta - 1)\eta, \\ &N_3 = (1 - 2\xi - 2\eta)(1 - \xi - \eta), \\ &N_4 = 4\xi\eta, \qquad N_5 = 4\eta(1 - \xi - \eta), \\ &N_6 = 4\xi(1 - \xi - \eta). \end{aligned} \tag{6.51}$$

6.3. QUADRILATERAL ELEMENTS

For the analysis of structural problems of complex shapes involving curved boundaries or surfaces, simple triangular or rectangular elements are not sufficient. This has led to the development of the so-called isoparametric quadrilateral elements that can have unparalleled edges. These elements are widely used in two-dimensional stress analysis, and analysis of plates and shells. The geometry of the quadrilateral element is nearly as good as that of the triangular element for modeling irregular boundaries. Further the use of quadrilateral elements allows one to use lesser number of elements to model any structure from geometrical point of view.

In this section, the element stiffness matrix and equivalent nodal load vector are derived for four noded isoparametric quadrilateral element, and the interpolation models for eight noded and nine noded isoparametric quadrilateral elements are developed. The formulation of element matrix and load vector for eight noded and nine noded quadrilateral elements is very similar to that for four noded quadrilateral element.

6.3.1 Four noded quadrilateral element

For four noded isoparametric quadrilateral element (Figure 6.8), the shape functions, N_α, which are bilinear in ξ and η, are defined as

$$N_\alpha(\xi, \eta) = \frac{1}{4}(1 + \xi_\alpha \xi)(1 + \eta_\alpha \eta), \tag{6.52}$$

where $(\xi_\alpha, \eta_\alpha)$ are the natural coordinates of the αth node.

Admissible displacement fields can be defined uniquely over the element by the nodal displacement components as

$$u_x(\xi,\eta) = \sum_{\alpha} N_\alpha(\xi,\eta) u_{x\alpha}$$

$$u_y(\xi,\eta) = \sum_{\alpha} N_\alpha(\xi,\eta) u_{y\alpha}, \tag{6.53}$$

where $(u_{x\alpha}, u_{y\alpha})$ is the displacement at the αth node and summation is over all four nodes.

The element nodal displacements are represented by the vector

$$\{U^e\} = [u_{x1}, u_{y1}, \;\cdots,\; u_{x4}, u_{y4}]^T. \tag{6.54}$$

In matrix notation, we can write Eq. (6.53) as

$$\{u\} = \begin{Bmatrix} u_x \\ u_y \end{Bmatrix} = [[N_1]\;\;[N_2]\;\;[N_3]\;\;[N_4]]\{U^e\} = [N]\{U^e\}, \tag{6.55}$$

where a typical shape function sub-matrix $[N_r]$ is given by

$$[N_r] = \begin{bmatrix} N_r & 0 \\ 0 & N_r \end{bmatrix}. \tag{6.56}$$

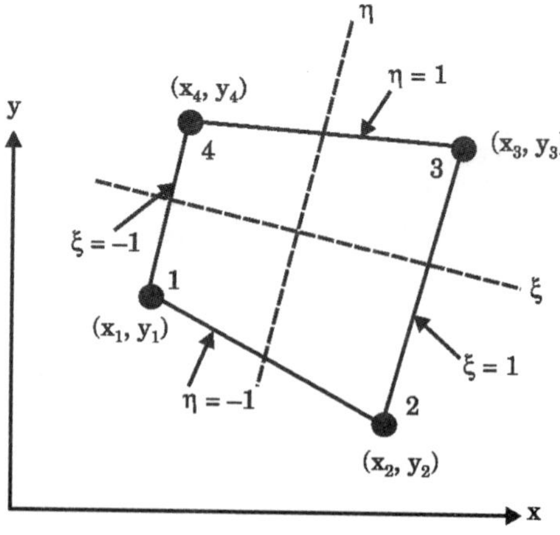

Fig. 6.8 Four-noded isoparametric quadrilateral element

In the isoparametric formulation, we use the same shape functions N_α to also express the coordinates of a point within the element in terms of nodal coordinates. Thus, coordinate transformation from the natural coordinate (ξ, η) to the global coordinates (x, y) is defined by

$$x(\xi,\eta) = \sum_{\alpha} N_{\alpha}(\xi,\eta)x_{\alpha}$$

$$y(\xi,\eta) = \sum_{\alpha} N_{\alpha}(\xi,\eta)y_{\alpha}, \tag{6.57}$$

where (x_{α}, y_{α}) are the global coordinates of the αth node.

Incompatible displacement modes

One of the drawbacks of this element is its inability to represent certain simple stress gradients since the interpolation functions defining the variation of displacements do not contain all the quadratic terms. Hence this element is not adequate in simulating flexural response. To overcome this difficulty, incompatible modes, proposed by Wilson *et al.* (1973) and later modified by Taylor *et al.* (1976), are added to the element shape functions. These modes are represented by the functions of the type

$$P_1 = \left(1-\xi^2\right), \quad P_2 = \left(1-\eta^2\right). \tag{6.58}$$

It may be noted that the functions P_i are chosen such that they are zero at the four nodes in order to maintain the displacement compatibility at the nodes and are of the same form as required to adequately represent bending deformation.

So, the displacement variation for an four-noded quadrilateral element with incompatible modes can be expressed as

$$\{u\} = [N]\{U^e\} + [P]\{a\}, \tag{6.59}$$

where $[P]$ is a 2×4 matrix given as

$$[P] = \begin{bmatrix} P_1 & 0 & P_2 & 0 \\ 0 & P_1 & 0 & P_2 \end{bmatrix}. \tag{6.60}$$

In Eq. (6.59), we have

$$\{a\} = \begin{bmatrix} a_{x1} & a_{y1} & a_{x2} & a_{y2} \end{bmatrix}^T, \tag{6.61}$$

where $a_{x\alpha}$ and $a_{y\alpha}$ are the α-*th* variables of incompatible modes, in x and y directions respectively.

Strain-displacement matrix

For 2D deformation, the strain components can be written as

$$\{\varepsilon\} = \begin{Bmatrix} \varepsilon_x \\ \varepsilon_y \\ \gamma_{xy} \end{Bmatrix} = \begin{bmatrix} \dfrac{\partial}{\partial x} & 0 \\ 0 & \dfrac{\partial}{\partial y} \\ \dfrac{\partial}{\partial y} & \dfrac{\partial}{\partial x} \end{bmatrix} \begin{Bmatrix} u_x \\ u_y \end{Bmatrix}. \tag{6.62}$$

Using Eq. (6.59), Eq. (6.62) can be written in matrix form as

$$\{\varepsilon\} = [N']\{U^e\} + [P']\{a\} = [B]\begin{Bmatrix} \{U^e\} \\ \{a\} \end{Bmatrix}, \tag{6.63}$$

where

$$[N'] = [[N_1'] \;\; [N_2'] \;\; [N_3'] \;\; [N_4']]$$

and

$$[P'] = [[P_1'] \;\; [P_2']]. \tag{6.64}$$

In the foregoing equation, typical submatrices $[N_r']$ and $[P_r']$ are defined as

$$[N_r'] = \begin{bmatrix} \dfrac{\partial N_r}{\partial x} & 0 & \dfrac{\partial N_r}{\partial y} \\ 0 & \dfrac{\partial N_r}{\partial y} & \dfrac{\partial N_r}{\partial x} \end{bmatrix}^T \tag{6.65}$$

and

$$[P_r'] = \begin{bmatrix} \dfrac{\partial P_r}{\partial x} & 0 & \dfrac{\partial P_r}{\partial y} \\ 0 & \dfrac{\partial P_r}{\partial y} & \dfrac{\partial P_r}{\partial x} \end{bmatrix}^T . \tag{6.66}$$

Element stiffness matrix

The element stiffness matrix is evaluated for this element as

$$\left[K^e\right] = h\int_{A^e}[B]^T[C][B]dA = h\int_{-1}^{+1}\int_{-1}^{+1}[B]^T[C][B]\det[J]d\xi d\eta, \quad (6.67)$$

in which h is the thickness of the element.

The stiffness matrix as given by Eq. (6.67) is of size 12 × 12 and includes coefficients pertaining to incompatible modes. However, these terms can be eliminated by using the standard static condensation procedure discussed below and the condensed stiffness matrix will be of the order 8 × 8 pertaining to the nodal degrees of freedom.

It is to be noted that the quantities $[B]$ and $\det[J]$ in the above integral (Eq. (6.67)) are involved functions of ξ and η, and thus the integration has to be performed numerically such as using Gaussian Quadrature.

Static condensation procedure

The stiffness matrix $[K^e]$ can be partitioned to correspond to element nodal displacements $\{U^e\}$ and to incompatible mode variables $\{a\}$. Noting that there are no external forces associated with $\{a\}$, the element equation of equilibrium can be written as

$$\begin{bmatrix}\left[K^e_{uu}\right] & \left[K^e_{ua}\right]\\ \left[K^e_{au}\right] & \left[K^e_{aa}\right]\end{bmatrix}\begin{Bmatrix}\{U^e\}\\ \{a\}\end{Bmatrix} = \begin{Bmatrix}\{F^e\}\\ \{0\}\end{Bmatrix}. \quad (6.68)$$

From second equation, we get

$$\left[K^e_{au}\right]\{U^e\} + \left[K^e_{aa}\right]\{a\} = 0,$$

i.e.

$$\{a\} = -\left[K^e_{aa}\right]^{-1}\left[K^e_{au}\right]\{U^e\}. \quad (6.69)$$

Putting $\{a\}$ from the above equation into the first equation of Eq. (6.68), we obtain

$$\left(\left[K^e_{uu}\right] - \left[K^e_{ua}\right]\left[K^e_{aa}\right]^{-1}\left[K^e_{au}\right]\right)\{U^e\} = \{F^e\}. \quad (6.70)$$

The matrix inside the parentheses is the required condensed matrix.

Element force vectors

Body force : A body force $\{b\} = [b_x \quad b_y]^T$ which is a distributed force per unit volume, contributes to the global load vector $\{F\}$. The (8×1) element body force vector is given by

$$\left\{F_b^e\right\} = h\int_{A^e} [N]^T \{b\}\, dA = h\int_{-1}^{+1}\int_{-1}^{+1} [N]^T \{b\} \det[J]\; d\xi d\eta. \qquad (6.71)$$

As with the stiffness matrix, the body force vector has also to be evaluated by numerical integration.

Traction force : Let us now consider uniformly distributed traction force on the boundary surface. The boundary surface of a quadrilateral element is a line. Assume that a constant traction force $\{t\} = [t_x \quad t_y]^T$, a force per unit area, is applied on a line of the quadrilateral element. If S_F^e is the boundary surface on which traction is applied, formed by element nodes, say 1 and 2, then the element traction force vector $\left\{F_s^e\right\}$ is given by

$$\left\{F_s^e\right\} = \int_{S_F^e} \left[N^s\right]^T \{t\}\, dS, \qquad (6.72)$$

in which $[N^s]$ is the surface interpolation matrix pertaining to edge $1-2$ and can be obtained by setting $\eta = -1$ in the coefficients of the shape function matrix $[N]$. Thus we get

$$\left[N^s\right] = \begin{bmatrix} N_1 & 0 & N_2 & 0 & 0 & 0 & 0 & 0 \\ 0 & N_1 & 0 & N_2 & 0 & 0 & 0 & 0 \end{bmatrix}. \qquad (6.73)$$

If we assume thickness of the element, h to be constant throughout the element, $dS = h dl$ where dl is the differential length along the edge. It can be seen from Eq. (A1.23) that

$$dl = \left[\left(\frac{\partial x}{\partial \xi}\right)^2 + \left(\frac{\partial y}{\partial \xi}\right)^2\right]^{1/2} d\xi, \qquad (6.74)$$

where $\frac{\partial x}{\partial \xi}$ and $\frac{\partial y}{\partial \xi}$ are the elements of the Jacobian matrix $[J]$ of 2D coordinate transformation. Using the transformation relation (Eq. (6.57)) and substituting $\eta = -1$ along the edge, we get

$$\frac{\partial x}{\partial \xi} = \frac{x_2 - x_1}{2} \text{ and } \frac{\partial y}{\partial \xi} = \frac{y_2 - y_1}{2}. \tag{6.75}$$

Substituting above values in Eq. (6.74), we obtain

$$dl = \frac{l_{1-2}}{2} d\xi, \tag{6.76}$$

where l_{1-2} is the length of the edge 1 – 2.

Hence, the integral of Eq. (6.72) reduces to

$$\left\{F_s^e\right\} = \frac{hl_{1-2}}{2} \int_{-1}^{+1} \left[N^s\right]^T \{t\} \, d\xi. \tag{6.77}$$

The above integral is a line integral and in this case, it can be evaluated explicitly. However in general, it becomes necessary to perform the integration numerically.

In many practical cases, an element edge will be subjected to a distributed loading in a normal and tangential direction prescribed to it as shown in Figure 6.9. These distributed forces need not be constant but can vary independently along the element edge. Since the element is a linear one, at best a linear distribution of these forces can be accommodated. The variation will be defined by prescribing the normal and tangential values at the two nodal points forming the element edge. In order to be consistent a tangential load is assumed to be positive if it acts in an anticlockwise direction with respect to the loaded element and a normal load is assumed to be positive if it acts in a direction into the element. The linear variation of tangential and normal traction force at any point on the line 1 – 2 can be represented by the following interpolation function:

$$t_t = N_1 t_{t1} + N_2 t_{t2}$$

$$t_n = N_1 t_{n1} + N_2 t_{n2}, \tag{6.78}$$

where $\left(t_{t\alpha}, t_{n\alpha}\right)$ are the tangential and normal traction forces at the αth node of the element and N_1 and N_2 are the two shape functions defining a linear variation along the element edge.

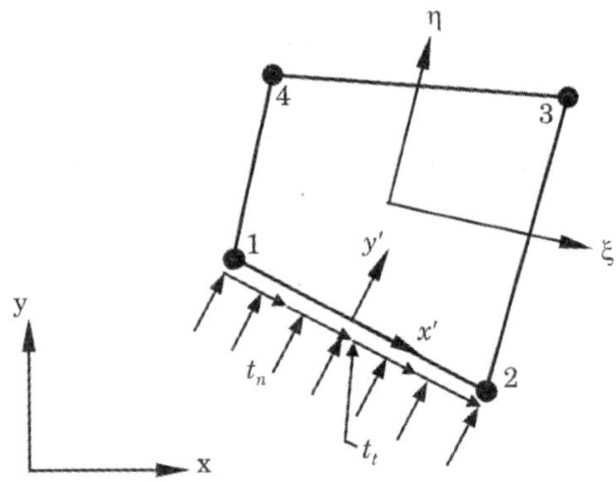

Fig. 6.9 Normal and tangential loads applied to a four-noded quadrilateral element

In order to determine the equivalent nodal forces using Eq. (6.77), the components of the traction forces given by Eq. (6.78) acting in the x and y direction must first be established.

Let $\hat{i},\hat{j},\hat{k}$ be unit vectors along global x, y and z directions, $\bar{e}_1,\bar{e}_2$ vectors along ξ and η directions and $\bar{l}_1,\bar{l}_2$ vectors along local x' and y' directions defined for an element edge as depicted in Figure 6.9. The vectors along the global and natural coordinate directions are related by the Jacobian matrix $[J]$ as

$$\begin{Bmatrix} e_1 \\ e_2 \end{Bmatrix} = \begin{bmatrix} \dfrac{\partial x}{\partial \xi} & \dfrac{\partial y}{\partial \xi} \\ \dfrac{\partial x}{\partial \eta} & \dfrac{\partial y}{\partial \eta} \end{bmatrix} \begin{Bmatrix} i \\ j \end{Bmatrix}. \tag{6.79}$$

Depending on the edge, the vector $\bar{l}_i$ will be identical to one of the vectors along the natural coordinate axes. For the edge 1 – 2, $\bar{l}_1$ is identical to $\bar{e}_1$. With respect to the figure, $\bar{l}_2$ can be given as

$$\bar{l}_2 = \frac{\bar{e}_1 \times \bar{e}_2}{|\bar{e}_1 \times \bar{e}_2|} \times \bar{l}_1 = \hat{k} \times \bar{e}_1$$

or

$$\bar{l}_2 = \begin{vmatrix} \hat{i} & \hat{j} & \hat{k} \\ 0 & 0 & 1 \\ \dfrac{\partial x}{\partial \xi} & \dfrac{\partial y}{\partial \xi} & 0 \end{vmatrix}.$$

Thus we get

$$\begin{Bmatrix} l_1 \\ l_2 \end{Bmatrix} = \begin{bmatrix} \dfrac{\partial x}{\partial \xi} & \dfrac{\partial y}{\partial \xi} \\ -\dfrac{\partial y}{\partial \xi} & \dfrac{\partial x}{\partial \xi} \end{bmatrix} \begin{Bmatrix} i \\ j \end{Bmatrix}. \tag{6.80}$$

By normalizing Eq. (6.80), we get unit vectors $\bar{l}_1^u, \bar{l}_2^u$ along local x' and y' directions, *i.e.*

$$\begin{Bmatrix} l_1^u \\ l_2^u \end{Bmatrix} = \frac{1}{\left[\left(\dfrac{\partial x}{\partial \xi}\right)^2 + \left(\dfrac{\partial y}{\partial \xi}\right)^2\right]^{1/2}} \begin{bmatrix} \dfrac{\partial x}{\partial \xi} & \dfrac{\partial y}{\partial \xi} \\ -\dfrac{\partial y}{\partial \xi} & \dfrac{\partial x}{\partial \xi} \end{bmatrix} \begin{Bmatrix} i \\ j \end{Bmatrix}. \tag{6.81}$$

Eq. (6.81) can be inverted to express unit vectors along global x and y directions in terms of unit vectors along local x' and y' directions. Thus we obtain

$$\begin{Bmatrix} i \\ j \end{Bmatrix} = \frac{1}{\left[\left(\dfrac{\partial x}{\partial \xi}\right)^2 + \left(\dfrac{\partial y}{\partial \xi}\right)^2\right]^{1/2}} \begin{bmatrix} \dfrac{\partial x}{\partial \xi} & -\dfrac{\partial y}{\partial \xi} \\ \dfrac{\partial y}{\partial \xi} & \dfrac{\partial x}{\partial \xi} \end{bmatrix} \begin{Bmatrix} l_1^u \\ l_2^u \end{Bmatrix}. \tag{6.82}$$

Thus, the components of the tangential and normal traction forces in the x and y direction can be evaluated using the coordinate transformation relation (Eq. (6.82)). Substituting these values into Eq. (6.77), we can write

$$\{F_s^e\} = h\int_{-1}^{+1} \left[N^s\right]^T \begin{bmatrix} \dfrac{\partial x}{\partial \xi} & -\dfrac{\partial y}{\partial \xi} \\ \dfrac{\partial y}{\partial \xi} & \dfrac{\partial x}{\partial \xi} \end{bmatrix} \begin{Bmatrix} t_t \\ t_n \end{Bmatrix} d\xi. \tag{6.83}$$

Stress calculation

From the known displacements, the strains are calculated at the Gauss points from strain-displacement matrix $[B]$ after eliminating the additional terms due to incompatible modes. Substituting $\{a\}$ from Eq. (6.69) into Eq. (6.63), we get

$$\{\varepsilon\} = \left(\left[N'\right] - \left[P'\right]\left[K^e_{aa}\right]^{-1}\left[K^e_{au}\right]\right)\left\{U^e\right\}. \tag{6.84}$$

These strains are extrapolated at the nodes of the element and a global smoothening of strain is performed for all the nodes of the finite element model by averaging of contributions from the neighboring finite elements. Finally the stresses in each element are obtained using the stress-strain relation for the material belonging to that element.

6.3.2. Nine-noded quadrilateral element

In this case (Figure 6.10), the quadratic interpolation functions can be written in terms of ξ and η as given below.

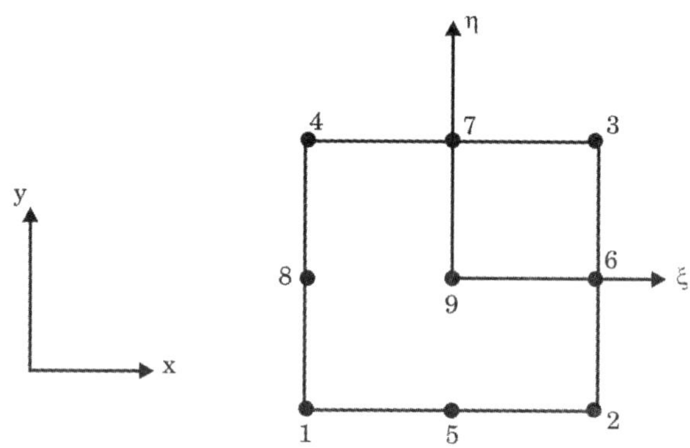

Fig. 6.10 Nine-noded isoparametric quadrilateral element

For corner nodes:

$$N_\alpha(\xi,\eta) = \frac{1}{4}(1+\xi_\alpha\xi)(1+\eta_\alpha\eta),\ \alpha = 1 \text{ to } 4;$$

for mid-side nodes:

$$N_5 = \frac{1}{2}\left(1-\xi^2\right)\eta(\eta-1)$$

$$N_6 = \frac{1}{2}\xi(1+\xi)\left(1-\eta^2\right)$$

$$N_7 = \frac{1}{2}\left(1-\xi^2\right)\eta(1+\eta)$$

$$N_8 = \frac{1}{2}\xi(\xi - 1)(1 - \eta^2);$$

and for centre node:

$$N_9 = (1 - \xi^2)(1 - \eta^2). \tag{6.85}$$

6.3.3. Eight-noded quadrilateral element

Eight noded quadrilateral element (Figure 6.11) falls into the category of serendipity elements. As discussed earlier, the interpolation functions for this element have to be constructed in the natural coordinate system using the procedure that employs all the criteria of these functions. One of such techniques is suggested in Zienkiewicz (1989).

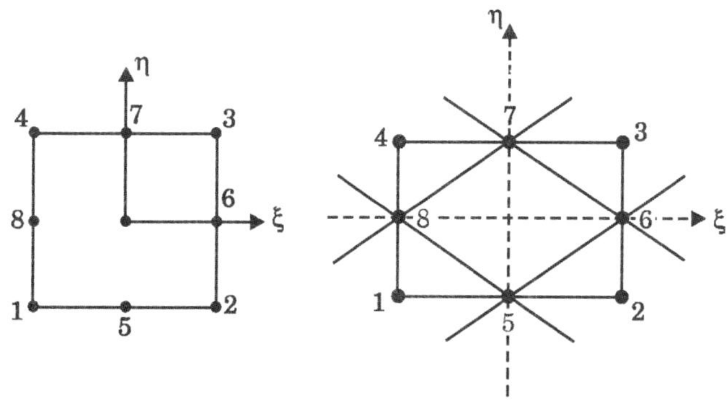

Fig. 6.11 Eight-noded isoparametric quadrilateral element

The interpolation function for node 1, N_1 should take on a value of unity at node 1 and zero at all other nodes i.e. 2, 3, ..., 8. In other words, N_1 should vanish on the sides 4-3 and 3-2 defined by the equations $1 - \eta = 0$ and $1 - \xi = 0$ respectively, and also on the line passing through nodes 8 and 5, and defined by the equation $1 + \xi + \eta = 0$. Therefore, N_1 is of the form:

$$N_1(\xi, \eta) = c(1 - \xi)(1 - \eta)(1 + \xi + \eta),$$

where c is a constant and is determined to be $-\frac{1}{4}$ by using the condition that $N_1(-1, -1) = 1$. Thus we have

$$N_1(\xi, \eta) = \frac{1}{4}(1 - \xi)(1 - \eta)(-\xi - \eta - 1).$$

A similar approach is used to determine interpolation functions for all other nodes. Finally we have

$$N_\alpha(\xi,\eta) = \frac{1}{4}(1+\xi_\alpha\xi)(1+\eta_\alpha\eta)(\xi_\alpha\xi+\eta_\alpha\eta-1), \ \alpha = 1 \text{ to } 4,$$

$$N_5 = \frac{1}{2}(1-\xi^2)(1-\eta),$$

$$N_6 = \frac{1}{2}(1+\xi)(1-\eta^2),$$

$$N_7 = \frac{1}{2}(1-\xi^2)(1+\eta)$$

and

$$N_8 = \frac{1}{2}(1-\xi)(1-\eta^2). \tag{6.86}$$

6.3.4. Axisymmetric element

For axisymmetric solids subjected to axisymmetric loading, due to the symmetry about the circumferential (θ) direction, the displacements are confined to the radial (r) and axial (z) directions only. Thus, as discussed in Chapter 2, any of the two-dimensional plane elements with little modifications can be used for the analysis of these solids. One such ring element is shown in Figure 6.12. For the axisymmetric case, the strain and stress components are given by Eq. (A2.33).

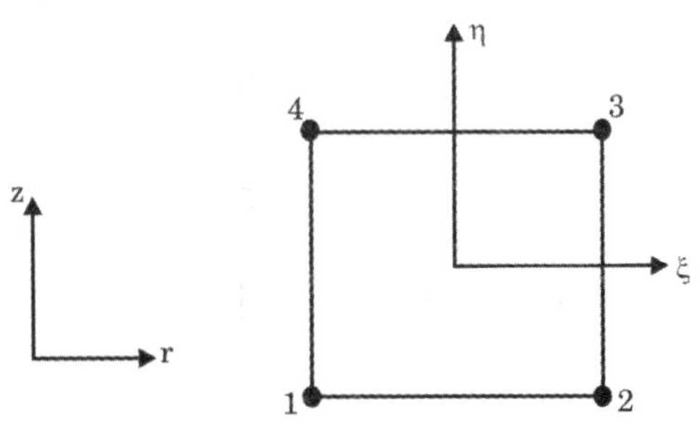

Fig. 6.12 Four-noded isoparametric quadrilateral element (Axisymmetric option)

The strain displacement relations can be rewritten as

$$\{\varepsilon\} = \begin{Bmatrix} \varepsilon_r \\ \varepsilon_z \\ \varepsilon_\theta \\ \gamma_{rz} \end{Bmatrix} = \begin{bmatrix} \dfrac{\partial}{\partial r} & 0 \\ 0 & \dfrac{\partial}{\partial z} \\ \dfrac{1}{r} & 0 \\ \dfrac{\partial}{\partial z} & \dfrac{\partial}{\partial r} \end{bmatrix} \begin{Bmatrix} u_r \\ u_z \end{Bmatrix}. \tag{6.87}$$

Using Eq. (6.87), strain displacement matrix [*B*] for a four-noded isoparametric quadrilateral element with axisymmetric option can be represented as

$$[B] = \begin{bmatrix} \dfrac{\partial N_1}{\partial r} & 0 & \dfrac{\partial N_2}{\partial r} & 0 & \dfrac{\partial N_3}{\partial r} & 0 & \dfrac{\partial N_4}{\partial r} & 0 \\ 0 & \dfrac{\partial N_1}{\partial z} & 0 & \dfrac{\partial N_2}{\partial z} & 0 & \dfrac{\partial N_3}{\partial z} & 0 & \dfrac{\partial N_4}{\partial z} \\ \dfrac{N_1}{r} & 0 & \dfrac{N_2}{r} & 0 & \dfrac{N_3}{r} & 0 & \dfrac{N_4}{r} & 0 \\ \dfrac{\partial N_1}{\partial z} & \dfrac{\partial N_1}{\partial r} & \dfrac{\partial N_2}{\partial z} & \dfrac{\partial N_2}{\partial r} & \dfrac{\partial N_3}{\partial z} & \dfrac{\partial N_3}{\partial r} & \dfrac{\partial N_4}{\partial z} & \dfrac{\partial N_4}{\partial r} \end{bmatrix}. \tag{6.88}$$

Element stiffness matrix

The element stiffness matrix can be derived for axisymmetric element as

$$\left[K^e\right] = \int_{V^e} [B]^T [C][B] dV = 2\pi \iint [B]^T [C][B] r dr dz$$

or

$$\left[K^e\right] = 2\pi \int_{-1}^{+1}\int_{-1}^{+1} [B]^T [C][B] r \det[J] d\xi d\eta, \tag{6.89}$$

where *r* is taken corresponding to the Gauss integration points when a quadrature rule is employed in evaluating the above integral.

Element force vectors

The element force vectors for axisymmetric elements can be evaluated in the same manner as that described in 2D case except that the volume and surface integrations have to be carried out over the whole ring of material.

6.4. EXAMPLE PROBLEMS

6.4.1 Example 1: A tapered cantilever beam

A tapered cantilever beam subjected to a concentrated load of 15 kN at the free end is considered as shown in Figure 6.13. With reference to the figure dimensions are: L = 1.25 m, a = 80 mm and h = 50 mm. Material property is: Young's modulus, $E = 2.1\times10^5$ N/mm^2.

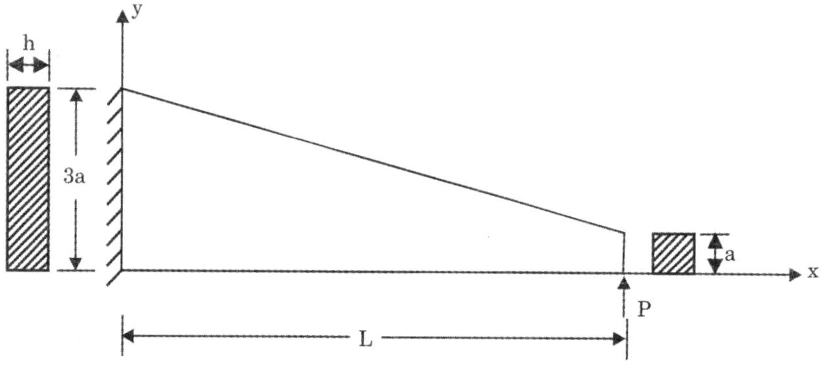

Fig. 6.13 A tapered cantilever beam subjected to a concentrated load

The beam is analyzed using four-noded quadrilateral elements and the number of elements taken is 10 × 8. The FEM results of vertical deflections along the length of the beam are compared with theoretical results (Krishnamoorthy, 1994), as depicted in Table 6.1. Since the structure is subjected to predominantly bending, it is expected that the quadrilateral element with incompatible modes gives better results than the element without incompatible modes.

Table 6.1 Comparison of results for a tapered cantilever beam

Distance from the free end (mm)	*Theoretical value of deflection (mm)*	*SB-FEM results (mm)*	
		Quadrilateral element without incompatible modes	*Quadrilateral element with incompatible modes*
250	0.052	0.048	0.054
500	0.223	0.197	0.230
750	0.543	0.465	0.555
1000	1.040	0.863	1.060
1250	1.714	1.369	1.742

6.4.2. Example 2: A curved beam subjected to a shear load

A curved beam subjected to a unit shear load at the free end (Figure 6.14) is analyzed as a plane stress problem using 2D plane elements. The dimensions are: Inner radius, a = 4.12; outer radius, b = 4.32; thickness, t = 0.1; arc = 90 degree. The beam is made of material with Young's modulus $E = 1.0 \times 10^7$ and Poisson's ratio $\nu = 0.25$. Consistent units are chosen here.

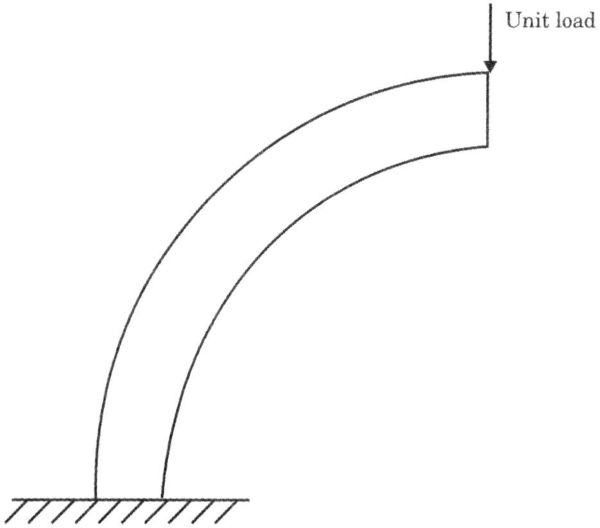

Fig. 6.14 A curved beam subjected to a unit shear load

The exact solution for the vertical deflection at the tip is given by (Timoshenko and Goodier, 1934)

$$v_{tip} = \frac{Pt\pi\left(a^2 + b^2\right)}{E\left[\left(a^2 - b^2\right) + \left(a^2 + b^2\right)\log\left(b / a\right)\right]},$$

where P is the load acting at the free end face per unit thickness of the beam. For the problem stated above, $v_{up} = 0.0886$ (upto 4 decimal places).

The beam is modeled with three different element types, *i.e.* 4-noded quadrilateral (with incompatible modes), 8-noded quadrilateral and 9-noded quadrilateral, and the results obtained for the vertical deflection at the tip are 0.0879, 0.0863 and 0.0864 units respectively. It is clear that the FEM results compare well with the theoretical value.

6.4.3. Example 3: A square plate with a circular hole

A square plate of side L = 300 mm with a circular hole of diameter d = 20 mm at the centre is subjected to a uniform tension of 100 N/mm^2 on its vertical edges (Figure 6.15). Thickness of the plate, h is 25 mm and the material properties are: $E = 2.0 \times 10^5$ N/mm^2, $\nu = 0.3$. This stress concentration problem is well treated analytically in standard text books (*e.g.*, Roark and Young, 1989).

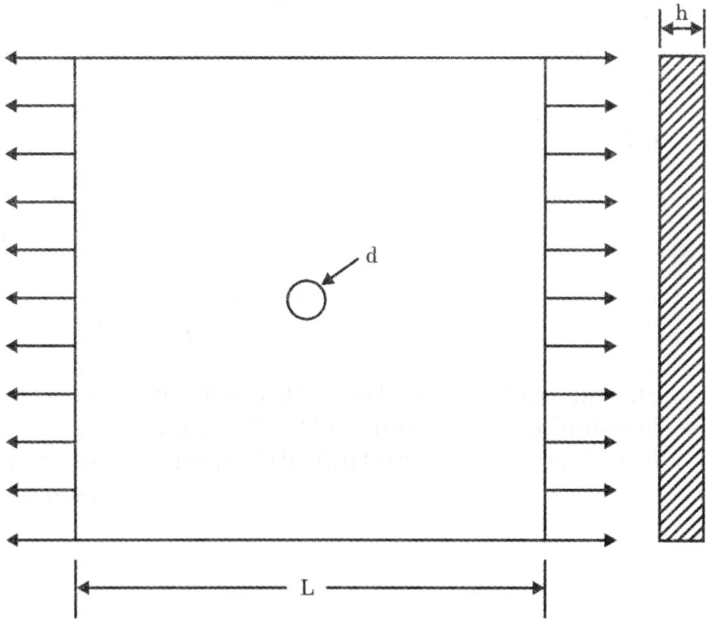

Fig. 6.15 Tensile loading of a square plate with a circular hole

The FEM analysis is done using three different element types, i.e. 4-noded, 8-noded and 9-noded quadrilaterals. Because of the symmetry of the problem, it is sufficient to model only one quadrant of the structure (Figure 6.16). The SB-FEM command file for this problem is listed in Appendix 8. The variation of stress component σ_x throughout the structure, determined using 4-noded quadrilateral elements is shown in Figure 6.17 as a contour plot. The values of σ_y and σ_x at selected points A and B (Figure 6.16) are shown in Table 6.2 along with theoretical values (Roark and Young, 1989). It is found that the present FEM results are in a reasonable agreement with the theoretical results. It may be expected that the results will improve with refining the mesh particularly near the hole of the plate.

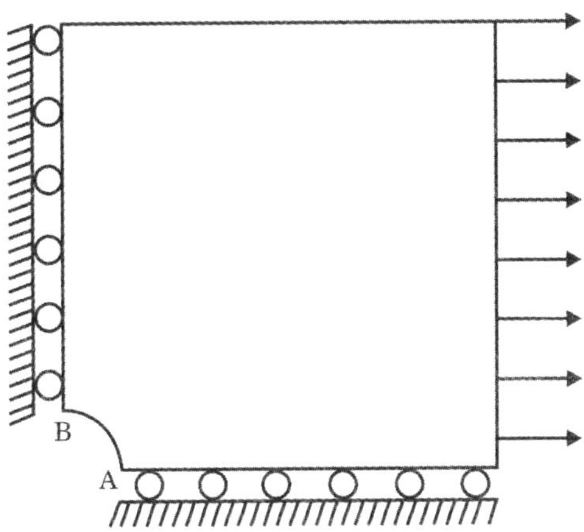

Fig. 6.16 Quarter symmetry

Table 6.2 Comparison of results for a square plate with a circular hole.

	Theoretical result	*FEM solution*		
		4-noded quadrilateral elements	*8-noded quadrilateral elements*	*9-noded quadrilateral elements*
σ_x at B, N/mm^2	300	292.64	318.92	313.73
σ_y at A, N/mm^2	–100	–108.31	–148.85	–145.25

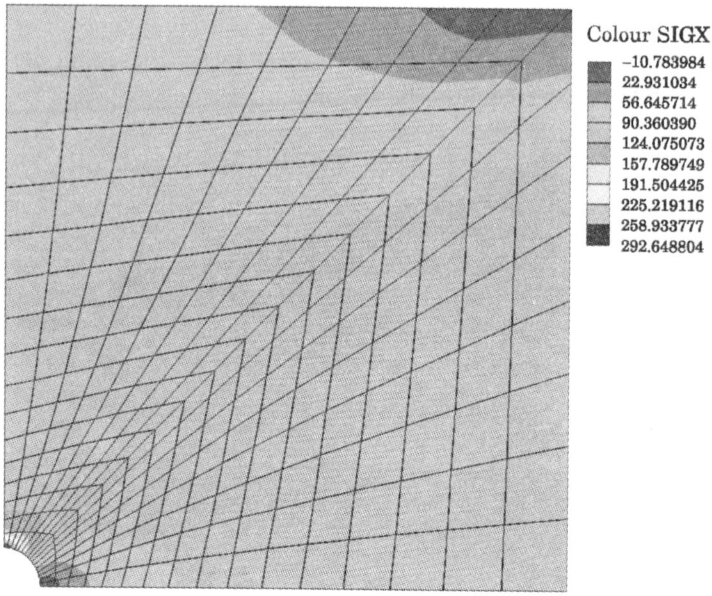

Fig. 6.17 Contour plot of σ_x

6.4.4. Example 4: A thick-walled cylinder

A long thick-walled cylinder (Figure 6.18) of inner radius, $a = 4$ units and outer radius, $b = 8$ units, is subjected to an internal pressure $p = 30000$ units. It is desired to calculate radial displacement, radial stress, axial stress and hoop stress distributions in the cylinder. Length of the cylinder is arbitrarily taken. Material properties data are: $E = 30\times10^6$ units, $\nu = 0.3$. Units are consistently chosen. This problem can be treated as an axisymmetric one and the desired results are u_r, σ_r, σ_z and σ_θ respectively. Considering a small thickness in the loaded area as depicted in Figure 6.18, the finite element representation for axisymmetric analysis is shown in Figure 6.19. The finite element results along with the analytical solution (Timoshenko, 1956) are given in Table 6.3. Good agreement between the two results can be observed.

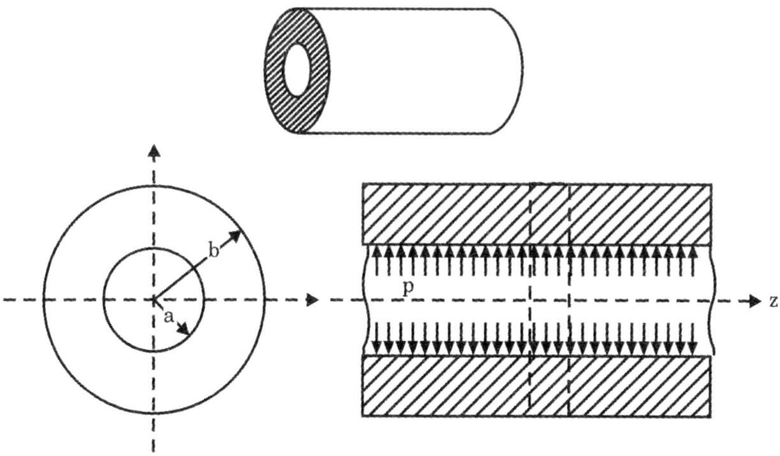

Fig. 6.18 A long thick-walled cylinder subjected to an internal pressure

Table 6.3 Comparison of results for thick-walled cylinder.

	Analytical solution, units	*FEM solution, units*		
		4-noded quadrilateral elements	*8-noded quadrilateral elements*	*9-noded quadrilateral elements*
Radial displacement at radius, a	0.0078666	0.007865	0.007867	0.007867
Radial stress at radius, a	–30000	–29032	–29543	–29543
Radial stress at radius, (a + b)/2	–7778	–7776	–7668	–7668
Radial stress at radius, b	0	–192	38	38
Hoop stress at radius, a	50000	50281	50181	50181
Hoop stress at radius, (a + b)/2	27778	27752	27824	27824
Hoop stress at radius, b	20000	19909	20016	20016

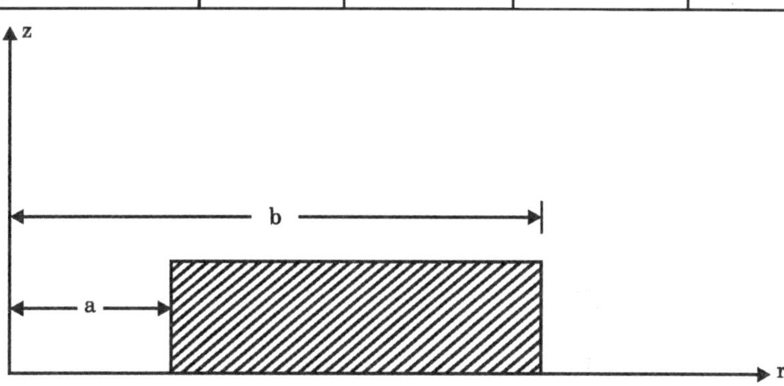

Fig. 6.19 Thick-walled cylinder - axisymmetric analysis

7

Analysis of Plates

7.1. INTRODUCTION

The word 'plate' is used to indicate bodies that are bounded by two parallel planes whose lateral dimensions are large compared to the separation between these planes. Geometrically, plates are similar to plane elastic bodies; however, plates are loaded transversely *i.e.* perpendicular to the plane of the plate.

7.2. FORCES AND MOMENTS ACTING ON THE PLATE

The Figure 7.1 below depicts forces and moments acting on a plate under transverse distributed loading of intensity $q\ (x, y)$ per unit area. Positive θ_x and θ_y are the cross-sectional rotations (rotations of a line normal to the mid-surface) around the y- and x-axes, respectively.

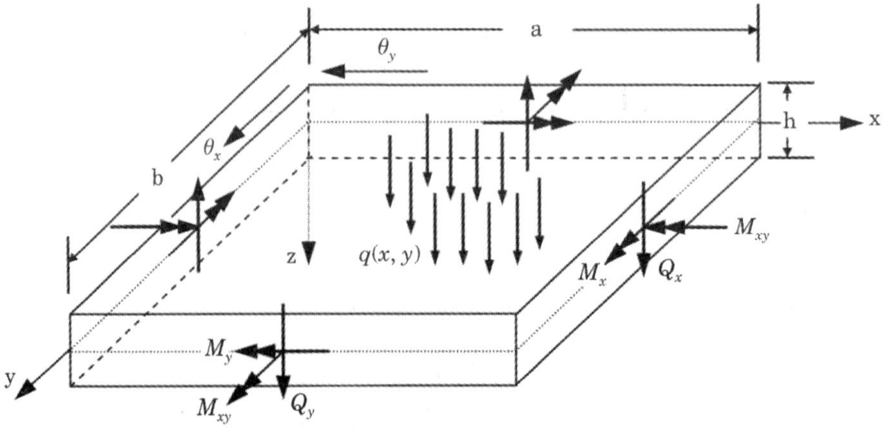

Fig. 7.1 Forces and moments acting on a plate

Different normal and shear stresses developed in the plate are shown in Figure 7.2.

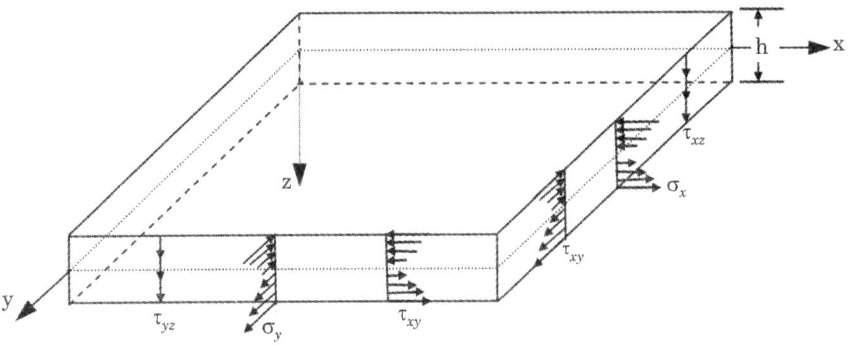

Fig. 7.2 Normal and shear stresses developed in the plate

In the case of plates, it is convenient to regard the generalized stress resultant vector $\{\bar{\sigma}\}$ instead of stress vector $\{\sigma\}$. $\{\bar{\sigma}\}$ is given by

$$\{\bar{\sigma}\} = \begin{bmatrix} M_x & M_y & M_{xy} & Q_{yz} & Q_{xz} \end{bmatrix}^T \tag{7.1}$$

or

$$\{\bar{\sigma}\} = \left[\int_{-h/2}^{h/2} \begin{pmatrix} \sigma_x z & \sigma_y z & \tau_{xy} z & \tau_{yz} & \tau_{xz} \end{pmatrix} dz \right]^T. \tag{7.2}$$

In the above equation, M_x, M_y, M_{xy} are internal moment resultants (moments per unit length) and Q_{yz}, Q_{xz} (also referred to briefly as Q_y and Q_x respectively) are transverse shear stress resultants (shear forces per unit length). Among the moment resultants, M_x and M_y are referred to as bending moments and M_{xy} is referred to as twisting moment.

7.3. THIN PLATE THEORY

Classical thin plate theory (also referred to as Kirchhoff's theory) is based upon the assumption that a straight line normal to the mid-surface of undeformed plate remains straight and normal to the deformed mid-surface after loading (Figure 7.3). This assumption implies that there is no transverse shear deformation effect, that is

$$\gamma_{yz} = \gamma_{xz} = 0. \tag{7.3}$$

Further, the plate is assumed inextensible in lateral direction *i.e.* $\varepsilon_z = 0$.

Let u, v and w be the displacements at any point (x, y, z). The slopes $\frac{\partial w}{\partial x}$ and $\frac{\partial w}{\partial y}$ in Kirchhoff's plates are set equal to the rotations of the normal during deformation. Thus, variation of the displacement u and v across the thickness can be expressed in terms of the displacement w as

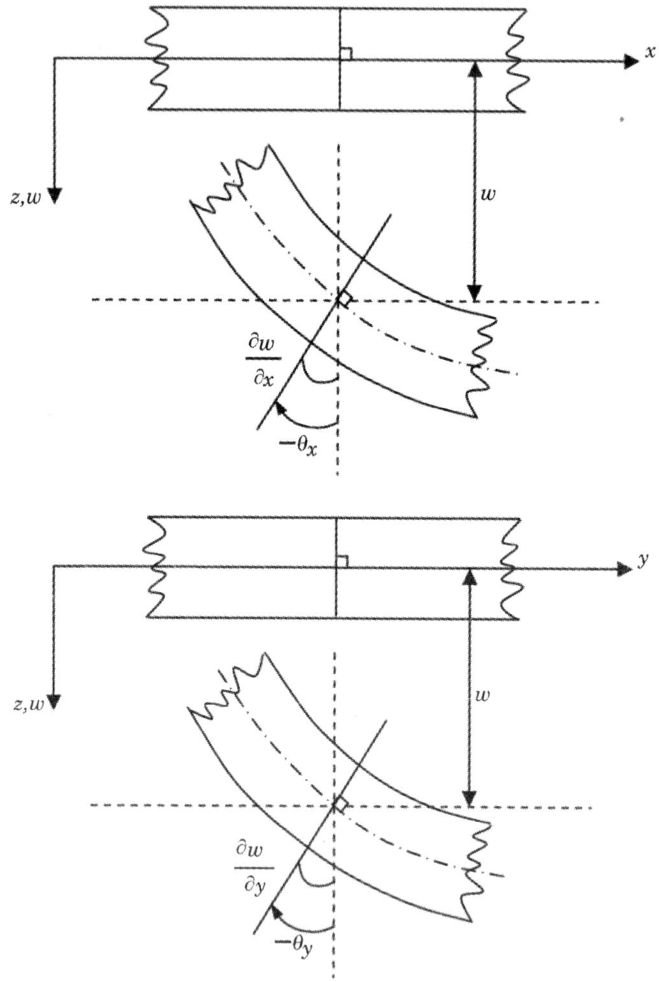

Fig. 7.3 Kirchhoff's plate

$$u = z\theta_x = -z\frac{\partial w}{\partial x}$$

$$v = z\theta_y = -z\frac{\partial w}{\partial y}. \qquad (7.4)$$

The strain distribution is give by

$$\varepsilon_x = \frac{\partial u}{\partial x} = -z\frac{\partial^2 w}{\partial x^2} = z\kappa_x,$$

$$\varepsilon_y = \frac{\partial v}{\partial y} = -z\frac{\partial^2 w}{\partial y^2} = z\kappa_y$$

and $$\gamma_{xy} = \frac{\partial u}{\partial y} + \frac{\partial v}{\partial x} = -2z\frac{\partial^2 w}{\partial x \partial y} = z\kappa_{xy}. \qquad (7.5)$$

It is to be noted here that there is no stretch of the mid-surface due to the deflection of the plate.

The constitutive law for plane stress is given as

$$\begin{Bmatrix} \sigma_x \\ \sigma_y \\ \tau_{xy} \end{Bmatrix} = \frac{E}{1-\upsilon^2}\begin{bmatrix} 1 & \upsilon & 0 \\ \upsilon & 1 & 0 \\ 0 & 0 & \dfrac{(1-\upsilon)}{2} \end{bmatrix}\begin{Bmatrix} \varepsilon_x \\ \varepsilon_y \\ \gamma_{xy} \end{Bmatrix}. \qquad (7.6)$$

Using Eqs. (7.5) and (7.6) into Eq. (7.2), we get

$$\begin{Bmatrix} M_x \\ M_y \\ M_{xy} \end{Bmatrix} = \frac{Eh^3}{12(1-\upsilon^2)}\begin{bmatrix} 1 & \upsilon & 0 \\ \upsilon & 1 & 0 \\ 0 & 0 & \dfrac{(1-\upsilon)}{2} \end{bmatrix}\begin{Bmatrix} \kappa_x \\ \kappa_y \\ \kappa_{xy} \end{Bmatrix}. \qquad (7.7)$$

Thus in the case of isotropic plates, we get the familiar relationship as

$$M_x = -D\left(\frac{\partial^2 w}{\partial x^2} + \upsilon\frac{\partial^2 w}{\partial y^2}\right),$$

$$M_y = -D\left(\frac{\partial^2 w}{\partial y^2} + \upsilon\frac{\partial^2 w}{\partial x^2}\right)$$

and $$M_{xy} = -D(1-\nu)\frac{\partial^2 w}{\partial x \partial y}, \tag{7.8}$$

where $D = \dfrac{Eh^3}{12\left(1-\upsilon^2\right)}$.

We observe that the solution of thin plate flexural problem depends entirely on the selection of a single component w, the transverse displacement.

7.3.1. Governing equation

It is now worthwhile to derive governing equation in terms of transverse displacement, w. Equilibrium equations are obtained by free body diagram shown in Figure 7.4. Moment equilibriums about y and x-axes and force equilibrium along z-axis yield after neglecting the higher order terms

$$\frac{\partial M_x}{\partial x} + \frac{\partial M_{xy}}{\partial y} - Q_x = 0,$$

$$\frac{\partial M_{xy}}{\partial x} + \frac{\partial M_y}{\partial y} - Q_y = 0$$

and $$\frac{\partial Q_x}{\partial x} + \frac{\partial Q_y}{\partial y} + q(x,y) = 0. \tag{7.9}$$

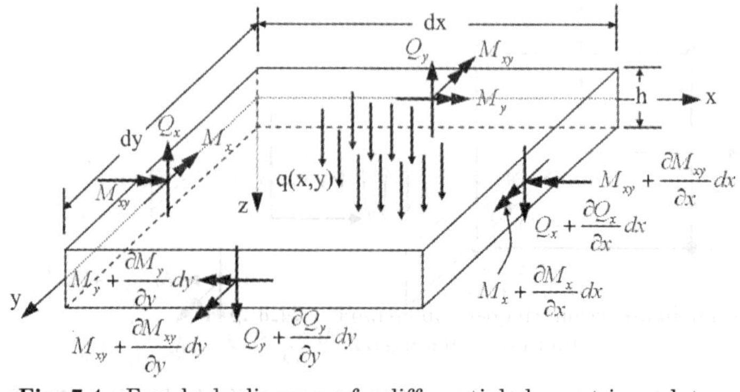

Fig. 7.4 Free body diagram of a differential element in a plate

By eliminating the shear forces from the foregoing equation, we obtain the following governing equation:

$$D\left(\frac{\partial^4 w}{\partial x^4}+2\frac{\partial^2 w}{\partial x^2}\frac{\partial^2 w}{\partial y^2}+\frac{\partial^4 w}{\partial y^4}\right)=q. \tag{7.10}$$

7.4. SHEAR DEFORMATION IN PLATES

In thin plate theory, the transverse shear deformation is neglected and deformation is completely described by a function $w(x,y)$ where w is the lateral displacement perpendicular to the plate. If the thickness of the plate is not thin, *e.g.* $(h/L >= 1/10)$ (L = characteristic length of the plate), then the effect of transverse shear deformation must be considered. Mindlin's theory has been found to be quite useful to consider shear deformation in plates and is widely adopted in finite element formulation.

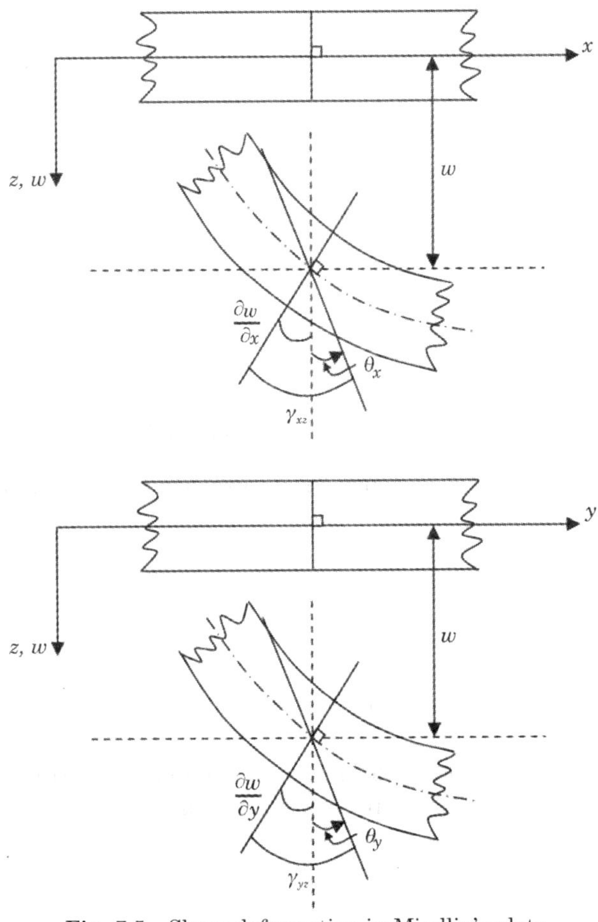

Fig. 7.5 Shear deformation in Mindlin's plate

Mindlin's approximation is that straight lines originally normal to the midsurface before deformation remain straight but not necessarily normal to the deformed midsurface, *i.e.* the average rotation of the section may be taken as the rotation in which normals remain perpendicular to the midsurface plus an additional rotation due to transverse shear (Figure 7.5).

Referring to the Figure 7.5, the rotations of a line normal to the mid-surface can be expressed as

$$\theta_x = -\frac{\partial w}{\partial x} + \gamma_{xz},$$

$$\theta_y = -\frac{\partial w}{\partial y} + \gamma_{yz}. \tag{7.11}$$

Here γ_{xz} and γ_{yz} are the average shear deformations. The above equations are based on the assumption that the total rotations θ_x and θ_y are small and the transverse strain ε_z is negligible. Note that if we impose the conditions that shear strains are zero, we can recover the relations applied in the thin plate theory.

The displacement components at a point located at a distance z from the midplane are

$$\begin{aligned} u &= z\theta_x \\ v &= z\theta_y. \end{aligned} \tag{7.12}$$

The strain distribution is give by

$$\varepsilon_x = \frac{\partial u}{\partial x} = z\frac{\partial \theta_x}{\partial x} = z\kappa_x,$$

$$\varepsilon_y = \frac{\partial v}{\partial y} = z\frac{\partial \theta_y}{\partial y} = z\kappa_y,$$

$$\gamma_{xy} = \frac{\partial u}{\partial y} + \frac{\partial v}{\partial x} = z\left(\frac{\partial \theta_x}{\partial y} + \frac{\partial \theta_y}{\partial x}\right) = z\kappa_{xy},$$

$$\gamma_{yz} = \theta_y + \frac{\partial w}{\partial y}$$

and $$\gamma_{xz} = \theta_x + \frac{\partial w}{\partial x}. \tag{7.13}$$

Here κ_x, κ_y and κ_{xy} are changes in curvature. κ_{xy} is referred to as a twist.

7.5. FINITE ELEMENT FORMULATION

Two-dimensional quadrilateral elements with isoparametric shape functions are used for the plate element formulation. Because the transverse displacement w and slopes θ_x and θ_y are independent in a shear deformable plate element, we need shape functions to interpolate them separately. For a quadrilateral plate element with n nodes, transverse displacement and slopes are interpolated as

$$w = \sum_{i=1}^{n} N_i w_i,$$

$$\theta_x = \sum_{i=1}^{n} N_i \theta_{xi}$$

and $$\theta_y = \sum_{i=1}^{n} N_i \theta_{yi}. \tag{7.14}$$

Here w_i are the nodal vertical displacements, and θ_{xi} and θ_{yi} are the nodal rotation terms.

The geometry of this isoparametric element is interpolated by the expression

$$x(\xi,\eta) = \sum_{\alpha} N_\alpha(\xi,\eta) x_\alpha$$

$$y(\xi,\eta) = \sum_{\alpha} N_\alpha(\xi,\eta) y_\alpha, \tag{7.15}$$

where (x_α, y_α) are the global coordinates of the α th node of the element.

Eq. (7.14) can be represented in matrix form as

$$\{u\} = \begin{Bmatrix} w \\ \theta_x \\ \theta_y \end{Bmatrix} = [N]\{U^e\}, \tag{7.16}$$

where $$[N]=[[N_1]\ [N_2]\ \dots\ [N_n]] \tag{7.17}$$

and $$\{U^e\}=\begin{bmatrix} w_1 & \theta_{x1} & \theta_{y1} & \cdots & w_n & \theta_{xn} & \theta_{yn}\end{bmatrix}^T. \tag{7.18}$$

A typical sub-matrix $[N_r]$ in Eq. (7.17) is given as

$$[N_r]=\begin{bmatrix} N_r & 0 & 0 \\ 0 & N_r & 0 \\ 0 & 0 & N_r \end{bmatrix}. \tag{7.19}$$

As in the case of beams, it is convenient here to express the consitutive relation and finite element model in terms of stress-resultants.

If we set

$$\{\bar{\varepsilon}\}=\begin{bmatrix}\kappa_x & \kappa_y & \kappa_{xy} & \gamma_{yz} & \gamma_{xz}\end{bmatrix}^T, \tag{7.20}$$

we have the following constitutive relation:

$$\{\bar{\sigma}\}=\left[\bar{D}\right]\{\bar{\varepsilon}\}, \tag{7.21}$$

where $\left[\bar{D}\right]$ is the stiffness matrix of size 5 × 5 and is given by

$$\left[\bar{D}\right]=\begin{bmatrix} D_{11} & D_{12} & D_{14} & 0 & 0 \\ D_{12} & D_{22} & D_{24} & 0 & 0 \\ D_{14} & D_{24} & D_{44} & 0 & 0 \\ 0 & 0 & 0 & S_{55} & S_{56} \\ 0 & 0 & 0 & S_{56} & S_{66} \end{bmatrix}. \tag{7.22}$$

In the above equation, the stiffness coefficients are defined by

$$D_{ij}=C_{ij}\int_{-h/2}^{h/2} z^2 dz=\frac{h^3}{12}C_{ij} \quad (i,j=1,2,4),$$

and $$S_{pq}=\alpha C_{pq}\int_{-h/2}^{h/2} dz=\alpha h C_{pq} \quad (p,q=5,6). \tag{7.23}$$

Here C_{ij} $(i,j=1,2,4,5,6)$ are the elastic constants and α is the shear correction factor.

Now, the strain-displacement relation can be written as

$$\{\varepsilon\} = [B]\{U^e\}, \tag{7.24}$$

in which $[B]$ matrix is evaluated as

$$[B] = [[B_1] \;\; [B_2] \;\; \cdots \;\; [B_n]]^T, \tag{7.25}$$

where $[B_r]$, $r = 1$ to 4 are strain-displacement submatrices of size 5×3 and are defined as

$$[B_r] = \begin{bmatrix} 0 & 0 & 0 & \frac{\partial N_r}{\partial y} & \frac{\partial N_r}{\partial x} \\ \frac{\partial N_r}{\partial x} & 0 & \frac{\partial N_r}{\partial y} & 0 & N_r \\ 0 & \frac{\partial N_r}{\partial y} & \frac{\partial N_r}{\partial x} & N_r & 0 \end{bmatrix}^T .$$

The principle of virtual work in this case can be written as

$$\delta I = \int_A \{\delta\bar{\varepsilon}\}^T \{\bar{\sigma}\}\, dA - \int_A \{\delta u^s\}^T \{t\}\, dA = 0, \tag{7.26}$$

where traction vector is given as $\{t\} = \begin{Bmatrix} q \\ 0 \\ 0 \end{Bmatrix}$.

Rewriting Eq. (7.26) as a sum of integration over the areas of all finite elements and substituting for variations of the displacements and strains within each element, we obtain

$$\sum_e \{\delta U^e\}^T \left(\int_{A^e} [B]^T \{\bar{\sigma}\}\, dA\right) = \sum_e \{\delta U^e\}^T \left(\int_{A^e} [N^s]^T \{t\}\, dA\right). \tag{7.27}$$

Sinceis $\{\delta U\}$ is arbitrary, Eq. (7.27) can be written as the following equation of equilibrium:

$$\left[\sum_e \left(\int_{A^e} [B]^T [\bar{D}][B] dA\right)\right]\{U\} = \sum_e \left(\int_{A^e} [N^s]^T \{t\}\, dA\right). \tag{7.28}$$

7.5.1 Element stiffness matrix

From Eq. (7.28) the stiffness matrix for an element is given as

$$[K^e] = \int_{A^e} [B]^T [\bar{D}][B] dA = \int_{-1}^{+1}\int_{-1}^{+1} [B]^T [\bar{D}][B] \det[J]\, d\xi\, d\eta, \tag{7.29}$$

in which the relation $dA = dx\,dy = \det[J]\,d\xi d\eta$ is used.

In the foregoing equation, the coefficients of Jacobian matrix $[J]$ as well as the Cartesian shape function derivatives appearing in $[B]$ matrix can be evaluated using the coordinate transformation relation (Eq. (7.15)). Putting these values in the equation, the integral therein may be evaluated numerically using Gaussian Quadrature.

When exact numerical integration is used with plate elements based on Mindlin theory, very disappointing results are obtained in application to thin plates. As the plate thickness becomes small, the influence of the shear terms tends to dominate the numerical solution thus causing imposition of the constraints of zero shear deformations ($\gamma_{yz} = \gamma_{xz} = 0$). This phenomenon known as shear locking leads to an overstiff element. As discussed in the case of beams in Chapter 5, shear locking is alleviated by separating the bending and shear terms in the overall element stiffness matrix and using selective integration procedure (Gallagher, 1976). For example, in the case of four noded quadrilateral plate element, the bending stiffness contribution is evaluated using 2 × 2 Gaussian quadrature and the shear stiffness contribution is calculated using 1 × 1 rule.

7.5.2. Element nodal force vector

The distributed lateral pressure load applied on the plate element can be represented in terms of nodal forces by making use of the consistent nodal force formulae given in Eq. (7.28). The integral can be written as

$$\left\{F_s^e\right\} = \int_{A^e} \left[N^s\right]^T \{t\}\, dA = \int_{-1}^{+1}\int_{-1}^{-1} \left[N^s\right]^T \{t\} \det[J]\, d\xi d\eta. \tag{7.30}$$

Substituting for $[N^s]$ which is same as $[N]$, we can calculate element nodal force vector corresponding to twelve element degrees of freedom.

7.6. EXAMPLE PROBLEMS

A square plate of side L having four edges clamped or simply-supported (hinged) and subjected to a uniform distributed load q or a concentrated force Q at the centre of the plate, is investigated. For this simple geometry, the analytical solutions for maximum deflection at the centre can be calculated using Timoshenko theory (Timoshenko *et. al.*, 1959) and are given in Table 7.1 for the different cases.

Table 7.1 Theoretical results for maximum deflection at the centre.

	Clamped	*Simply-supported*
Under uniform load q	$0.00126qL^4/D$	$0.00406qL^4/D$
Under concentrated force Q	$0.00560QL^2/D$	$0.0116QL^2/D$

Here D is as defined in Eq. (7.8).

7.6.1 Example 1: Clamped square plate

A square thick plate of size 2175 mm × 2175 mm is clamped on all of its edges (Figure 7.6). The plate is 40 mm thick and is subjected to two different loading cases: (a) unit point load (UPL) at the centre and (b) uniformly distributed load of unit intensity (UUDL). Material properties data are: E = 2.0×10^5 N/mm^2, $\nu = 0.3$.

The problem is investigated using 8-noded plate elements based on the plate theory. Taking advantage of symmetry, only one quarter of the plate is analyzed with suitable symmetric boundary conditions along the lines of symmetry. The SB-FEM command file is listed in Appendix 8. The present results of maximum deflection at the centre of the plate for the two loading cases are compared with analytical results in Table 7.2. Good agreements can be observed between the two results.

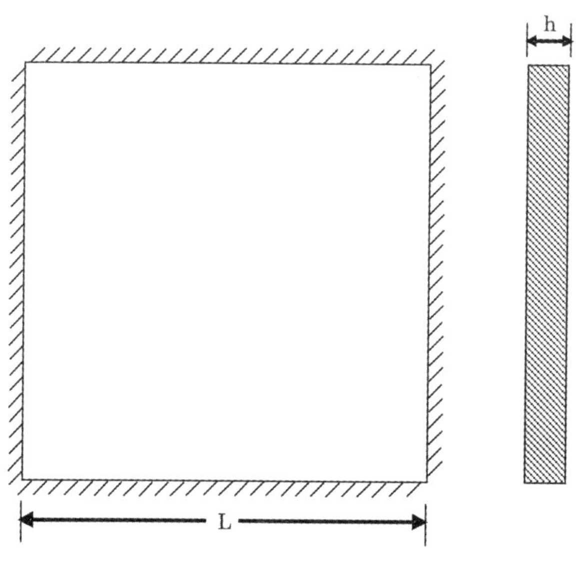

Fig. 7.6 Problem of a clamped square plate

Table 7.2 Comparison of results for a clamped square plate.

Loading case	*Displacement at the centre of the plate, mm*	
	Analytical results	*SB-FEM results*
UPL	2.26×10^{-5}	2.3×10^{-5}
UUDL	24.06	24.29

7.6.2. Example 2: Simply-supported square plate

The above plate is also analyzed with simply-supported boundary conditions at the four edges. The SB-FEM results of maximum deflection are depicted in Table 7.3 along with analytical results, for the two loading cases. Again, a good agreement can be observed between the two results.

Table 7.3 Comparison of results for a simply-supported square plate.

Loading case	*Displacement at the centre of the plate, mm*	
	Analytical results	*SB-FEM results*
UPL	4.68×10^{-5}	4.7×10^{-5}
UUDL	77.53	77.67

8

Three-dimensional Stress Analysis

8.1. INTRODUCTION

Most engineering problems are three dimensional. Solid or three-dimensional elements enable the solution of three-dimensional stress analysis problems. These elements belong to tetrahedron, triangular prism and hexahedron families. When a force is applied on a three dimensional body occupying volume V and surface area S, the body deforms and the deformation at a point $\{x\} = [x, y, z]^T$ is given by the three components of its displacement in the x-, y- and z-directions, *i.e.*

$$\{u\} = [u, v, w]^T \tag{8.1}$$

The stresses and strains are given by

$$\{\sigma\} = \begin{bmatrix} \sigma_x & \sigma_y & \sigma_z & \tau_{xy} & \tau_{yz} & \tau_{xz} \end{bmatrix}^T, \tag{8.2}$$

$$\{\varepsilon\} = \begin{bmatrix} \varepsilon_x & \varepsilon_y & \varepsilon_z & \gamma_{xy} & \gamma_{yz} & \gamma_{xz} \end{bmatrix}^T. \tag{8.3}$$

The stress-strain relations are given by

$$\{\sigma\} = [C]\{\varepsilon\}, \tag{8.4}$$

where $[C]$ is a (6×6) symmetric material matrix as discussed in Appendix 2.

The strain-displacement relations are given by

$$\{\varepsilon\} = \left[\frac{\partial u}{\partial x}, \frac{\partial v}{\partial y}, \frac{\partial w}{\partial z}, \frac{\partial u}{\partial y} + \frac{\partial v}{\partial x}, \frac{\partial v}{\partial z} + \frac{\partial w}{\partial y}, \frac{\partial w}{\partial x} + \frac{\partial u}{\partial z}\right]^T. \tag{8.5}$$

8.2. TETRAHEDRAL ELEMENTS

The tetrahedral family of elements exhibits properties similar to those of the triangle family in two-dimensional analysis. This family includes 4 noded linear tetrahedron, 10 noded quadratic tetrahedron and 20 noded cubic tetrahedron (Zienkiewicz, 1977). The formulation of element matrices and nodal load vector for a linear tetrahedral element is described in this section.

8.2.1. Linear tetrahedral element

This is the simplest tetrahedral element with four nodes, one at each vertex (Figure 8.1). A consistent node numbering must be used for the element. Here, staring with a vertex (1) of the element, nodes are numbered around the element face (2-3-4) opposite to the vertex, in the counter-clockwise direction when viewed from the starting vertex.

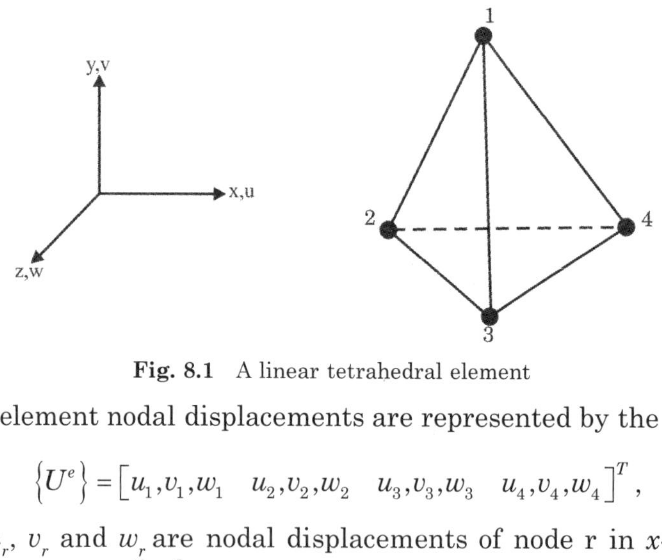

Fig. 8.1 A linear tetrahedral element

The element nodal displacements are represented by the vector

$$\{U^e\} = [u_1, v_1, w_1 \quad u_2, v_2, w_2 \quad u_3, v_3, w_3 \quad u_4, v_4, w_4]^T, \tag{8.6}$$

where u_r, v_r and w_r are nodal displacements of node r in x-, y- and z-directions respectively.

The displacements u, v and w are assumed to be linear functions within the element such that at any internal point $P(x, y, z)$, we have

$$u = a_0 + a_1 x + a_2 y + a_3 z,$$

$$v = b_0 + b_1 x + b_2 y + b_3 z$$

and

$$w = c_0 + c_1 x + c_2 y + c_3 z, \tag{8.7}$$

where a_i, b_i, c_i ($i = 0,1,2,3$) are constants.

Displacements given by Eq. (8.7) should satisfy the following equations at each node:

$$\left.\begin{aligned} u_r &= a_0 + a_1 x_r + a_2 y_r + a_3 z_r \\ v_r &= b_0 + b_1 x_r + b_2 y_r + b_3 z_r \\ w_r &= c_0 + c_1 x_r + c_2 y_r + c_3 z_r \end{aligned}\right\} \quad (r = 1,2,3,4). \tag{8.8}$$

Solving these equations, the coefficients a_i, b_i, and c_i can be determined in terms of displacements and coordinates of nodes.

Putting values of these coefficients into Eq. (8.7) and rearranging the terms we obtain

$$u(x,y,z) = \sum_{i=1}^{4} N_i(x,y,z)u_i,$$

$$v(x,y,z) = \sum_{i=1}^{4} N_i(x,y,z)v_i$$

and
$$w(x,y,z) = \sum_{i=1}^{4} N_i(x,y,z)w_i. \tag{8.9}$$

In the above equation, N_i are defined as shape function and have the values

$$N_i = \frac{1}{6V}\left(\alpha_{i1} + \alpha_{i2}x + \alpha_{i3}y + \alpha_{i4}z\right), \tag{8.10}$$

where V is the volume of the tetrahedron given by

$$V = \frac{1}{6}\det\begin{bmatrix} 1 & x_1 & y_1 & z_1 \\ 1 & x_2 & y_2 & z_2 \\ 1 & x_3 & y_3 & z_3 \\ 1 & x_4 & y_4 & z_4 \end{bmatrix} = \frac{1}{6}\begin{vmatrix} 1 & x_1 & y_1 & z_1 \\ 1 & x_2 & y_2 & z_2 \\ 1 & x_3 & y_3 & z_3 \\ 1 & x_4 & y_4 & z_4 \end{vmatrix} \tag{8.11}$$

and α_{ij} are cofactors of the matrix (in Eq. (8.11)) in ith row and jth column. For example,

$$\alpha_{11}=(-1)^{1+1}\begin{vmatrix} x_2 & y_2 & z_2 \\ x_3 & y_3 & z_3 \\ x_4 & y_4 & z_4 \end{vmatrix}, \quad \alpha_{12}=(-1)^{1+2}\begin{vmatrix} 1 & y_2 & z_2 \\ 1 & y_3 & z_3 \\ 1 & y_4 & z_4 \end{vmatrix},$$

$$\alpha_{13}=(-1)^{1+3}\begin{vmatrix} 1 & x_2 & z_2 \\ 1 & x_3 & z_3 \\ 1 & x_4 & z_4 \end{vmatrix} \text{ and } \alpha_{14}=(-1)^{1+4}\begin{vmatrix} 1 & x_2 & y_2 \\ 1 & x_3 & y_3 \\ 1 & x_4 & y_4 \end{vmatrix}. \tag{8.12}$$

In matrix notation, we can write Eq. (8.9) as

$$\{u\}=\begin{Bmatrix} u \\ v \\ w \end{Bmatrix}=[N]\{U^e\}, \tag{8.13}$$

where shape function matrix $[N]$ contains the appropriate N_i terms to satisfy Eq. (8.13) so that

$$[N]=[[N_1]\ \ [N_2]\ \ [N_3]\ \ [N_4]], \tag{8.14}$$

where a typical sub-matrix $[N_r]$ is given as

$$[N_r]=\begin{bmatrix} N_r & 0 & 0 \\ 0 & N_r & 0 \\ 0 & 0 & N_r \end{bmatrix}. \tag{8.15}$$

The Eq. (8.10) can also be written as

$$N_1=\frac{V_1}{V},\ N_2=\frac{V_2}{V},\ N_3=\frac{V_3}{V},\ N_4=\frac{V_4}{V}, \tag{8.16}$$

where $V_i (i = 1,2,3,4)$ is the volume of tetrahedron formed by face opposite to node i and the point P (Figure 8.2). The ratios of the volumes of these sub-tetrahedrons to the volume of the full tetrahedron are referred to as volume coordinates.

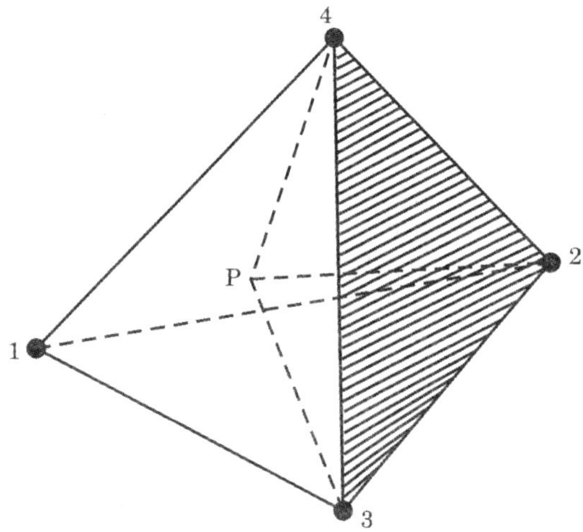

Fig. 8.2 Volume coordinate system of the tetrahedral element

Now, strain-displacement relation can be evaluated as

$$\{\varepsilon\} = \begin{bmatrix} \frac{\partial}{\partial x} & 0 & 0 \\ 0 & \frac{\partial}{\partial y} & 0 \\ 0 & 0 & \frac{\partial}{\partial z} \\ \frac{\partial}{\partial y} & \frac{\partial}{\partial x} & 0 \\ 0 & \frac{\partial}{\partial z} & \frac{\partial}{\partial y} \\ \frac{\partial}{\partial z} & 0 & \frac{\partial}{\partial x} \end{bmatrix} \begin{Bmatrix} u \\ v \\ w \end{Bmatrix}. \tag{8.17}$$

Substituting from Eq. (8.13) into the above equation, we obtain

$$\{\varepsilon\} = [B]\{U^e\}, \tag{8.18}$$

in which $[B]$ matrix is evaluated as

$$[B] = \left[[B_1] \; [B_2] \; [B_3] \; [B_4]\right]^T, \tag{8.19}$$

where each sub-matrix is defined as

$$[B_r] = \begin{bmatrix} \frac{\partial N_r}{\partial x} & 0 & 0 & \frac{\partial N_r}{\partial y} & 0 & \frac{\partial N_r}{\partial z} \\ 0 & \frac{\partial N_r}{\partial y} & 0 & \frac{\partial N_r}{\partial x} & \frac{\partial N_r}{\partial z} & 0 \\ 0 & 0 & \frac{\partial N_r}{\partial z} & 0 & \frac{\partial N_r}{\partial y} & \frac{\partial N_r}{\partial x} \end{bmatrix}^T$$

$$= \frac{1}{6V}\begin{bmatrix} \alpha_{r2} & 0 & 0 & \alpha_{r3} & 0 & \alpha_{r4} \\ 0 & \alpha_{r3} & 0 & \alpha_{r2} & \alpha_{r4} & 0 \\ 0 & 0 & \alpha_{r4} & 0 & \alpha_{r3} & \alpha_{r2} \end{bmatrix}^T. \tag{8.20}$$

Interpolation model in terms of local (natural) coordinates

We now introduce natural coordinates (ξ, η, ζ) on the tetrahedron as shown in Figure 8.3. We choose the unit right tetrahedron as the master element. Then the shape functions for this linear element can be represented as

$$N_1 = \xi,\ N_2 = \eta,\ N_3 = \zeta,\ N_4 = 1 - \xi - \eta - \zeta. \tag{8.21}$$

We have two coordinate systems for the element: the global coordinates (*x*, *y*, *z*) and the natural coordinates (ξ, η, ζ). The relation between the two can be given as

$$x(\xi,\eta,\zeta) = \sum_\alpha N_\alpha(\xi,\eta,\zeta)\, x_\alpha,$$

$$y(\xi,\eta,\zeta) = \sum_\alpha N_\alpha(\xi,\eta,\zeta)\, y_\alpha$$

and

$$z(\xi,\eta,\zeta) = \sum_\alpha N_\alpha(\xi,\eta,\zeta)\, z_\alpha, \tag{8.22}$$

where summation is over all four nodes.

Writing in matrix form, we have

$$\begin{Bmatrix} 1 \\ x \\ y \\ z \end{Bmatrix} = \begin{bmatrix} 1 & 1 & 1 & 1 \\ x_1 & x_2 & x_3 & x_4 \\ y_1 & y_2 & y_3 & y_4 \\ z_1 & z_2 & z_3 & z_4 \end{bmatrix} \begin{Bmatrix} N_1 \\ N_2 \\ N_3 \\ N_4 \end{Bmatrix}. \tag{8.23}$$

Explicit inversion of the above matrix results in the same values for N_α as given in Eq. (8.10).

If, using the approach similar to 2D triangular element, we compute the strain-displacement matrix $[B]$ in terms of natural coordinates, we arrive at the same equation as in Eq. (8.20).

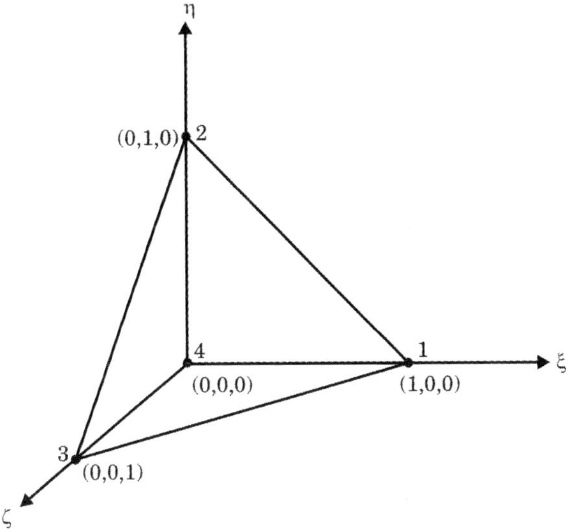

Fig. 8.3 Natural coordinates for a tetrahedron

Element stiffness matrix

The usual expression for the element stiffness matrix for this element is given as

$$\left[K^e\right] = \int_{V^e} [B]^T [C][B]\, dV. \tag{8.24}$$

Assuming that the elasticity matrix $[C]$ does not vary over the element, the integrand in Eq. (8.24) is constant because matrix $[B]$ as given by Eq. (8.20) is constant. Thus, the element stiffness matrix can be explicitly evaluated as

$$\left[K^e\right] = V\left([B]^T [C][B]\right). \tag{8.25}$$

Element force vectors

Body force : Owing to a body force $\{b\} = \left[b_x, b_y, b_z\right]^T$ which is a distributed force per unit volume, the element body force vector is given by

$$\left\{F_b^e\right\} = \int_{V^e} [N]^T \{b\} dV = \int_0^1 \int_0^{1-\xi} \int_0^{1-\xi-\eta} [N]^T \{b\} \det[J] d\zeta d\eta d\xi. \qquad (8.26)$$

Using the approach similar to its 2D counterpart *i.e.* the linear triangular element, the above integrand can be evaluated explicitly.

Not surprisingly, it can once more be found as in the case of linear triangle that if the body forces are constant the nodal components of the total resultant are distributed in four equal parts, *i.e.*,

$$\left\{F_b^e\right\} = \frac{V}{4}\begin{bmatrix} b_x & b_y & b_z & b_x & b_y & b_z & b_x & b_y & b_z & b_x & b_y & b_z \end{bmatrix}^T. \qquad (8.27)$$

Like triangle, there is also a useful formula for integration of quantities defined as polynomials of natural coordinates over the tetrahedral region such as in Eq. (8.26). The formula is

$$\begin{aligned} \int_{V^e} \xi^a \eta^b \zeta^c dV &= \int_0^1 \int_0^{1-\xi} \int_0^{1-\xi-\eta} \xi^a \eta^b \zeta^c \det[J] d\zeta d\eta d\xi \\ &= \frac{a!b!c!}{(a+b+c+3)!} 6V \end{aligned} \qquad (8.28)$$

Traction force : Let us now consider uniformly distributed traction force on the boundary surface. The boundary surface of a tetrahedral is a triangle. Assume that a constant traction force $\{t\} = \left[t_x, t_y, t_z\right]^T$, a force per unit area, is applied on a triangle formed by element nodes 1, 2 and 3 of the tetrahedral element. If S_{123} is the area of the face on which traction is applied, then the element traction force vector $\left\{F_s^e\right\}$ is given by

$$\left\{F_s^e\right\} = \int_{S_{123}} \left[N^s\right]^T \{t\} dS, \qquad (8.29)$$

in which $[N^s]$ is the surface interpolation matrix pertaining to the triangular face $1-2-3$ (also referred to as face 4, *i.e.* the face opposite to the 4th corner) and can be derived by imposing $N_4 = 0$ (which is equivalent to $\zeta = 1 - \xi - \eta$) into the shape function matrix $[N]$. Thus, we have

$$\left[N^s\right] = \left[\left[N_1\right] \ \left[N_2\right] \ \left[N_3\right] \ [0]\right], \qquad (8.30)$$

where [0] is a 3×3 zero matrix.

Since the traction force $\{t\}$ is assumed to be constant over the surface, it can be seen that the element traction forces are equally distributed among the three nodes 1, 2 and 3, which define the loaded face.

Thus, we get

$$\{F_s^e\} = \frac{A_{123}}{3}\begin{bmatrix} t_x & t_y & t_z & t_x & t_y & t_z & t_x & t_y & t_z & 0 & 0 & 0 \end{bmatrix}^T, \quad (8.31)$$

in which A_{123} is the area of the face 1–2–3 of the element and can be calculated as described below.

Since the equation of the ith face in terms of natural coordinates is $N_i = 0$, it follows immediately from Eq. (8.10) that its Cartesian equation is

$$\alpha_{i1} + \alpha_{i2}x + \alpha_{i3}y + \alpha_{i4}z = 0. \quad (8.32)$$

This can be transformed to the normal equation of a plane by scaling, *i.e.*

$$\frac{1}{S_i}\left(\alpha_{i1} + \alpha_{i2}x + \alpha_{i3}y + \alpha_{i4}z\right) = a_i x + b_i y + c_i z + d_i = 0,$$

in which $S_i = +\sqrt{\alpha_{i2}^2 + \alpha_{i3}^2 + \alpha_{i4}^2}$. Thus, $a_i = \dfrac{\alpha_{i2}}{S_i}$, $b_i = \dfrac{\alpha_{i3}}{S_i}$ and $c_i = \dfrac{\alpha_{i4}}{S_i}$ are the direction cosines of the unit interior normal to the ith face and $d_i = \dfrac{\alpha_{i1}}{S_i}$. The distance from the ith corner to the ith face (i.e. face opposite to the corner) is thus given as

$$h_i = \left|a_i x_i + b_i y_i + c_i z_i + d_i\right|. \quad (8.33)$$

h_i are also called tetrahedron altitudes. Now, expressing volume of the tetrahedron as $V = A_i h_i$, the area of the ith face A_i can be calculated.

Point load/Concentrated force: Let a point or concentrated force $\{P\} = \left[P_x, P_y, P_z\right]^T$ is acting on the body. Point loads are considered in the usual manner by having a structural node at that point and simply adding to the global load vector.

8.3. EIGHT-NODED BRICK ELEMENT

Let us consider a general three-dimensional isoparametric hexahedral/ brick element (Figure 8.4). We use indicial notation here for the sake of convenience. The local nodes are numbered as 1, 2, 3, 4, 5, 6, 7, 8 as shown, and (x_{1i}, x_{2i}, x_{3i}) are the coordinates of node i. In the hexahedral elements, a consistent node numbering scheme must be followed for defining the connectivity.

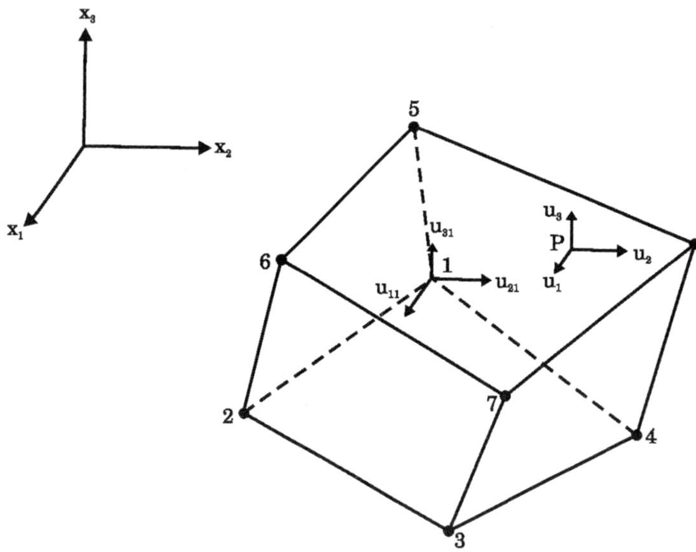

Fig. 8.4 A general 3-D eight-noded brick element

The displacement of an interior point P located at (x_1, x_2, x_3) is represented as

$$\{u\} = \left[u_1(x_1, x_2, x_3) \quad u_2(x_1, x_2, x_3) \quad u_3(x_1, x_2, x_3)\right]^T. \tag{8.34}$$

The element nodal displacements are represented by the vector

$$\{U^e\} = \left[u_{11}, u_{21}, u_{31}, \quad \cdots, \quad u_{18}, u_{28}, u_{38}\right]^T, \tag{8.35}$$

where u_{1r}, u_{2r} and u_{3r} are nodal displacements of node r.

8.3.1. Interpolation function

The Lagrange shape functions defining the geometry and variation of displacements for eight-noded brick element are given by (see Section 2.3)

$$N_i = \frac{1}{8}(1+\xi_i\xi)(1+\eta_i\eta)(1+\zeta_i\zeta) \quad i = 1 \text{ to } 8, \tag{8.36}$$

where (ξ_i, η_i, ζ_i) represents the coordinates of node i of the element in the natural coordinate (ξ, η, ζ) system.

Like four-noded 2D quadrilateral element, this element is also not adequate in representing flexural response because the shape functions given by Eq. (8.36) do not contain quadratic terms. To overcome this difficulty, incompatible modes (Wilson *et. al.*, 1973 and Taylor *et. al.*, 1976) are added to the brick element shape functions. These additional modes are represented by the functions of the type

$$P_1 = (1-\xi^2), P_2 = (1-\eta^2), P_3 = (1-\zeta^2). \tag{8.37}$$

It may be noted again that the functions P_i are zero at the eight nodes in order to maintain the displacement compatibility at the nodes and are of the same form as required to adequately represent bending deformation.

So, the displacement variation for an eight-noded solid element with incompatible modes can be expressed as

$$\{u\} = \sum_{r=1}^{8}[N_r]\{u_r\} + \sum_{s=1}^{3}[P_s]\{a_s\}, \tag{8.38}$$

where
$$[N_r] = \begin{bmatrix} N_r & 0 & 0 \\ 0 & N_r & 0 \\ 0 & 0 & N_r \end{bmatrix}, \tag{8.39}$$

$$[P_s] = \begin{bmatrix} P_s & 0 & 0 \\ 0 & P_s & 0 \\ 0 & 0 & P_s \end{bmatrix}, \tag{8.40}$$

$$\{u_r\} = [u_{1r} \quad u_{2r} \quad u_{3r}]^T \tag{8.41}$$

and
$$\{a_s\} = [a_{1s} \quad a_{2s} \quad a_{3s}]^T. \tag{8.42}$$

In the foregoing equation, a_{1s}, a_{2s} and a_{3s} are s-th variables of incompatible modes, in x_1, x_2 and x_3 directions respectively.

Eq. (8.38) can also be expressed as

$$\{u\} = [N]\{U^e\} + [P]\{a\}, \tag{8.43}$$

where

$$[N] = [[N_1] \;\cdots\; [N_8]],$$

$$[P] = [[P_1] \;\; [P_2] \;\; [P_3]] \tag{8.44}$$

and

$$\{a\} = [a_{11}, a_{21}, a_{31}, \;\; a_{12}, a_{22}, a_{32}, \;\; a_{13}, a_{23}, a_{33}]^T.$$

8.3.2. Strain-displacement matrix

Using Eq. (8.38), the strain-displacement relation (see Eq. (2.23)) can be written as

$$\{\varepsilon\} = \sum_{r=1}^{8}[N_r']\{u_r\} + \sum_{s=1}^{3}[P_s']\{a_s\}, \tag{8.45}$$

where $[N_r']$, r = 1 to 8 and $[P_s']$, s = 1 to 3 are strain-displacement submatrices of size 6 × 3 and are defined as

$$[N_r'] = \begin{bmatrix} \frac{\partial N_r}{\partial x_1} & 0 & 0 & \frac{\partial N_r}{\partial x_2} & 0 & \frac{\partial N_r}{\partial x_3} \\ 0 & \frac{\partial N_r}{\partial x_2} & 0 & \frac{\partial N_r}{\partial x_1} & \frac{\partial N_r}{\partial x_3} & 0 \\ 0 & 0 & \frac{\partial N_r}{\partial x_3} & 0 & \frac{\partial N_r}{\partial x_2} & \frac{\partial N_r}{\partial x_1} \end{bmatrix}^T \tag{8.46}$$

and

$$[P_s'] = \begin{bmatrix} \frac{\partial P_s}{\partial x_1} & 0 & 0 & \frac{\partial P_s}{\partial x_2} & 0 & \frac{\partial P_s}{\partial x_3} \\ 0 & \frac{\partial P_s}{\partial x_2} & 0 & \frac{\partial P_s}{\partial x_1} & \frac{\partial P_s}{\partial x_3} & 0 \\ 0 & 0 & \frac{\partial P_s}{\partial x_3} & 0 & \frac{\partial P_s}{\partial x_2} & \frac{\partial P_s}{\partial x_1} \end{bmatrix}^T. \tag{8.47}$$

Eq. (8.45) can also be written as

$$\{\varepsilon\} = [N']\{U^e\} + [P']\{a\} = [B]\begin{Bmatrix} \{U^e\} \\ \{a\} \end{Bmatrix}, \tag{8.48}$$

where

$$[N'] = \left[[N_1'] \quad \cdots \quad [N_8']\right] \text{ and } [P'] = \left[[P_1'] \quad \cdots \quad [P_3']\right].$$

$[B] = \left[[N'] \quad [P']\right]$ is the strain-displacement matrix of size 6 × 33.

8.3.3. Element stiffness matrix

The element stiffness matrix is evaluated for this element as

$$\begin{aligned}[K^e] &= \int_{V^e} [B]^T [C][B] dV \\ &= \int_{-1}^{+1}\int_{-1}^{+1}\int_{-1}^{+1} [B]^T [C][B] \det[J]\ d\xi d\eta d\zeta,\end{aligned} \tag{8.49}$$

where

$$dV = dx_1\ dx_2\ dx_3 = \det[J] d\xi d\eta d\zeta.$$

The stiffness matrix as given by Eq. (8.49) is of size 33 × 33 and includes coefficients pertaining to incompatible modes. However, these terms can be eliminated by using the static condensation procedure as discussed in Section 6.3 for four-noded 2D quadrilateral element and thus reduced stiffness matrix will be of the order 24 × 24 pertaining to the nodal degrees of freedom.

The quantities $[B]$ and $\det[J]$ in the above integral are involved functions of ξ, η and ζ and the integration has to be performed numerically.

8.3.4 Element force vectors

Body force

A body force $\{b\} = [b_1, b_2, b_3]^T$ contributes to the global load vector $\{F\}$. The (24 × 1) element body force vector is given by

$$\{F_b^e\} = \int_{V^e} [N]^T \{b\} dV = \int_{-1}^{+1}\int_{-1}^{+1}\int_{-1}^{+1} [N]^T \{b\} \det[J]\ d\xi d\eta d\zeta. \tag{8.50}$$

As with the stiffness matrix, the body force vector has also to be evaluated by numerical integration.

Traction force

Let us now consider uniformly distributed traction force on the boundary surface. The boundary surface of a hexahedral is a quadrilateral. Assume that a constant traction force $\{t\} = [t_1, t_2, t_3]^T$, a force per unit area, is applied on a quadrilateral of the hexahedral element. Without loss of generality, if S_F^e is the boundary surface on which traction is applied, formed by element nodes 1, 2, 3, and 4, then the element traction force vector $\{F_s^e\}$ is given by

$$\{F_s^e\} = \int_{S_F^e} \left[N^s\right]^T \{t\}\, dS, \tag{8.51}$$

in which $[N^s]$ is the surface interpolation matrix pertaining to the face $1-2-3-4$ and can be obtained by setting $\zeta = -1$ in the coefficients of the shape function matrix $[N]$. Further, the surface integral can be evaluated by using the expression given by Eq. (A1.22).

8.3.5. Stress calculation

From the known displacements, the strains are calculated at the Gauss points from strain-displacement matrix $[B]$ after eliminating the additional terms due to incompatible modes as described in the case of four-noded 2D quadrilateral element. These strains are extrapolated at the nodes of the element and a global smoothening of strain is performed for all nodes of the finite element model by averaging of contributions from the neighboring finite elements. Finally the stresses in each element are obtained using the stress-strain relation for the material belonging to the element.

8.4. TWENTY-NODED BRICK ELEMENT

This element belongs to the serendipity family. On the master cube (Figure 8.5), the isoparametric shape functions defining the geometry and variation of displacements for this element are given by

$$N_i = \frac{1}{8}(1+\xi_i\xi)(1+\eta_i\eta)(1+\zeta_i\zeta)(\xi_i\xi+\eta_i\eta+\zeta_i\zeta-2) \quad i = 1 \text{ to } 8$$

$$N_i = \frac{1}{4}(1-\xi^2)(1+\eta_i\eta)(1+\zeta_i\zeta) \quad i = 9,11,13,15$$

$$N_i = \frac{1}{4}\left(1-\eta^2\right)\left(1+\xi_i\xi\right)\left(1+\zeta_i\zeta\right) \quad i = 10,12,14,16$$

$$N_i = \frac{1}{4}\left(1-\zeta^2\right)\left(1+\xi_i\xi\right)\left(1+\eta_i\eta\right) \quad i = 17,18,19,20, \tag{8.52}$$

where (ξ_i, η_i, ζ_i) represents the coordinates of node i of the element in the natural coordinate (ξ, η, ζ) system.

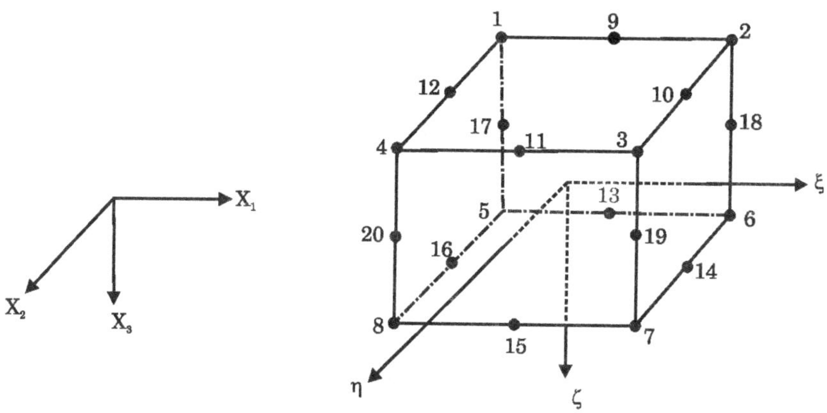

Fig. 8.5 Twenty-noded isoparametric brick element

The formulation of element matrix and nodal load vector for this element is very similar to that for the eight noded isoparametric brick element without incompatible modes.

8.4.1. Stress calculation

From the known displacements, the strains are calculated at reduced integration Gaussian points 2 × 2 × 2 inside elements, since they have most precise values at these points. These strains are extrapolated at the vertex nodes of the element using trilinear extrapolation in the local coordinate system. Values for midside nodes are calculated as an average between values for two vertex nodes. Then a global smoothing is performed for all nodes of the finite element model by averaging of contributions from the neighboring finite elements. Finally the stresses in each element are obtained using the stress-strain relation for the material belonging to that element.

8.5. LAYERED BRICK ELEMENTS

In the three-dimensional analysis of composites, layered version of three-dimensional isoparametric brick element is used to account for changes in material properties and orientation of the plies inside the element. Layered eight-noded and twenty-noded brick elements are depicted in Figures (8.6) and (8.7) respectively.

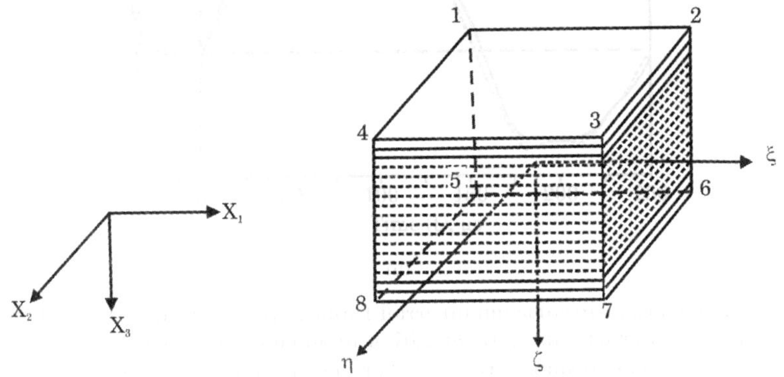

Fig. 8.6. Layered version of eight-noded isoparametric brick element.

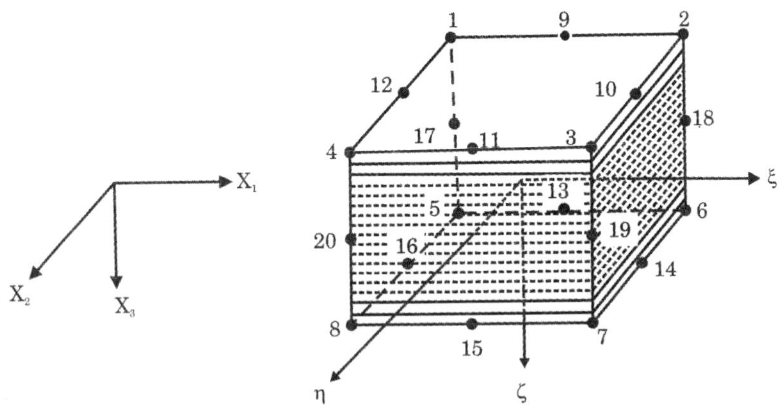

Fig. 8.7 Layered version of twenty-noded isoparametric brick element

Element stiffness matrices for these elements are given as

$$\begin{aligned}\left[K^e\right] &= \int_{-1}^{+1}\int_{-1}^{+1}\int_{-1}^{+1}[B]^T[C][B]\ \det[J]\ d\xi d\eta d\zeta \\ &= \int_{-1}^{+1}\int_{-1}^{+1}\int_{-1}^{+1}[B]^T[T]^T\left[\bar{C}\right][T][B]\ \det[J]\ d\xi d\eta d\zeta,\end{aligned} \tag{8.53}$$

in which the constitutive relation for composites is used (See Appendix 3). Here $\left[\bar{C}\right]$ is the material property matrix relating the strains within the ply to the stresses in ply coordinate system and [*C*] is the material property matrix transformed to the global directions. [*T*] is the transformation matrix relating the strains in the ply principal directions to those in the global reference axis.

The elasticity matrix [*C*] in the above equation depends on the material properties and the orientations of the plies through the thickness of the element. When the material properties and ply orientations are same through the thickness of the element, numerical integration of this equation using Gaussian quadrature can be carried out at element level as given below,

$$\begin{aligned}\left[K^e\right] &= \int_{-1}^{1}\int_{-1}^{1}\int_{-1}^{1}\left[f(\xi,\eta,\zeta)\right]d\xi d\eta d\zeta \\ &= \sum_{K=1}^{ng}\sum_{J=1}^{ng}\sum_{I=1}^{ng}\left[f(\xi_I,\eta_J,\zeta_K)\right]w_I w_J w_K,\end{aligned} \tag{8.54}$$

in which $\left[f(\xi,\eta,\zeta)\right]$ represents the integrand matrix in Eq. (8.53) and ng is the number of Gauss points in each direction.

However, when the material properties and ply orientations are non-constant throughout the element thickness, it is no longer possible to perform the volume integration using a minimum number of Gauss points. Rather the integration must be performed using several points from ply to ply through the element thickness. The procedure adopted is to assume the element thickness to be divided into a number of sections equal to the number of plies through the element thickness, with the ply boundary delineating each section. The integration along ζ direction is then to be carried out by splitting the limits through each ply.

Thus Eq. (8.54) becomes

$$\left[K^e\right] = \sum_{r=1}^{nply}\left(\int_{\zeta_{r(bot)}}^{\zeta_{r(top)}}\int_{-1}^{1}\int_{-1}^{1}\left[f(\xi,\eta,\zeta)\right]d\xi d\eta d\zeta\right), \tag{8.55}$$

where the limit of integration in ζ direction in rth ply is from $\zeta_{r(bot)}$ (value of ζ at the bottom of *r*th ply) to $\zeta_{r(top)}$ (value of ζ at the top of the *r*th ply).

The integration along ζ direction in the above equation may be carried out either by numerical integration or analytically. If numerical quadrature is used, the limit of integration should be from –1 to 1 in each ply in order to apply the known coefficients of the Gauss integration rule. This is achieved by suitably modifying the variable ζ to $\zeta^{(r)}$ in the

rth ply such that $\zeta^{(r)}$ varies from –1 to 1 in that ply (Kumar, 1998). The change of variable is obtained from the following transformation (Figure 8.8):

$$\zeta = -1 + \frac{2}{h}\left[\sum_{j=1}^{r-1} h_j + \left(\frac{1+\zeta^{(r)}}{2}\right) h_r\right], \tag{8.56}$$

where h_r is the thickness of the rth ply and h is the full thickness of the element at the point.

From Eq. (8.56), we obtain

$$d\zeta = \frac{h_r}{h} d\zeta^{(r)}. \tag{8.57}$$

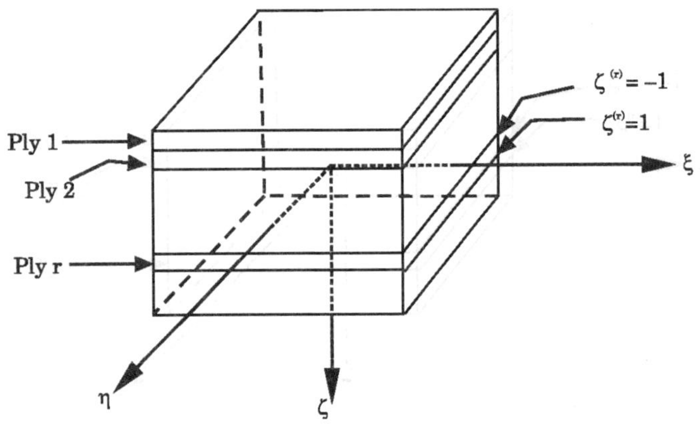

Fig. 8.8 Transformation of ζ coordinate from element to ply

Eq. (8.55) may then be written as

$$\begin{aligned}\left[K^e\right] &= \sum_{r=1}^{nply}\left(\int_{-1}^{1}\int_{-1}^{1}\int_{-1}^{1}\left[f\left(\xi,\eta,\zeta^{(r)}\right)\right]\frac{h_r}{h} d\xi d\eta d\zeta^{(r)}\right) \\ &= \sum_{r=1}^{nply}\left(\frac{h_r}{h}\sum_{K=1}^{ng}\sum_{J=1}^{ng}\sum_{I=1}^{ng}\left[f\left(\xi_I,\eta_J,\zeta_K^{(r)}\right)\right] w_I w_J w_K\right).\end{aligned} \tag{8.58}$$

8.5.1 Stress calculation

From the known displacements, the strains are calculated at the Gauss points from strain-displacement matrix $[B]$. These strains are extrapolated at the nodes of the element and a global smoothening of strain is performed. Since the element may contain several plies of different orientations, the strains in each individual ply are calculated from element strains using interpolation procedure. The strains in each ply are then transformed in material axes system using standard formula for transformation of strains. Finally the stresses in each ply are obtained using the stress-strain relation for the ply in material axes system.

8.6. EXAMPLE PROBLEMS

8.6.1. Example 1: A tapered aluminum alloy bar

Here, we consider the problem of tapered aluminum alloy bar solved in Chapter 4. The axial load of 10000 units applied to the free end of the bar is converted to an equivalent surface load. The problem is solved using both 3D 8 noded and 20 noded brick elements. The SB-FEM input file listing is shown in Appendix 8.

The FEM results for maximum axial deflection and axial stress at mid-length of the bar are shown in Table 8.1 and are found to be in good agreement with the theoretical results (Harris, 1959).

Table 8.1 Comparison of results for a tapered aluminum alloy bar.

	Theoretical results (Harris, 1959)	*SB-FEM results*	
		8-noded brick elements	*20-noded brick elements*
Maximum axial deflection	0.0048077 units	0.004815 units	0.004817 units
Axial stress at mid-length	4444 units	4468 units	4441 units

8.6.2. Example 2: A curved beam subjected to a shear load

The problem of a curved beam subjected to a shear load at the free end is solved using both 3D 8 noded and 20 noded brick elements. The geometrical, material properties and loading data of the beam are same as taken in Chapter 6. Number of elements chosen for the model are 512 in case of 8 noded elements and 64 in case of 20 noded elements. The vertical deflection at the top of the free end is calculated as 0.08760 and 0.08405 units respectively by the use of 8 noded and 20 noded brick elements. It is clear that both FEM results compare well with the theoretical value while the agreement is better in case of 8 noded brick elements.

8.6.3. Example 3: Long thick-walled cylinder

A long thick-walled cylinder subjected to an internal pressure is considered. This problem is same as that solved in Chapter 6. Because of symmetry only a quarter of the cylinder is analyzed (Figure 8.9). The cylinder segment is modeled using both 3D 8 noded and 20 noded brick elements.

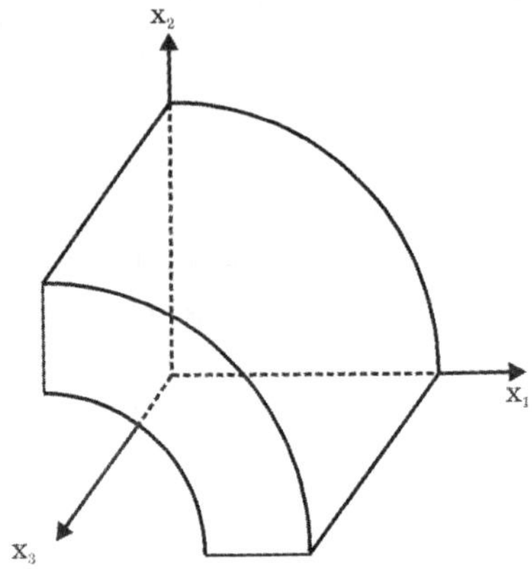

Fig. 8.9 Thick-walled cylinder - quarter symmetry

Since this problem is solved in Cartesian coordinates instead of cylindrical coordinates, the radial displacement, radial stress and hoop stress distributions (defined in cylindrical coordinates) may be calculated to be same as the distributions of u_1, σ_1 and σ_2 along the horizontal line (line in x_1 direction) with origin located at the centerline of the cylinder. The finite element results obtained by the use of both the elements are given in Table 8.2 along with the analytical results (Timoshenko, 1956). Good agreement between the two results can be seen, particularly those for displacements.

Table 8.2 Comparison of results for thick-walled cylinder.

	Analytical solution, units	*SB-FEM results, units*	
		8-noded brick elements	*20-noded brick elements*
Radial displacement at radius, a	0.0078666	0.007823	0.007863
Radial stress at radius, a	–30000	–21419	–25344
Radial stress at radius, $(a + b)/2$	–7778	–7814	–5738
Radial stress at radius, b	0	–1319	865
Hoop stress at radius, a	50000	51386	51240
Hoop stress at radius, $(a + b)/2$	27778	27192	28782
Hoop stress at radius, b	20000	19278	20404

8.6.4. Example 4: Composite laminated and sandwich plates

A symmetrically laminated cross-ply plate and a sandwich plate, both with simply-supported boundary conditions are considered (Figure 8.10).

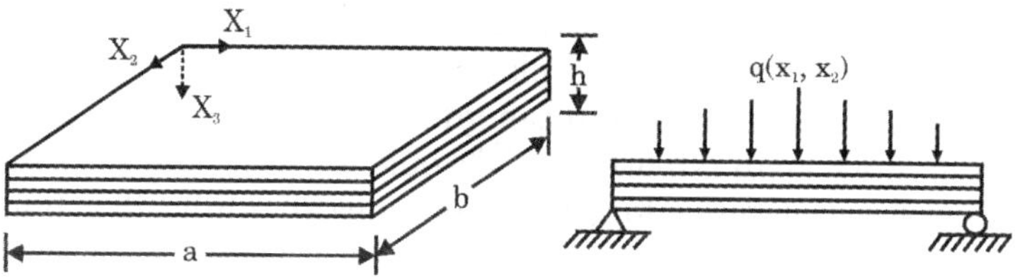

Fig. 8.10. Problem description of a laminated plate simply-supported at its edges and subjected to a transverse sinusoidal loading.

The material properties considered for the laminated plate in ply coordinate system are (Pagano, 1970) $E_x = 25 \times 10^6$; $E_y = E_z = 1 \times 10^6$; $G_{xy} = G_{xz} = 0.5 \times 10^6$; $G_{yz} = 0.2 \times 10^6$; $\nu_{xy} = \nu_{yz} = \nu_{xz} = 0.25$, and each ply thickness is $h/4$, in which h is the laminate thickness taken as 0.02. For sandwich plate, the material properties of face sheets are same as for laminated plate with the thickness of each face sheet as $h/10$ (plate thickness, $h = 0.02$) and fibre orientation in the face sheet as 0°. The properties of the middle core material are (Pagano, 1970) $E_x = E_y = 0.04 \times 10^6$; $E_z = 0.5 \times 10^6$; $G_{xy} = 0.016 \times 10^6$; $G_{yz} = G_{xz} = 0.06 \times 10^6$; $\upsilon_{xy} = \upsilon_{yz} = \upsilon_{xz} = 0.25$ and thickness is $0.8h$. Consistent units are assumed.

The plates are square with $a = b = 10$h and are subjected to a transverse sinusoidal loading (applied at the top faces) of a form represented by $q(x_1, x_2) = q_0 \sin\left(\frac{\pi x_1}{a}\right) \sin\left(\frac{\pi x_2}{b}\right)$. The transverse deflection (u_3) and bending stresses (σ_1 and σ_2) presented are non-dimensionalised using the following multiplying factors:

$$m_1 = \frac{100h^3 E_y}{q_0 a^4} \text{ and } m_2 = \frac{h^2}{q_0 a^2},$$

where E_y corresponds to the top layer/face sheet.

Results of the analysis are presented in Table 8.3 along with analytical results available in literature. The present results are in reasonable agreement with those obtained by 3D elasticity solution (Pagano, 1970) and by Pandya and Kant (1987).

Table 8.3 Deflection and stresses in a square simply supported four layer cross-ply and sandwich plate under sinusoidal transverse loading.

Symmetrically laminated plate (0/90/90/0)			
	$u_3 \times m_1$ *at location* $(a/2, b/2, h/2)$	$\sigma_1 \times m_2$ *at* $(a/2, b/2, h)$	$\sigma_2 \times m_2$ *at* $(a/2, b/2, 3h/4)$
SB-FEM result	0.7362	0.8304	0.3977
Classical plate theory (Stavsky, 1961 and Pagano, 1970)	0.4312	0.5390	0.2690
3-D Elasticity (Pagano, 1970)	0.7370	0.5590	0.4010
Pandya and Kant (1987)	0.7070	0.5358	0.3893
Sandwich plate			
	$u_3 \times m_1$ *at location* $(a/2, b/2, h/2)$	$\sigma_1 \times m_2$ *at* $(a/2, b/2, h)$	$\sigma_2 \times m_2$ *at* $(a/2, b/2, h)$
SB-FEM result	2.232	1.0819	0.1017
Classical plate theory (Stavsky, 1961 and Pagano, 1970)	—	1.0970	0.0543
3-D Elasticity (Pagano, 1970)	—	1.1520	0.1099
Pandya and Kant (1987)	2.0179	1.1140	0.1077

9

Transient Dynamic Analysis

9.1. INTRODUCTION

Transient dynamics analysis, sometimes called 'Time-history analysis', is a technique used to determine the dynamic response of a structure under the action of any general time dependant loading. This type of analysis is used to determine the time-varying displacements, strains, stresses and forces as it responds to any combination of static, transient and harmonic loads. The time-scale of loading is such that inertial or damping effects of the structure are considered to be important. Cases where such effects play a major role are under step or impulse loading conditions, where there is a sharp load change in a fraction of time. If inertia effects are negligible for the loading conditions being considered, a static analysis may be used instead. A transient analysis is more involved than the static analysis and requires a good understanding of the dynamic behavior of the structure.

9.2 GENERAL DERIVATION OF FINITE ELEMENT DYNAMIC EQUILIBRIUM EQUATIONS

Finite element dynamic equilibrium equation is derived using the Hamilton's principle. The energy functional for this principle is the Lagrangian which includes the kinetic energy in addition to the strain energy and potential energy due to external loads, and is given by

$$L = T - U + W \tag{9.1}$$

where,

T = kinetic energy ,

U = strain energy and

$-W$ = potential energy due to applied loads (negative of the work done).

These energy terms are given as

$$T = \int_V \tfrac{1}{2}\rho\{\dot{u}\}^T\{\dot{u}\}\,dV$$

$$U = \int_V \tfrac{1}{2}\{\varepsilon\}^T\{\sigma\}\,dV = \int_V \tfrac{1}{2}\{\varepsilon\}^T[C]\{\varepsilon\}\,dV$$

and
$$W = \int_V \{u\}^T\{b\}\,dV + \int_{S_F}\{u^s\}^T\{t\}\,dS + \sum_I \{u^I\}^T\{f_c^I\}, \tag{9.2}$$

where V and S represent volume and surface of the body. ρ is the mass density of the material. $\{u\}$ and $\{\dot{u}\}$ are vectors of displacement and velocity respectively. The superscript s over u denotes that surface displacements are considered and the superscript I denotes the displacements at the point where the concentrated forces $\{f_c^I\}$ are applied. All other vectors such as $\{b\}$, $\{t\}$, $\{\varepsilon\}$ and $\{\sigma\}$, and matrix $[C]$ have their usual meanings same as in the static analysis.

After substituting all terms in Eq. (9.1), we have

$$L = \int_V \tfrac{1}{2}\rho\{\dot{u}\}^T\{\dot{u}\}\,dV - \int_V \tfrac{1}{2}\{\varepsilon\}^T[C]\{\varepsilon\}\,dV + \int_V \{u\}^T\{b\}\,dV + \int_{S_F}\{u^s\}^T\{t\}\,dS + \sum_I \{u^I\}^T\{f_c^I\}. \tag{9.3}$$

The Hamilton's principle states that the variation of the Lagrangian during any time interval t_1 to t_2 must be equal to zero, *i.e.*

$$\delta\int_{t_1}^{t_2} L\,dt = 0 \tag{9.4}$$

Substituting from Eq. (9.3) into Eq. (9.4), and applying the variational principle, we obtain

$$\int_{t_1}^{t_2}\Big(\int_V \rho\{\delta\dot{u}\}^T\{\dot{u}\}\,dV - \int_V \{\delta\varepsilon\}^T[C]\{\varepsilon\}\,dV + \int_V \{\delta u\}^T\{b\}\,dV + \int_{S_F}\{\delta u^s\}^T\{t\}\,dS + \sum_I \{\delta u^I\}^T\{f_c^I\}\Big)\,dt = 0. \tag{9.5}$$

Now, integration of the first term by parts with respect to time gives

$$\int_{t_1}^{t_2} \int_V \rho \{\delta \dot{u}\}^T \{\dot{u}\} \, dV dt = \int_V \rho \Big[\{\delta u\}^T \{\dot{u}\} \Big]_{t_1}^{t_2} dV - \int_{t_1}^{t_2} \int_V \rho \{\delta u\}^T \{\ddot{u}\} \, dV dt \tag{9.6}$$

According to Hamilton's principle, the displacement configuration must satisfy the conditions given at time t_1 and t_2. Hence, $\{\delta u(t_1)\} = \{\delta u(t_2)\} = \{0\}$ and the first term of right hand side of Eq. (9.6) vanishes.

Substitution of second right hand side term of Eq. (9.6) in Eq. (9.5) yields

$$\begin{aligned} \int_{t_1}^{t_2} \Big(&-\int_V \rho \{\delta u\}^T \{\ddot{u}\} \, dV - \int_V \{\delta \varepsilon\}^T [C] \{\varepsilon\} \, dV + \int_V \{\delta u\}^T \{b\} \, dV \\ &+ \int_{S_F} \{\delta u^s\}^T \{t\} \, dS + \sum_I \{\delta u^I\}^T \{f_c^I\} \Big) dt = 0 . \end{aligned} \tag{9.7}$$

It follows that

$$\begin{aligned} &-\int_V \rho \{\delta u\}^T \{\ddot{u}\} \, dV - \int_V \{\delta \varepsilon\}^T [C] \{\varepsilon\} \, dV + \int_V \{\delta u\}^T \{b\} \, dV \\ &+ \int_{S_F} \{\delta u^s\}^T \{t\} \, dS + \sum_I \{\delta u^I\}^T \{f_c^I\} = 0 , \end{aligned} \tag{9.8}$$

or, more explicitly,

$$\begin{aligned} &\int_V \rho \{\delta u\}^T \{\ddot{u}\} \, dV + \int_V \{\delta \varepsilon\}^T [C] \{\varepsilon\} \, dV \\ &= \int_V \{\delta u\}^T \{b\} \, dV + \int_{S_F} \{\delta u^s\}^T \{t\} \, dS + \sum_I \{\delta u^I\}^T \{f_c^I\} . \end{aligned} \tag{9.9}$$

Now, the principle can be used effectively as a mechanism to generate finite element equations that govern the response of a structure or continuum.

As discussed earlier for static analysis, the nodal displacements pertaining to a particular element e, $\{U^e\}$ and the global displacements $\{U\}$ are related by a Boolean connectivity matrix $[L^e]$ as

$$\{U^e\} = [L^e]\{U\}. \tag{9.10}$$

Assuming that the element accelerations within each element e are interpolated in the same way as the element displacements, the displacements and accelerations within the element are related to the nodal displacements $\{U^e\}$ and nodal accelerations $\{\ddot{U}^e\}$ by its shape function matrix $[N]$ so that

$$\{u\} = [N]\{U^e\}$$

and
$$\{\ddot{u}\} = [N]\{\ddot{U}^e\}. \tag{9.11}$$

The corresponding element strains can now be evaluated as

$$\{\varepsilon\} = [B]\{U^e\}, \tag{9.12}$$

where $[B]$ is the strain displacement matrix; the rows of $[B]$ are obtained by appropriately differentiating and combining rows of the matrix $[N]$.

Using the assumption on the displacements within each finite element, the equilibrium equations corresponding to the nodal point displacements of the assemblage of finite elements can now be derived. First, Eq. (9.9) is rewritten as a sum of integration over the volume and areas of all finite elements, *i.e.*

$$\begin{aligned}&\sum_e \int_{V^e} \rho\{\delta u\}^T \{\ddot{u}\}\,dV + \sum_e \int_{V^e} \{\delta\varepsilon\}^T [C]\{\varepsilon\}\,dV \\ &\quad = \sum_e \int_{V^e} \{\delta u\}^T \{b\}\,dV + \sum_e \int_{S_F^e} \{\delta u^s\}^T \{t\}\,dS + \sum_I \{\delta u^I\}^T \{f_c^I\},\end{aligned} \tag{9.13}$$

where $e = 1,2, \ldots, E$ and E = number of elements.

Now, substituting into Eq. (9.13) for the element displacements and strains, using Eqs. (9.11) and (9.12), we obtain

$$\begin{aligned}&\sum_e \{\delta U^e\}^T \left[\left(\int_{V^e} \rho[N]^T [N]dV\right)\{\ddot{U}^e\} + \left(\int_{V^e} [B]^T [C][B]dV\right)\{U^e\}\right] \\ &\quad = \sum_e \{\delta U^e\}^T \left[\left(\int_{V^e} [N]^T \{b\}\,dV\right) + \left(\int_{S_F^e} \left[N^s\right]^T \{t\}\,dS\right)\right] + \{\delta U\}^T \{F_c\},\end{aligned} \tag{9.14}$$

where the surface displacement interpolation matrices $[N^s]$ are obtained from the volume displacement interpolation matrices $[N]$ by substituting element surface coordinates. $\{F_c\}$ is a vector of externally applied forces to the nodes of the element assemblage. It should be noted that the *ith* component in $\{F_c\}$ is the concentrated nodal forces corresponding to the ith displacement component in $\{U\}$.

Now the terms in Eq. (9.14) can be defined as follows:

Element mass matrix,

$$\left[M^e\right] = \int_{V^e} \rho [N]^T [N] dV \tag{9.15}$$

Element stiffness matrix,

$$\left[K^e\right] = \int_{V^e} [B]^T [C][B] dV \tag{9.16}$$

Element body force vector,

$$\left\{F_b^e\right\} = \int_{V^e} [N]^T \{b\} dV \tag{9.17}$$

Element surface force vector,

$$\left\{F_s^e\right\} = \int_{S_F^e} \left[N^s\right]^T \{t\} dS. \tag{9.18}$$

Substituting Eq. (9.10) and (9.15-9.18) into Eq. (9.14) and rearranging, we have

$$\begin{aligned}\{\delta U\}^T &\left[\left(\sum_e \left[L^e\right]^T \left[M^e\right]\left[L^e\right]\right)\{\ddot{U}\} + \left(\sum_e \left[L^e\right]^T \left[K^e\right]\left[L^e\right]\right)\{U\}\right] \\ &= \{\delta U\}^T \left[\sum_e \left[L^e\right]^T \left\{F_b^e\right\} + \sum_e \left[L^e\right]^T \left\{F_s^e\right\} + \left\{F_c\right\}\right].\end{aligned} \tag{9.19}$$

Since $\{\delta U\}$ is arbitrary, Eq. (9.19) can be written as

$$[M]\{\ddot{U}\} + [K]\{U\} = \{F_b\} + \{F_s\} + \{F_c\}, \tag{9.20}$$

where,

Global mass matrix,

$$[M] = \sum_e \left[L^e\right]^T \left[M^e\right]\left[L^e\right] \tag{9.21}$$

Global stiffness matrix,

$$[K] = \sum_e \left[L^e\right]^T \left[K^e\right]\left[L^e\right] \tag{9.22}$$

Global body force vector,

$$\{F_b\} = \sum_e [L^e]^T \{F_b^e\} \tag{9.23}$$

Global surface force vector,

$$\{F_s\} = \sum_e [L^e]^T \{F_s^e\}. \tag{9.24}$$

Representing the sum of all load vectors in Eq. (9.20) as $\{F\}$, we obtain the following governing equation:

$$[M]\{\ddot{U}\} + [K]\{U\} = \{F\}. \tag{9.25}$$

9.3. NUMERICAL SOLUTION OF FINITE ELEMENT EQUATIONS

Several methods are available for the solution of the finite element equation of motion (Eq. (9.25)). All the methods can be divided into two classes: direct integration methods and the mode superposition method. In direct integration methods, Eq. (9.25) is integrated numerically by using a step-by-step procedure (Brusha and Nigro, 1980). The term direct denotes that no transformation of the equations into a different form is used prior to numerical integration. These methods calculate the solution at the next required time step from a knowledge of the solution until the previous time step. The computational work involved in the direct integration methods is proportional to the number of time steps used in the analysis. Hence in general, the use of direct integration methods is expected to be efficient when the response over only a relatively short duration (involving few time steps) is required. However, if the integration has to be performed for many time steps, it may be more effective to transform Eq. (9.25) into a form in which the step-by-step solution is less costly. The mode superposition or normal mode method is a technique wherein Eq. (9.25) is first transformed into a convenient form before integration is performed. The mode superposition method is well described in Rao (2004).

9.3.1. Direct integration methods

The direct integration methods are based on the following ideas:

(a) Instead of trying to find a solution $\{U(t)\}$ that satisfies Eq. (9.25) for any time t, we can try to satisfy Eq. (9.25) only at discrete time intervals Δt apart.

(b) Within any time interval, the nature of variation of displacement $\{U\}$, and acceleration $\{\ddot{U}\}$ can be assumed in a suitable manner.

Here, the time interval Δt and the nature of variation of $\{U\}$ and $\{\ddot{U}\}$ within any Δt are chosen by considering factors such as accuracy, stability, and cost of solution. The central difference, Houbolt, Wilson and Newmark methods fall under the category of direct methods (Wood *et al.*, 1980 and Wood, 1984). In the central difference method, the solution at time $t + \Delta t$ is based on using the equilibrium conditions at time t. For this reason, this procedure is called an explicit integration method. This method does not require the assembly of the global stiffness matrix and is very ideal for certain categories of transient dynamic problems. Explicit methods require very small time steps to ensure numerical stability and therefore, should be used carefully. On the other hand, the Houbolt, Wilson and Newmark methods use the equilibrium conditions at time $t + \Delta t$ and are called implicit integration methods. The implicit methods generally have the benefits with regard to numerical stability and are therefore widely used for several types of problems. The most popular of these is the Newmark's method which is discussed next.

Newmark direct Integration Method

Newmark integration method is an extension of the linear acceleration method in which a linear variation of acceleration from time t to $t + \Delta t$ is assumed. From knowledge of the displacement, velocity and acceleration at time t_n (time at nth time step), the method calculates the velocity and displacement at time t_{n+1} (time at $(n+1)$th time step, *i.e.* $t_n + \Delta t$) using the following equations:

$$\{\dot{U}_{n+1}\} = \{\dot{U}_n\} + \left[(1-\alpha)\{\ddot{U}_n\} + \alpha\{\ddot{U}_{n+1}\}\right]\Delta t \tag{9.26}$$

$$\{U_{n+1}\} = \{U_n\} + \{\dot{U}_n\}\Delta t + \left[\left(\frac{1}{2}-\beta\right)\{\ddot{U}_n\} + \beta\{\ddot{U}_{n+1}\}\right](\Delta t)^2, \tag{9.27}$$

where Δt is the time step, n is the step number, and the parameters α and β can be determined to obtain integration accuracy and stability. For $\alpha = 0.5$ and $\beta = 0.25$, this method is unconditionally stable for any time step.

Since the primary aim is the computation of displacements $\{U_{n+1}\}$, the governing equation (Eq. (9.25)) is evaluated at time t_{n+1} as

$$[M]\{\ddot{U}_{n+1}\}+[K]\{U_{n+1}\}=\{F_{n+1}\}. \tag{9.28}$$

The solution for the displacements at time t_{n+1} is obtained by first rearranging Eq. (9.27), such that

$$\{\ddot{U}_{n+1}\}=a_0\left(\{U_{n+1}\}-\{U_n\}\right)-a_1\{\dot{U}_n\}-a_2\{\ddot{U}_n\}, \tag{9.29}$$

where

$$a_0=\frac{1}{\beta(\Delta t)^2},\quad a_1=\frac{1}{\beta(\Delta t)},\quad a_2=\frac{1-2\beta}{2\beta}.$$

Eq. (9.29) is now substituted into Eq. (9.28) to form

$$\left[\hat{K}\right]\{U_{n+1}\}=\left\{\hat{F}_{n+1}\right\}, \tag{9.30}$$

where $\left[\hat{K}\right]$ and $\left\{\hat{F}_{n+1}\right\}$ can be termed as effective stiffness matrix and effective load vector respectively at time t_{n+1} and are given as

$$\left[\hat{K}\right]=[K]+a_0[M] \tag{9.31}$$

and

$$\left\{\hat{F}_{n+1}\right\}=\{F_{n+1}\}+[M]\left(a_0\{U_n\}+a_1\{\dot{U}_n\}+a_2\{\ddot{U}_n\}\right). \tag{9.32}$$

Once the displacements $\{U_{n+1}\}$ at time t_{n+1} are obtained by solving Eq. (9.30), the velocities $\{\dot{U}_{n+1}\}$ and accelerations $\{\ddot{U}_{n+1}\}$ are computed using Eqs. (9.26) and (9.29) respectively.

Inherent to the Newmark method is that the values of $\{U_0\}$, $\{\dot{U}_0\}$ and $\{\ddot{U}_0\}$ at the start of the transient must be known. Non-zero initial conditions are input by performing a static analysis load step(s) prior to the start of the transient itself.

9.4 CONSISTENT AND LUMPED MASS MATRICES

The mass matrix derived using Eq. (9.15) is called the "consistent" mass matrix of the element. It is called consistent because the same shape functions are used to describe the variation of displacement and acceleration over the element in formulating the stiffness and mass matrices. It is useful to note that several dynamic problems have been solved with simpler forms of mass matrices which are convenient and economical. The simplest form of mass matrix that can be used is that obtained by placing point (concentrated) masses at node points in the directions of the assumed displacement degrees of freedom. The concentrated masses refer to translational and rotational inertia of the element and are calculated by assuming that the material within the mean locations on either side of the particular displacement behaves like a rigid body while the remainder of the element does not participate in the motion. Thus, this assumption excludes the dynamic coupling that exists between the element displacements, and hence the resulting element mass matrix is purely diagonal and is called the "'lumped" mass matrix.

Among the mass lumping schemes, the simplest is the equal lumping of mass in which the mass is lumped equally to all nodes of the element. As an example, consider a three-dimensional eight noded brick element as described in Chapter 8. The consistent mass matrix given by Eq. (9.15) can be expressed for this element as

$$\left[M^e\right]=\int_{-1}^{+1}\int_{-1}^{+1}\int_{-1}^{+1}[N]^T\rho[N]\ \det[J]\ d\xi d\eta d\zeta, \qquad (9.33)$$

which can be evaluated numerically. This matrix, in general, is fully populated. On the other hand, the lumped mass matrix may be established by lumping equal parts of the total element mass to its nodal points and can be written as

$$\left[M^e\right]=\frac{\rho^e V^e}{8}[I], \qquad (9.34)$$

where ρ^e and V^e are the material density and volume of the element. $[I]$ is a 24×24 unit matrix.

If we take 8 noded layered brick element, the consistent mass matrix is again evaluated using Eq. (9.33). However, the integration in the thickness direction must be performed from ply to ply in the case when the material density varies over different plies. Alternatively, the lumped mass matrix can be used which may be calculated as

$$\left[M^e\right]=\frac{\sum_{i=1}^{n_{ply}}\rho_i V_i^e}{8}[I], \qquad (9.35)$$

where ρ_i and V_i^e are the material density and volume of the *ith* ply of the element. n_{ply} represents number of plies in the element. $[I]$ is again a 24 × 24 unit matrix.

9.5. EXAMPLE PROBLEMS

Transient dynamic finite element analysis is performed for problems of isotropic and composite plates subjected to dynamic (impact) loading using three dimensional hexahedral elements. The problem description of a laminated composite plate subjected to transverse low velocity non-penetrating impact by a foreign object is shown in Figure 9.1.

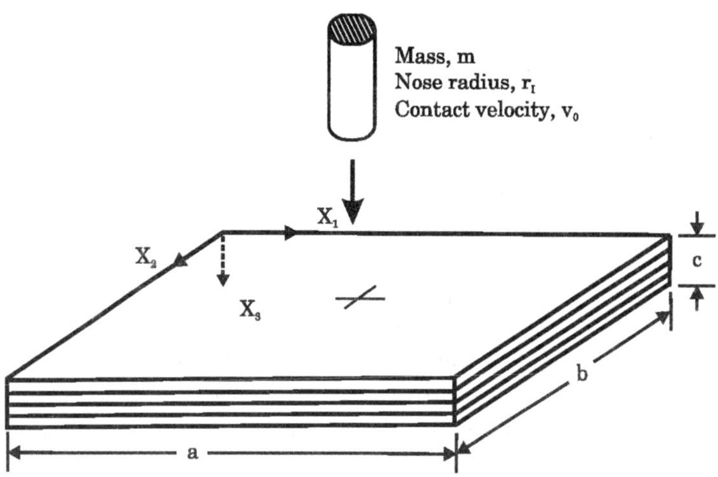

Fig. 9.1 Problem description of transverse impact on a rectangular laminated plate by a foreign object

9.5.1 Impact analysis

When a composite laminate is impacted by a mass, local deformation takes place in the contact region and in most cases must be accounted for in the analysis if the contact force history is to be predicted accurately. A practical approach consists of determining the relationship between contact force F^c with the indentation depth α (the change in the distance between the centre of the impactor's nose and the mid-surface of the plate).

Contact law

Isotropic plate. For isotropic plate, the contact force is given by the following classical Hertzian law (Goldsmith, 1960):

$$F^c = k\alpha^{1.5}, \tag{9.36}$$

in which the constant k is given by the expression

$$k = \frac{4}{3}\sqrt{r_I}\frac{1}{\left[\left(1-\nu_I^2\right)/E_I + \left(1-\nu_T^2\right)/E_T\right]},$$

in which r_I, E_I and ν_I are the local nose radius, the Young's modulus and the Poisson's ratio of the impactor, respectively. E_T and ν_T are the Young's modulus and the Poisson's ratio of the target (plate), respectively.

Composite plate. For composite plate, the modified version of Hertzian contact law proposed by Yang and Sun (1982) based on static indentation tests can be used. During loading, the contact force is given by the following relation:

$$F^c = k\alpha^{1.5}, \tag{9.37}$$

where k is the modified constant of the Hertz contact theory and is given by

$$k = \frac{4}{3}\sqrt{r_I}\frac{1}{\left[\left(1-\nu_I^2\right)/E_I + 1/E_y\right]},$$

in which E_y is the transverse modulus normal to the fibre direction in the uppermost composite layer.

Upon unloading, the contact force is simulated by the following relation as proposed by Yang and Sun (1982)

$$F^c = F_m^c\left(\frac{\alpha - \alpha_0}{\alpha_m - \alpha_0}\right)^{2.5}, \tag{9.38}$$

where F_m^c is the maximum contact force just before unloading, α_m is the maximum indentation corresponding to F_m^c, and α_0 is the permanent indentation during this loading-unloading process. The permanent indentation can be determined from the following expressions:

$$\alpha_0 = 0 \qquad \text{when } \alpha_m < \alpha_{cr}$$

$$\alpha_0 = \alpha_m \left[1 - \left(\frac{\alpha_{cr}}{\alpha_m} \right)^{2/5} \right] \quad \text{when } \alpha_m \geq \alpha_{cr}, \tag{9.39}$$

where α_{cr} is the critical indentation and is approximately 8.03×10^{-2} mm for graphite/epoxy (Yang and Sun, 1982).

Solution procedure

In order to implement the non-linear contact law in a linear analysis, the following procedure is adopted (Pradhan and Kumar, 2000; Kumar *et. al.*, 2007 and Kumar, 2011b). Since the contact area is generally small in comparison with the dimensions of the plate, a point load representing the resultant contact force is assumed.

The applied load vector $\{F_{n+1}\}$ in Eq. (9.32) consists of time-varying contact force F^c_{n+1} at $(n+1)$th time-step and can be written as

$$\left\{F_{n+1}\right\} = F^c_{n+1} \left\{I\right\}, \tag{9.40}$$

where $\{I\}$ is a column vector whose component corresponding to the node at the contact point and in the direction of contact force is unity, all other components being zeroes. Contact force F^c_{n+1} is not known at the beginning of each time-step and can be calculated using the following procedure.

The displacement vector $\left\{U_{n+1}\right\}$ in Eq. (9.30) can be decomposed into two parts $\left\{U^A_{n+1}\right\}$ and $\left\{U^M_{n+1}\right\}$, so that

$$\left[\hat{K}_{n+1}\right]\left\{U^A_{n+1}\right\} = \left\{F_{n+1}\right\} = F^c_{n+1} \left\{I\right\} \tag{9.41}$$

and

$$\left[\hat{K}_{n+1}\right]\left\{U^M_{n+1}\right\} = \left\{F^M_{n+1}\right\}, \tag{9.42}$$

where $\left\{F^M_{n+1}\right\} = [M]\left(a_0 \left\{U_n\right\} + a_1 \left\{\dot{U}_n\right\} + a_2 \left\{\ddot{U}_n\right\}\right)$ is used.

For a unit contact force, Eq. (9.41) becomes

$$\left[\hat{K}_{n+1}\right]\left\{U^i_{n+1}\right\} = \left\{I\right\}, \tag{9.43}$$

where $\left\{U^i_{n+1}\right\}$ is the displacement caused by the unit contact force.

From Eqs. (9.41) and (9.43), it is seen that

$$\left\{U_{n+1}^{A}\right\} = F_{n+1}^{c}\left\{U_{n+1}^{i}\right\}. \tag{9.44}$$

The contact force F_{n+1}^{c} can be written in a general form of contact law as

$$F_{n+1}^{c} = \phi\left(\alpha_{n+1}\right) = \phi\left(d_{n+1} - w_{n+1}\right) = \phi\left(d_n + \dot{d}_n \Delta t - \frac{1}{4}\frac{F_n^c + F_{n+1}^c}{m}\left(\Delta t\right)^2 - w_{n+1}^{M} - F_{n+1}^{c} w_{n+1}^{i}\right), \tag{9.45}$$

where m is the mass of the impactor, and d_{n+1} is the displacement of the centre point of the impactor at *(n*+1)th time-step and is calculated by applying Newmark's method (with $\alpha = 0.5$ and $\beta = 0.25$) to equation of motion of impactor as shown in the above equation. w_{n+1} is the displacement of the mid-surface of the plate at the impact point in the direction of impact and is evaluated as $\left(w_{n+1}^{M} + F_{n+1}^{c} w_{n+1}^{i}\right)$ using Eqs. (9.42) and (9.44), w_{n+1}^{M} and w_{n+1}^{i} being the components of the displacement vectors $\left\{U_{n+1}^{M}\right\}$ and $\left\{U_{n+1}^{i}\right\}$ corresponding to the node at the centre of the mid-surface of the plate and in the direction of impact. These displacement vectors are first calculated at each time step using Eqs. (9.42) and (9.43), since the terms on the right hand side of these equations are known at the beginning of *(n*+1)th time-step. Having known all other terms in Eq. (9.45), the contact force F_{n+1}^{c} can be evaluated by solving the equation using Newton–Raphson root-finding algorithm. The displacement vector $\left\{U_{n+1}^{A}\right\}$ is now evaluated from Eq. (9.44). Addition of $\left\{U_{n+1}^{A}\right\}$ and $\left\{U_{n+1}^{M}\right\}$ gives the total displacement vector $\left\{U_{n+1}\right\}$. This procedure is repeated for each time step.

9.5.2 Example 1: Impact on an isotropic plate

An example of an isotropic rectangular steel plate of length 200 mm, width 200 mm and thickness 8 mm simply supported on its edges is considered. The plate is subjected to an impact induced by a 20 mm diameter steel ball with initial velocity 1 m/s.

The SB-FEM command listing for this problem is shown in Appendix 8. The present FEM solutions for the contact force and displacements of the impactor and centre of the plate are compared with the analytical solutions (Goldsmith, 1960). As shown in Figure 9.2, an excellent agreement exists between the two solutions.

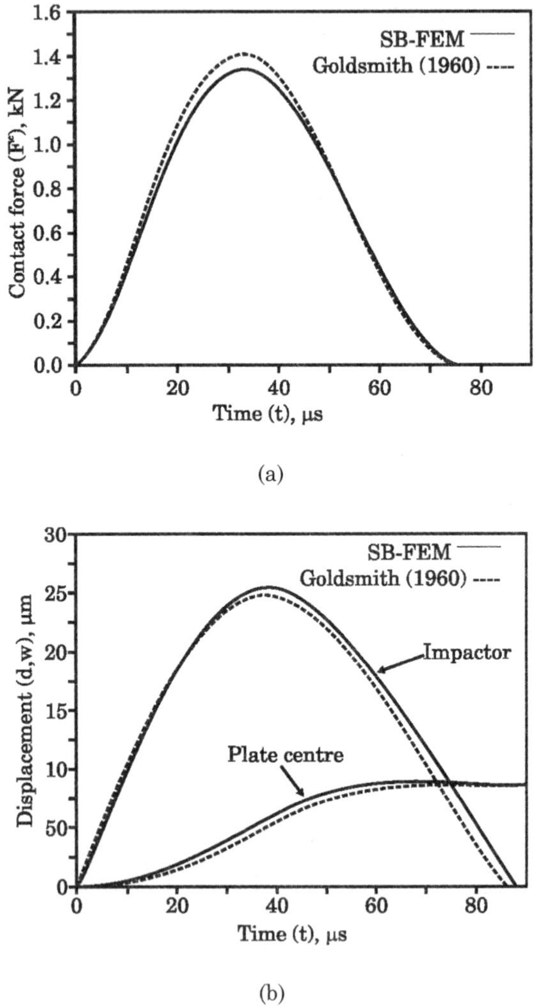

Fig. 9.2 Comparison of (a) contact force and (b) impactor and plate centre displacements in a 200 mm × 200 mm × 8 mm steel plate with simply-supported edges impacted by a 20 mm diameter steel sphere at a velocity of 1 m/s

9.5.3 Example 2: Impact on a graphite/epoxy composite plate

Impact response of a rectangular graphite/epoxy laminated plate with a ply orientation of $[0/\text{-}45/45/90]_{2S}$ is investigated. The plate of length

(a) 76.2 mm, width (b) 76.2 mm and thickness (c) 2.54 mm clamped on its edges is impacted by an aluminum sphere of 12.7 mm diameter traveling at 25.4 m/s. The material properties of fiberite T300/934 graphite/epoxy unidirectional ply are taken as: ply thickness = 0.15875 mm; density, ρ = 1536 kg/m^3; E_x = 145.4 GPa; $E_y = E_z$ = 9.998 GPa; $G_{xy} = G_{xz}$ = 5.689 GPa; G_{yz} = 3.845 GPa; $\nu_{xy} = \nu_{yz} = \nu_{xz}$ = 0.3. The time-varying results of contact force, impactor displacement and plate centre displacement are presented in Figure 9.3 along with the results reported by Wu and Chang (1989). A good agreement between the two results may be observed.

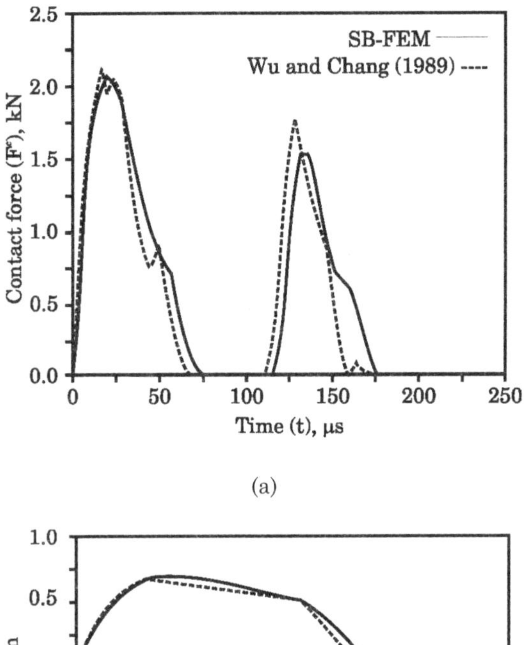

(a)

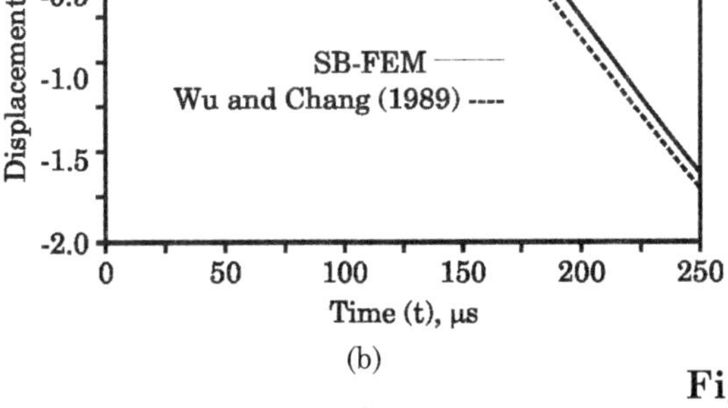

(b)

Fig. 9.3 : *contd.*

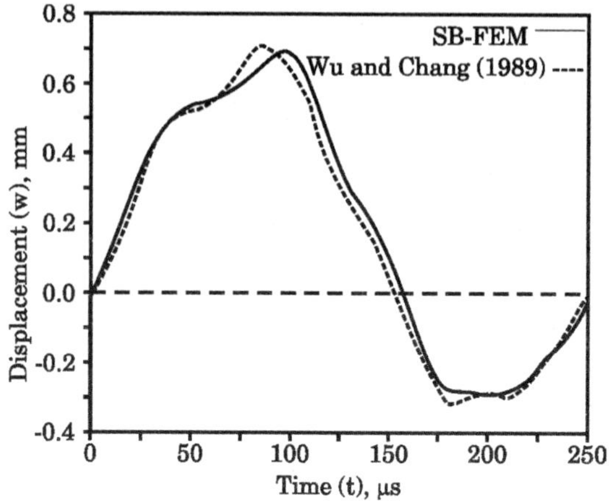

(c)

Fig. 9.3 Comparison of (a) contact force, (b) impactor displacement and (c) plate centre displacement in a 76.2 by 76.2 mm T300/934 graphite/epoxy laminated plate ($[0/-45/45/90]_{2S}$) with clamped edges and impacted by 12.7 mm diameter aluminum sphere at a velocity of 25.4 ms^{-1}

10

Geometrical Non-Linearity

10.1. INTRODUCTION

In all the structural problems studied so far, linear analyses were performed assuming that the analyzed model retained throughout the deformation process, whatever stiffness it possessed in its undeformed shape prior to loading. This assumption is valid when both displacements and strains developed in the structure are small and the geometrical equations (strain-displacement relations) as well as constitutive equations (stress-strain relations) are linear. However, in several problems the stiffness of the structure changes during the deformation process depending upon changes in geometry, material properties and restraints. If changes in stiffness come only from changes in shape as in the case of structure undergoing large displacements and large rotations, nonlinear behavior is defined as geometric nonlinearity. In many cases large displacements may be associated with large strains; however, in several other cases large displacements may occur without causing large strains for which linear stress-strain relationship may be assumed.

In the geometric nonlinear analysis, we use the principle of virtual work as in a linear analysis but now include the possibility that the strain-displacement relationship is non-linear. We then proceed in a completely analogous manner to establish the non-linear finite element equilibrium equations that govern the non-linear response of the structure. For non-linear analysis, the total load is applied in a number of increments since it may no longer be possible to directly obtain a stress distribution which equilibrates a given set of external loads. Incrementing loads requires definition of time (pseudo or actual) varying loads and definition of time increments (time points) in a solution set. In the solution procedure, a linear prediction of the nonlinear response is made within each increment and subsequent iterative operations are performed in order to restore equilibrium. Such a solution procedure is commonly referred to as an incremental-iterative method. The stiffness matrix [K] of the finite element equations must be updated as the nonlinear solution progresses through the iterative process. Since the solution here is obtained in an iterative manner, geometric nonlinearity may often be combined with material nonlinearity such as small strain plasticity (Zienkiewicz, 1977). The coupling is particularly simple if an incremental elasticity matrix may be established.

Two basically different formulations can be employed for the description of large deformation problems: (a) total Lagrangian approach and (b) updated Lagrangian approach. In a total Lagrangian approach, the current (2nd Piola-Kirchhoff) stress and (Green-Lagrange) strain fields are referred to the original geometric configuration and the displacement field gives the current configuration of the system in relation to its initial position. In the updated Lagrangian method, the current configuration of the system is used to define the current (Cauchy) stress and (Almansi) strain state. Here the geometry of the structure is successively updated during the incremental process and the stress and strain fields are referred to the last evaluated configuration. The stress and strain measures employed in the two approaches above are defined in such a way as to be work conjugates; *i.e.* an increment of strain multiplied by the current stress must be a proper infinitesimal incremental work quantity per unit volume of the material.

In this chapter, non-linear finite element formulation for large deformation analysis is developed based on the total Lagrangian approach. In the formulation, strains are considered to be reasonably small, so that the logarithmic strain measure and the engineering strain measure practically coincide and the standard elasticity relations used previously can be assumed to be valid. Extension of geometrical nonlinearity to transient dynamic analysis is also accomplished.

10.2. NON-LINEAR STRAIN-DISPLACEMENT RELATIONS

The general three-dimensional non-linear (large displacement) Green-Lagrange strain-displacement relationship can be written as (Sokolnikoff, 1956)

$$\varepsilon_1 = \frac{\partial u_1}{\partial x_1} + \frac{1}{2}\left[\left(\frac{\partial u_1}{\partial x_1}\right)^2 + \left(\frac{\partial u_2}{\partial x_1}\right)^2 + \left(\frac{\partial u_3}{\partial x_1}\right)^2\right],$$

$$\varepsilon_2 = \frac{\partial u_2}{\partial x_2} + \frac{1}{2}\left[\left(\frac{\partial u_1}{\partial x_2}\right)^2 + \left(\frac{\partial u_2}{\partial x_2}\right)^2 + \left(\frac{\partial u_3}{\partial x_2}\right)^2\right],$$

$$\varepsilon_3 = \frac{\partial u_3}{\partial x_3} + \frac{1}{2}\left[\left(\frac{\partial u_1}{\partial x_3}\right)^2 + \left(\frac{\partial u_2}{\partial x_3}\right)^2 + \left(\frac{\partial u_3}{\partial x_3}\right)^2\right],$$

$$\gamma_{12} = \frac{\partial u_1}{\partial x_2} + \frac{\partial u_2}{\partial x_1} + \frac{\partial u_1}{\partial x_1}\frac{\partial u_1}{\partial x_2} + \frac{\partial u_2}{\partial x_1}\frac{\partial u_2}{\partial x_2} + \frac{\partial u_3}{\partial x_1}\frac{\partial u_3}{\partial x_2},$$

$$\gamma_{23} = \frac{\partial u_2}{\partial x_3} + \frac{\partial u_3}{\partial x_2} + \frac{\partial u_1}{\partial x_2}\frac{\partial u_1}{\partial x_3} + \frac{\partial u_2}{\partial x_2}\frac{\partial u_2}{\partial x_3} + \frac{\partial u_3}{\partial x_2}\frac{\partial u_3}{\partial x_3}$$

and

$$\gamma_{13} = \frac{\partial u_1}{\partial x_3} + \frac{\partial u_3}{\partial x_1} + \frac{\partial u_1}{\partial x_1}\frac{\partial u_1}{\partial x_3} + \frac{\partial u_2}{\partial x_1}\frac{\partial u_2}{\partial x_3} + \frac{\partial u_3}{\partial x_1}\frac{\partial u_3}{\partial x_3}, \qquad (10.1)$$

where all derivatives are evaluated with respect to the reference configuration.

10.3. LARGE DISPLACEMENT FINITE ELEMENT FORMULATION

In large displacement problem, strain at any point within an element consists of both linear and non-linear parts and can be written as

$$\{\varepsilon\} = \{\varepsilon_L\} + \{\varepsilon_{NL}\}. \qquad (10.2)$$

The linear part of the strain within each element e are related to the nodal displacements as

$$\{\varepsilon_L\} = [B_L]\{U^e\}, \qquad (10.3)$$

where $[B_L]$ is the usual strain-displacement matrix pertaining to the linear part of the Green's strain.

The non-linear part of the strain in Eq. (10.2) can be expressed as

$$\{\varepsilon_{NL}\} = \frac{1}{2}[A]\{g\}, \tag{10.4}$$

where $[A]$ is a matrix of slopes calculated from Eq. (10.1) as

$$[A] = \begin{bmatrix} \frac{\partial u_1}{\partial x_1} & \frac{\partial u_2}{\partial x_1} & \frac{\partial u_3}{\partial x_1} & 0 & 0 & 0 & 0 & 0 & 0 \\ 0 & 0 & 0 & \frac{\partial u_1}{\partial x_2} & \frac{\partial u_2}{\partial x_2} & \frac{\partial u_3}{\partial x_2} & 0 & 0 & 0 \\ 0 & 0 & 0 & 0 & 0 & 0 & \frac{\partial u_1}{\partial x_3} & \frac{\partial u_2}{\partial x_3} & \frac{\partial u_3}{\partial x_3} \\ \frac{\partial u_1}{\partial x_2} & \frac{\partial u_2}{\partial x_2} & \frac{\partial u_3}{\partial x_2} & \frac{\partial u_1}{\partial x_1} & \frac{\partial u_2}{\partial x_1} & \frac{\partial u_3}{\partial x_1} & 0 & 0 & 0 \\ 0 & 0 & 0 & \frac{\partial u_1}{\partial x_3} & \frac{\partial u_2}{\partial x_3} & \frac{\partial u_3}{\partial x_3} & \frac{\partial u_1}{\partial x_2} & \frac{\partial u_2}{\partial x_2} & \frac{\partial u_3}{\partial x_2} \\ \frac{\partial u_1}{\partial x_3} & \frac{\partial u_2}{\partial x_3} & \frac{\partial u_3}{\partial x_3} & 0 & 0 & 0 & \frac{\partial u_1}{\partial x_1} & \frac{\partial u_2}{\partial x_1} & \frac{\partial u_3}{\partial x_1} \end{bmatrix} \tag{10.5}$$

and $\{g\}$ is defined as

$$\{g\} = \begin{bmatrix} \frac{\partial u_1}{\partial x_1} & \frac{\partial u_2}{\partial x_1} & \frac{\partial u_3}{\partial x_1} & \frac{\partial u_1}{\partial x_2} & \frac{\partial u_2}{\partial x_2} & \frac{\partial u_3}{\partial x_2} & \frac{\partial u_1}{\partial x_3} & \frac{\partial u_2}{\partial x_3} & \frac{\partial u_3}{\partial x_3} \end{bmatrix}^T. \tag{10.6}$$

$\{\varepsilon_{NL}\}$ within each element e can be related to the nodal displacements as

$$\{\varepsilon_{NL}\} = \frac{1}{2}[A][G]\{U^e\} = \frac{1}{2}[B_{NL}]\{U^e\}, \tag{10.7}$$

where $[G]$ is a matrix of shape function derivatives defined by the relation

$$\{g\} = [G]\{U^e\} \tag{10.8}$$

and $\frac{1}{2}[B_{NL}]$ is the strain-displacement matrix contributed from the quadratic part of the Green's strain.

The total strain can thus be expressed as

$$\{\varepsilon\} = \left[[B_L] + \frac{1}{2}[B_{NL}]\right]\{U^e\} = [\bar{B}]\{U^e\}. \tag{10.9}$$

Now, the variation of strain is given by

$$\{\delta\varepsilon\} = \{\delta\varepsilon_L\} + \{\delta\varepsilon_{NL}\}. \tag{10.10}$$

$\{\delta\varepsilon_{NL}\}$ is given as

$$\{\delta\varepsilon_{NL}\} = \frac{1}{2}[A]\{\delta g\} + \frac{1}{2}[\delta A]\{g\}.$$

It can easily be shown that $[A]\{\delta g\} = [\delta A]\{g\}$ and thus

$$\{\delta\varepsilon_{NL}\} = [A]\{\delta g\}. \tag{10.11}$$

Thus, we have

$$\{\delta\varepsilon\} = [B_L]\{\delta U^e\} + [A][G]\{\delta U^e\} = [[B_L] + [B_{NL}]]\{\delta U^e\} = [B]\{\delta U^e\}. \tag{10.12}$$

10.3.1. Static analysis

Using the displacement interpolation functions and substituting from Eq. (10.12), Eq. (2.21) can be written as

$$\begin{aligned}&\sum_e \{\delta U^e\}^T \left(\int_{V^e} [B]^T \{\sigma\}\, dV\right)\\ &-\sum_e \{\delta U^e\}^T \left(\int_{V^e} [N]^T \{b\}\, dV + \int_{S_F^e} [N^s]^T \{t\}\, dS\right) = 0,\end{aligned} \tag{10.13}$$

where

$$\{\sigma\} = [C]\{\varepsilon\} = [C][\bar{B}]\{U^e\}. \tag{10.14}$$

In Eq. (10.14), [*C*] is the material property matrix relating the strains to the stresses in global coordinate system.

Eq. (10.13) can be rewritten after dropping $\{\delta U\}$ from each term as follows:

$$\sum_e \int_{V^e} [B]^T \{\sigma\} dV - \{F\} = 0, \tag{10.15}$$

where $\{F\}$ is the applied load vector given by

$$\{F\} = \sum_e \int_{V^e} [N]^T \{b\} dV + \sum_e \int_{S_F^e} \left[N^s\right]^T \{t\} dS. \tag{10.16}$$

The equation of equilibrium given by Eq. (10.15) can be written in a non-linear sense as

$$\{\Psi\} = \sum_e \int_{V^e} [B]^T \{\sigma\} dV - \{F\}, \tag{10.17}$$

where $\{\Psi\}$ is residual or out-of-balance force vector at a particular solution iteration.

Putting Eq. (10.14) into Eq. (10.17), a equilibrium equation can be written as

$$\{\Psi\} = \left(\sum_e \int_{V^e} [B]^T [C] \left[\bar{B}\right] dV\right) \{U\} - \{F\}. \tag{10.18}$$

The solution of above non-linear Eq. (10.18) may be achieved iteratively using direct iteration method. However, Newton-Raphson method is generally adopted. In this method, linearization of Eq. (10.18) is achieved by Taylor expansion in the neighborhood of a solution point, say $\{U\}$, so that

$$\left\{\Psi_{U+\delta U}\right\} = \left\{\Psi_U\right\} + \left\{\delta\Psi_U\right\} \approx \{0\}. \tag{10.19}$$

It is, therefore, necessary in this method to find relation between $\{\delta\Psi\}$ and $\{\delta U\}$. Taking differentiation of residual force (Eq. (10.17)) with respect to $\{U\}$, we have

$$\begin{aligned}\{\delta\Psi\} &= \sum_e \int_{V^e} [B]^T \{\delta\sigma\} dV + \sum_e \int_{V^e} [\delta B]^T \{\sigma\} dV \\ &= \sum_e \int_{V^e} [B]^T \{\delta\sigma\} dV + \sum_e \int_{V^e} [\delta B_{NL}]^T \{\sigma\} dV,\end{aligned} \tag{10.20}$$

where

$$\{\delta\sigma\} = [C]\{\delta\varepsilon\} = [C][B]\{\delta U^e\}. \tag{10.21}$$

Using Eq. (10.7), $[\delta B_{NL}]^T$ in Eq. (10.20) can be written as

$$[\delta B_{NL}]^T = [G]^T [\delta A]^T. \tag{10.22}$$

It can easily be shown that

$$[\delta A]^T \{\sigma\} = [S]\{\delta g\}, \tag{10.23}$$

where $[S]$ is a matrix of stress array and is given for 3D analysis as

$$[S] = \begin{bmatrix} \sigma_1[I] & \tau_{12}[I] & \tau_{13}[I] \\ & \sigma_2[I] & \tau_{23}[I] \\ sym. & & \sigma_3[I] \end{bmatrix}, \tag{10.24}$$

in which $[I]$ is an identity matrix of size 3 × 3.

Using Eqs. (10.21) to (10.23), Eq. (10.20) can be written as

$$\{\delta\Psi\} = \sum_e \left(\left(\int_{V^e} [B]^T [C][B] dV \right) \{\delta U^e\} \right) + \sum_e \left(\left(\int_{V^e} [G]^T [S][G] dV \right) \{\delta U^e\} \right)$$

or

$$[K^T]\{\delta U\} = \{\delta\Psi\}, \tag{10.25}$$

where tangent stiffness matrix $[K^T]$ is evaluated as

$$[K^T] = \sum_e [K^{T(e)}] = \sum_e \left(\int_{V^e} [B]^T [C][B] dV + \int_{V^e} [G]^T [S][G] dV \right). \tag{10.26}$$

Using Eq. (10.25) into Eq. (10.19) results in the following incremental equilibrium equation:

$$[K^T]\{\delta U\} = -\{\Psi_U\}. \tag{10.27}$$

Solution procedure

The various steps involved in the non-linear (large deformation) analysis using Newton-Raphson incremental iterative scheme are described below:

(a) The linear (small deformation) solution is obtained as a first approximation, say $\{U_0\}$.

(b) The unbalanced force vector $\left\{\Psi_{U_0}\right\}$ is found using Eq. (10.17).

(c) The tangent stiffness matrix $[K^T]$ is evaluated.

(d) Correction is computed as $\{\delta U\} = -\left[K^T\right]^{-1}\left\{\Psi_{U_0}\right\}$ and $\{U_0\}$ is updated as $\{U_0\} = \{U_0\} + \{\delta U\}$.

The above processes (steps b to d) are repeated until the equilibrium is established. This can be achieved by incorporating suitable convergence criteria. In many example problems, the convergence criteria used are (i) maximum displacement variation less than 2 % and (ii) residual force less than 2 %.

10.3.2. Transient dynamic analysis

Large displacement formulation can easily be extended to transient dynamic analysis. Using the displacement interpolation functions, substituting from Eq. (10.12) and rearranging, Eq. (9.13) can be written as

$$\sum_e \{\delta U_e\}^T\left[\int_{V^e}[B]^T\{\sigma\}\,dV + \int_{V^e}[N]^T\rho\{\ddot{u}\}\,dV\right] - \sum_e \{\delta U_e\}^T\left[\int_{V^e}[N]^T\{b\}\,dV + \int_{S_F^e}\left[N^s\right]^T\{t\}\,dS\right] - \{\delta U\}^T\{F_c\} = 0. \tag{10.28}$$

Eq. (10.28) can be rewritten after dropping $\{\delta U\}$ from each term, *i.e.*

$$\sum_e \int_{V^e}[B]^T\{\sigma\}\,dV + \sum_e \int_{V^e}[N]^T\rho\{\ddot{u}\}\,dV - \{F\} = 0, \tag{10.29}$$

where $\{F\}$ is the applied load vector given by

$$\{F\} = \sum_e\left[\int_{V^e}[N]^T\{b\}\,dV + \int_{S_F^e}\left[N^s\right]^T\{t\}\,dS\right] + \{F_c\}. \tag{10.30}$$

The equation of dynamic equilibrium given by Eq. (10.29) can be written at time t_{n+1} in a non-linear sense as

$$\{\Psi_{n+1}\} = \sum_e \int_{V^e}[B]^T\{\sigma_{n+1}\}\,dV + \sum_e \int_{V^e}[N]^T\rho_{n+1}\{\ddot{u}_{n+1}\}\,dV - \{F_{n+1}\}, \tag{10.31}$$

where $\{\Psi_{n+1}\}$ is residual or out-of-balance force vector at a particular solution iteration of $(n+1)$th time-step.

Here Newmark time integration method is used to integrate equilibrium equation step-by-step with respect to time. The nodal acceleration vectors $\{\ddot{U}_{n+1}\}$ can be expressed in terms of nodal displacements $\{U_{n+1}\}$as given in Eq. (9.29).

Eq. (9.29) along with Eq. (9.11) can be put into Eq. (10.31) to form

$$\{\Psi_{n+1}\} = \sum_e \int_{V^e} [B]^T \{\sigma_{n+1}\} dV +$$

$$\frac{1}{\beta(\Delta t)^2}\left(\sum_e \int_{V^e} [N]^T \rho_{n+1} [N] dV\right)\{U_{n+1}\} - \{F_{n+1}^M\} - \{F_{n+1}\}, \quad (10.32)$$

in which $\{F_{n+1}^M\}$ is given as

$$\{F_{n+1}^M\} = [M]\left(\frac{1}{\beta(\Delta t)^2}\{U_n\} + \frac{1}{\beta(\Delta t)}\{\dot{U}_n\} + \frac{1-2\beta}{2\beta}\{\ddot{U}_n\}\right). \quad (10.33)$$

In Eq. (10.33), $[M] = \sum_e [M^e] = \sum_e \int_{V^e} [N]^T \rho_{n+1} [N] dV$ is the global mass matrix. $\{F_{n+1}^M\}$ can be referred to as the load vector due to inertia term at time t_{n+1}.

Putting Eq. (10.14) into Eq. (10.32), an effective static equilibrium equation can be written as

$$\{\Psi_{n+1}\} = \left(\sum_e \int_{V^e} [B]^T [C][\bar{B}] dV + \frac{1}{\beta(\Delta t)^2}\left(\sum_e \int_{V^e} [N]^T \rho_{n+1} [N] dV\right)\right)\{U_{n+1}\} -$$

$$\{F_{n+1}^M\} - \{F_{n+1}\}. \quad (10.34)$$

Eq. (10.34) can be written as

$$\{\Psi_{n+1}\} = [\hat{K}]\{U_{n+1}\} - \{\hat{F}_{n+1}\}, \quad (10.35)$$

where

$$\left[\hat{K}\right]=\left[K\right]+\frac{1}{\beta(\Delta t)^2}\left[M\right], \tag{10.36}$$

$$\left\{\hat{F}_{n+1}\right\}=\left\{F_{n+1}^M\right\}+\left\{F_{n+1}\right\} \tag{10.37}$$

and

$$\left[K\right]=\sum_e\left[K^e\right]=\sum_e\int_{V^e}\left[B\right]^T\left[C\right]\left[\bar{B}\right]dV. \tag{10.38}$$

The solution of above non-linear Eq. (10.35) is achieved iteratively. As discussed in the case of static analysis, if Newton-Raphson incremental iterative scheme is adopted here, it is necessary to find relation between $\{\delta\Psi_{n+1}\}$ and $\{\delta U_{n+1}\}$.

Taking differentiation of residual force (Eq. (10.32)) with respect to $\{U_{n+1}\}$, we have

$$\left\{\delta\Psi_{n+1}\right\}=\sum_e\int_{V^e}\left[B\right]^T\left\{\delta\sigma_{n+1}\right\}dV+\sum_e\int_V\left[\delta B_{NL}\right]^T\left\{\sigma_{n+1}\right\}dV+ \frac{1}{\beta(\Delta t)^2}\left(\sum_e\int_{V^e}\left[N\right]^T\rho_{n+1}\left[N\right]dV\right)\left\{\delta U_{n+1}\right\}. \tag{10.39}$$

Using Eqs. (10.21) to (10.23), Eq. (10.39) can be written as

$$\left[\hat{K}^T\right]\left\{\delta U_{n+1}\right\}=\left\{\delta\Psi_{n+1}\right\}, \tag{10.40}$$

where effective tangent stiffness matrix $\left[\hat{K}^T\right]$ is given as

$$\left[\hat{K}^T\right]=\left[K^T\right]+\frac{1}{\beta(\Delta t)^2}\left[M\right]$$

and

$$\left[K^T\right]=\sum_e\left[K^{T(e)}\right]=\sum_e\left(\int_{V^e}\left[B\right]^T\left[C\right]\left[B\right]dV+\int_{V^e}\left[G\right]^T\left[S_{n+1}\right]\left[G\right]dV\right). \tag{10.41}$$

10.4. EXAMPLE PROBLEMS

10.4.1. Example 1: Large deformation of hinged-hinged and pinned-pinned beams

As an example, a benchmark problem of a simply-supported isotropic beam involving geometric nonlinearity as solved by Reddy (2004) is considered. A uniform beam of length $L = 100$, cross-section dimensions of 1×1, made of a material with $E = 30 \times 10^6$ is simply supported at both ends and subjected to a uniformly distributed load of intensity q per unit length. The units are consistently chosen, so that the exact deflection at the middle of the beam in linear bending theory is 0.5208, when $q = 1$. In linear bending theory, when the beam is assumed to undergo pure bending (*i.e.* there is no axial deformation); it is immaterial to consider whether the beam is allowed free movement in the axial direction at the supported ends. However, in non-linear bending, this is a crucial distinction. Hinged-hinged (HH) condition is designated for the case where there is no axial restraint at both ends and pinned-pinned (PP) condition for the one where there is full restraint. In the former case, inextensional bending occurs which is largely of a linear nature, and in the latter case, bending with extension exists. The HH case is ideal to test the consistency aspect of the problem. As a non-linear element formulation is being used, a correct model should be able to recover the purely linear bending response under increasing load. This is possible only if the element can ensure that the inextensional axial condition (*i.e.* there is no axial restraint at both ends, no axial force should develop) is consistently recovered throughout. The PP case is ideal to examine the significance of the consistency aspect of the problem where the non-linear action becomes important.

Both the cases are considered here. A non-linear analysis is performed to calculate the deflection under the load as q increases from 1 to 10. The solution is obtained using both eight noded brick elements with incompatible modes and twenty noded brick elements. The present results are compared with the results from Reddy (2004) in which two versions of beam element has been used. The first version which will have locking uses 2 points integration for bending energy and extensional energy while the second version which will be lock free uses 2 points integration for bending energy and 1 point integration for the extensional energy.

Eight noded brick element

Using symmetry, half of the beam is modeled with 8 × 1 × 1 elements. Table 10.1 shows the SB-FEM results obtained for the central deflection of the beam along with the literature results (Reddy, 2004). It is clear from the table that the non-linear model using eight noded brick elements is able to capture the behavior correctly under both H-H and P-P conditions.

Table 10.1 Central deflection of a simply supported beam under a uniformly distributed load (UDL) (8 noded brick element).

UDL per unit length, q	*Present solution*	*Reddy (2004)*	
	2×2×2 integration rule	*2×2 selective (energy-based) integration rule*	*2×1 selective (energy-based) integration rule*
Hinged-hinged (HH) condition			
1	0.517697	0.5108	0.5208
2	1.03519	0.9739	1.0417
3	1.55206	1.3764	1.5625
4	2.06791	1.7265	2.0833
5	2.58231	2.0351	2.6042
6	3.09489	2.3116	3.1250
7	3.60521	2.5630	3.6458
8	4.11301	2.7930	4.1667
9	4.61784	3.0060	4.6875
10	5.11938	3.2051	5.2083
Pinned-pinned (PP) condition			
1	0.367949	0.3669	0.3687
2	0.545474	0.5424	0.5466
3	0.664471	0.6601	0.6663
4	0.756362	0.7510	0.7591
5	0.832352	0.8263	0.8361
6	0.897760	0.8912	0.9027
7	0.955557	0.9485	0.9617
8	1.00758	1.0002	1.0150
9	1.05506	1.0473	1.0638
10	1.09885	1.0908	1.1089

Twenty noded brick element

Using symmetry, half of the beam is modeled with 4 × 1 × 1 elements and three different integration schemes *i.e.* standard 3 × 3 × 3 (3 gauss points in all directions), 3 × 3 × 2 (2 gauss points in thickness direction) and standard 14 points integration were employed. Table 10.2 depicts the results obtained for the central deflection of the beam using different integration schemes along with the literature results. Again, it is clear that the non-linear formulation with twenty noded brick elements is able to capture both the behaviors correctly.

Table 10.2 Central deflection of a simply supported beam under uniformly distributed load (UDL) (20 noded brick element).

UDL per unit length, q	*Present solution*			*Reddy (2004)*	
	3×3×3 integration rule	*3×3×2 integration rule*	*14 point integration rule*	*2×2 selective (energy-based) integration rule*	*2×1 selective (energy-based) integration rule*
Hinged-hinged (HH) condition					
1	0.514751	0.514751	0.513443	0.5108	0.5208
2	1.02881	1.02881	1.02620	0.9739	1.0417
3	1.54128	1.54128	1.53738	1.3764	1.5625
4	2.05131	2.05131	2.04612	1.7265	2.0833
5	2.55805	2.55806	2.55161	2.0351	2.6042
6	3.06073	3.06073	3.05304	2.3116	3.1250
7	3.55859	3.55859	3.54970	2.5630	3.6458
8	4.05098	4.05098	4.04091	2.7930	4.1667
9	4.53727	4.53727	4.52606	3.0060	4.6875
10	5.01692	5.01692	5.00460	3.2051	5.2083
Pinned-pinned (PP) condition					
1	0.366698	0.366698	0.366272	0.3669	0.3687
2	0.544018	0.544018	0.543636	0.5424	0.5466
3	0.662900	0.662900	0.662555	0.6601	0.6663
4	0.754704	0.754704	0.754387	0.7510	0.7591
5	0.830619	0.830620	0.830323	0.8263	0.8361
6	0.895961	0.895961	0.895682	0.8912	0.9027
7	0.953697	0.953698	0.953432	0.9485	0.9617
8	1.00567	1.00567	1.00541	1.0002	1.0150
9	1.05309	1.05309	1.05285	1.0473	1.0638
10	1.09683	1.09683	1.09660	1.0908	1.1089

10.4.2. Example 2: Clamped 16-ply symmetric laminate subjected to uniform load

The non-linear formulation discussed above can easily be extended to the analysis of laminated composite structures by using layered brick elements (Kumar, 2010a and Kumar, 2010b). A 16-ply $[\pm 45/0_2/\mp 45/90_2]_S$ graphite-epoxy square laminate of size $b = 254$ mm and thickness $h = 2.114$ mm is considered as investigated by Noor (1985). The plate is clamped on all edges and subject to a uniform load q. The material properties of graphite-epoxy unidirectional ply are taken as:

$$E_x = 131 \text{ GPa}; \; E_y = E_z = 13.03 \text{ GPa};$$

$$G_{xy} = G_{yz} = G_{xz} = 6.41 \text{ GPa}; \; \nu_{xy} = \nu_{yz} = \nu_{xz} = 0.38.$$

The SB-FEM input file is listed in Appendix 8. Eight-noded layered brick elements are used to model the laminate. Figure 10.1 gives the comparison between present analysis and the results by Noor (1985) for laminate central deflection of the mid-ply (w_{max}). An excellent agreement can be observed.

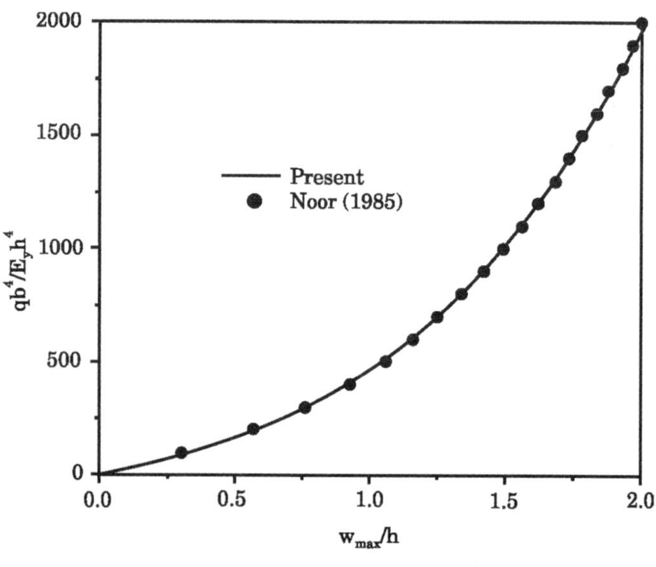

Fig. 10.1 Load-deflection curve for 16-ply $[\pm 45/0_2/\mp 45/90_2]_S$ square laminate (size $b = 254$ mm and thickness $h = 2.114$ mm) with clamped edges and subjected to a uniform load

10.4.3. Example 3: Clamped [0/90] asymmetric cross-ply cylindrical shell panel subjected to uniform load

As another example of laminated composite, the asymmetric cross-ply cylindrical shell panel investigated by Reddy and Chandrashekhara (1985) is considered. The radius and length of the cylindrical shell panel are 64.5 m and 12.9 m respectively. The angle subtended by the arc is 0.2 radian. The thickness of the shell is 64.5 mm. The material properties used are:

$E_x = 172$ GPa; $E_y = E_z = 7$ GPa;

$G_{xy} = G_{yz} = G_{xz} = 3.5$ GPa; $\nu_{xy} = \nu_{yz} = \nu_{xz} = 0.25.$

Eight-noded layered brick elements are used to model the laminated shell. The load-deflection curve of the non-linear solution is depicted in Figure 10.2 along with that reported by Reddy and Chandrashekhara (1985). Fairly good agreement can be seen between the two results.

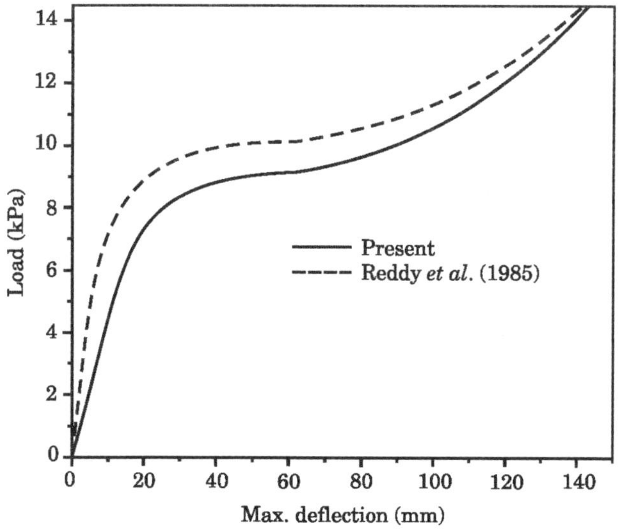

Fig. 10.2 Load-deflection curve for Clamped [0/90] asymmetric cross-ply cylindrical shell panel subjected to a uniform load

11

Non-linear Analysis of Shells

11.1. INTRODUCTION

Many areas of structural design require analysis of shells subjected to different types of loads. Analytical solutions to shell structures are limited in scope and in general are not applicable to arbitrary shapes, actual loading conditions, irregular stiffening and support conditions, cutouts, variation in mechanical properties and several other aspects of practical design. The finite element method has consequently been prominent in the analysis of such shells because of its capability in handling such complexities. Three distinct approaches in the finite element representation of shell structures have been employed (Hinton and Owen, 1984) based on (a) flat triangular or quadrilateral elements, (b) curved elements formulated on the basis of various shell theories and (c) elements derived from three-dimensional elements by the use of the degeneration methods.

The earliest numerical studies of shell problems involved faceting the shell by plane triangular plate elements, onto which a membrane stiffness was superimposed. Assembly of these elements gives a geometry which only approximates the actual shell surface. The results obtained were found to be satisfactory (Clough and Johnson, 1968; Backlund and Wennerstrom, 1974; Argyris *et al.*, 1980 and Bathe *et al.*, 1983), although several difficulties and shortcomings occurred from the application of flat elements to curved shells as pointed out by Gallagher (1976).

In the second approach a classical concept is employed whereby a shell theory is used as the starting point of the finite element formulation. Different shell theories have been developed from the three-dimensional field equations by incorporating various assumptions appropriate to the shell behavior. Based on the shell theories, a range of finite elements with different degrees of complexity have been formulated for deep and shallow shells (Gallagher, 1975 and Idelsohn, 1981).

The third approach employed in the formulation of shell elements avoids the complexities of fully general shell theories by discretizing directly the three-dimensional equations of continuum mechanics. Isoparametric shell elements were developed by degenerating the three-dimensional stress and strain conditions to shell behavior. By this approach the displacements and rotations of the shell midsurface are taken as degrees of freedom. This procedure was originally introduced by Ahmad *et al.* (1970) for the linear analysis of moderately thick shells, and later improved by Zienkiewicz *et al.* (1971) and Pawsey and Clough (1971).

In this chapter, finite element formulation for the analysis of shells based on Sanders shell theory (Sanders, 1959) is developed in which geometric nonlinearity is also incorporated. The formulation is also extended to layered anisotropic shells for the analysis of laminated composite structures (Kumar, 2008).

11.2. BASIC EQUATIONS

Finite element analysis of shells with general geometry takes a variety of formulations, depending upon the shell theory used. According to a first order shear deformation theory (FSDT), cross-sections normal to the reference plane (mid-surface) before deformation are assumed to remain planar but not necessarily normal to mid-surface after deformation. The displacement field $\{u\} = [u,v,w]^T$ for a doubly-curved shell (Figure 11.1), according to the FSDT is given by

$$u(x,y,z) = u_0(x,y) + z\theta_x(x,y)$$

$$v(x,y,z) = v_0(x,y) + z\theta_y(x,y)$$

$$w(x,y,z) = w_0(x,y), \tag{11.1}$$

where u is the membrane displacement in the x-direction, v is the membrane displacement in the y-direction and w is the out-of-plane transverse displacement in the z-direction. u_0, v_0 and w_0 are the midplane displacements. Positive θ_x and θ_y are the cross-sectional rotations around the y- and x-axes, respectively.

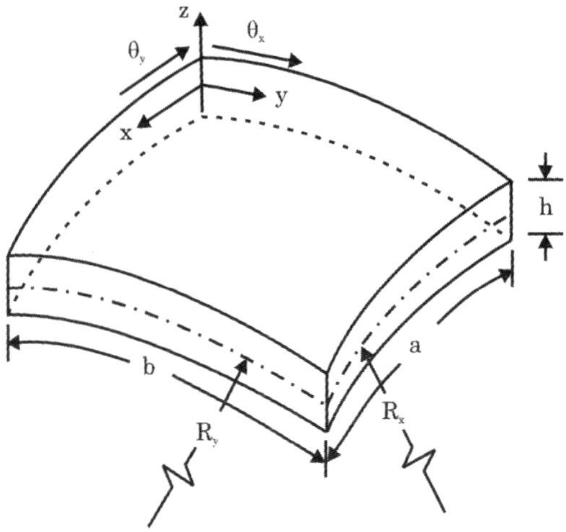

Fig. 11.1 A general doubly-curved shell

11.2.1 Strain-displacement relations

According to Sanders shell theory, the linear strain-displacement relations can be written as

$$\begin{aligned}
\varepsilon_{xL} &= \varepsilon_{xL}^0 + z\kappa_x \\
\varepsilon_{yL} &= \varepsilon_{yL}^0 + z\kappa_y \\
\gamma_{xyL} &= \gamma_{xyL}^0 + z\kappa_{xy} \\
\gamma_{yzL} &= \gamma_{yzL}^0 \\
\gamma_{xzL} &= \gamma_{xzL}^0 ,
\end{aligned} \tag{11.2}$$

where the linear mid-plane strains ε_{xL}^0, ε_{yL}^0, γ_{xyL}^0; the changes in curvature κ_x, κ_y, κ_{xy}; and the linear transverse shear strains γ_{yzL}^0, γ_{xzL}^0 are given by

$$\varepsilon_{xL}^0 = \frac{\partial u_0}{\partial x} + \frac{w_0}{R_x}$$

$$\varepsilon_{yL}^{0} = \frac{\partial v_0}{\partial y} + \frac{w_0}{R_y}$$

$$\gamma_{xyL}^{0} = \frac{\partial u_0}{\partial y} + \frac{\partial v_0}{\partial x}$$

$$\kappa_x = \frac{\partial \theta_x}{\partial x}$$

$$\kappa_y = \frac{\partial \theta_y}{\partial y}$$

$$\kappa_{xy} = \frac{\partial \theta_x}{\partial y} + \frac{\partial \theta_y}{\partial x} - c_R \left(\frac{\partial v_0}{\partial x} - \frac{\partial u_0}{\partial y} \right)$$

$$\gamma_{yzL}^{0} = \theta_y + \frac{\partial w_0}{\partial y} - \frac{v_0}{R_y}$$

$$\gamma_{xzL}^{0} = \theta_x + \frac{\partial w_0}{\partial x} - \frac{u_0}{R_x}. \tag{11.3}$$

In Eq. (11.3), R_x and R_y are the radii of curvature in the x- and y-directions respectively and

$$c_R = \frac{1}{2}\left(\frac{1}{R_x} - \frac{1}{R_y} \right).$$

Non-linear equations

The strain-displacement relations based on Sanders theory and incorporating geometrical nonlinearity (Stein, 1986) can be written as

$$\varepsilon_{xNL} = \varepsilon_{xNL}^{0} + z\kappa_x$$

$$\varepsilon_{yNL} = \varepsilon_{yNL}^{0} + z\kappa_y$$

$$\gamma_{xyNL} = \gamma_{xyNL}^{0} + z\kappa_{xy} \tag{11.4}$$

$$\gamma_{yzNL} = \gamma_{yzNL}^{0}$$

$$\gamma_{xzNL} = \gamma^0_{xzNL},$$

where the non-linear mid-plane strains $\varepsilon^0_{xNL}, \varepsilon^0_{yNL}, \gamma^0_{xyNL}$ and the transverse shear strains $\gamma^0_{yzNL}, \gamma^0_{xzNL}$ are given by

$$\varepsilon^0_{xNL} = \varepsilon^0_{xL} + \frac{1}{2}\left[\left(\frac{\partial u_0}{\partial x}\right)^2 + \left(\frac{\partial v_0}{\partial x}\right)^2 + \left(\frac{\partial w_0}{\partial x} - \frac{u_0}{R_x}\right)^2\right]$$

$$\varepsilon^0_{yNL} = \varepsilon^0_{yL} + \frac{1}{2}\left[\left(\frac{\partial u_0}{\partial y}\right)^2 + \left(\frac{\partial v_0}{\partial y}\right)^2 + \left(\frac{\partial w_0}{\partial y} - \frac{v_0}{R_y}\right)^2\right]$$

$$\gamma^0_{xyNL} = \gamma^0_{xyL} + \frac{\partial u_0}{\partial x}\frac{\partial u_0}{\partial y} + \frac{\partial v_0}{\partial x}\frac{\partial v_0}{\partial y} + \left(\frac{\partial w_0}{\partial x} - \frac{u_0}{R_x}\right)\left(\frac{\partial w_0}{\partial y} - \frac{v_0}{R_y}\right)$$

$$\gamma^0_{yzNL} = \gamma^0_{yzL}$$

$$\gamma^0_{xzNL} = \gamma^0_{xzL}. \tag{11.5}$$

In this equation, higher order terms for the derivatives of cross-sectional rotations and the non-linear terms in transverse shear strains have been neglected.

11.2.2 Constitutive equations

As in the case of beams and plates, it is convenient here to express the consitutive relation in terms of stress-resultants.

If we set degree of freedom vector as

$$\{\bar{u}\} = \begin{bmatrix} u_0 & v_0 & w_0 & \theta_x & \theta_y \end{bmatrix}^T \tag{11.6}$$

and strain components as

$$\{\bar{\varepsilon}\} = \begin{bmatrix} \varepsilon^0_x & \varepsilon^0_y & \gamma^0_{xy} & \kappa_x & \kappa_y & \kappa_{xy} & \gamma^0_{yz} & \gamma^0_{xz} \end{bmatrix}^T, \tag{11.7}$$

the strain-displacement relations (Eq. 11.5) can be written in matrix form as

$$\{\bar{\varepsilon}\} = [\Delta]\{\bar{u}\}. \tag{11.8}$$

Here [Δ] is the appropriate derivative operator matrix which also includes non-linear terms.

The constitutive equation can be written as

$$\{\bar{\sigma}\} = \left[\bar{D}\right]\{\bar{\varepsilon}\}, \tag{11.9}$$

where $\{\bar{\sigma}\}$ is the generalized stress resultant vector and $\left[\bar{D}\right]$ is the stiffness matrix. $\{\bar{\sigma}\}$ consists of membrane, bending and shear effects and is given as

$$\{\bar{\sigma}\} = \begin{bmatrix} N_x & N_y & N_{xy} & M_x & M_y & M_{xy} & Q_{yz} & Q_{xz} \end{bmatrix}^T$$

or

$$\{\bar{\sigma}\} = \left[\int_{-h/2}^{h/2} \begin{pmatrix} \sigma_x & \sigma_y & \tau_{xy} & \sigma_x z & \sigma_y z & \tau_{xy} z & \tau_{yz} & \tau_{xz} \end{pmatrix} dz\right]^T. \tag{11.10}$$

In the above equation, N_x, N_y, N_{xy} are internal in-plane (membrane) stress resultants (membrane forces per unit length); M_x, M_y, M_{xy} are internal moment resultants (moments per unit length) and Q_{yz}, Q_{xz} are transverse shear stress resultants (shear forces per unit length).

Matrix $\left[\bar{D}\right]$ is an 8 × 8 matrix and can be written as

$$\left[\bar{D}\right] = \begin{bmatrix} A_{11} & A_{12} & A_{14} & B_{11} & B_{12} & B_{14} & 0 & 0 \\ A_{12} & A_{22} & A_{24} & B_{12} & B_{22} & B_{24} & 0 & 0 \\ A_{14} & A_{24} & A_{44} & B_{14} & B_{24} & B_{44} & 0 & 0 \\ B_{11} & B_{12} & B_{14} & D_{11} & D_{12} & D_{14} & 0 & 0 \\ B_{12} & B_{22} & B_{24} & D_{12} & D_{22} & D_{24} & 0 & 0 \\ B_{14} & B_{24} & B_{44} & D_{14} & D_{24} & D_{44} & 0 & 0 \\ 0 & 0 & 0 & 0 & 0 & 0 & S_{55} & S_{56} \\ 0 & 0 & 0 & 0 & 0 & 0 & S_{56} & S_{66} \end{bmatrix}, \tag{11.11}$$

in which the extensional, bending-stretching coupling and bending stiffness coefficients are defined by

$$\left(A_{ij}, B_{ij}, D_{ij}\right) = C_{ij} \int_{-h/2}^{h/2} \left(1, z, z^2\right) dz = C_{ij}\left(h, 0, \frac{h^3}{12}\right) \quad (i, j = 1,2,4) \tag{11.12}$$

and shear stiffness coefficients are given as

$$S_{pq} = \alpha C_{pq} \int_{-h/2}^{h/2} dz = \alpha h C_{pq} \quad (p,q = 5,6). \tag{11.13}$$

Here C_{ij} $(i,j = 1,2,4,5,6)$ are the elastic constants relating the strains to the stresses and α is the shear correction factor.

In case of laminated shell consisting of N number of plies (layers), we have the following relations:

$$\left(A_{ij}, B_{ij}, D_{ij}\right) = \sum_{k=1}^{N} \int_{z_{k-1}}^{z_k} C_{ij}^k \left(1, z, z^2\right) dz \quad (i,j = 1,2,4)$$

and

$$S_{pq} = \sum_{k=1}^{N} \int_{z_{k-1}}^{z_k} \alpha C_{pq}^k dz \quad (p,q = 5,6). \tag{11.14}$$

In Eq. (11.14), C_{ij}^k $(i,j = 1,2,4,5,6)$ are the transformed elastic constants for the kth layer and given by the relation (see Appendix 3)

$$[C] = [T]^T [\bar{C}][T]. \tag{11.15}$$

In the above equation, $[T]$ is the transformation matrix relating the strains in the ply principal directions to those in the shell reference axis and $[\bar{C}]$ is the elasticity matrix relating the strains within the ply to the stresses in the ply coordinate system.

11.3 FINITE ELEMENT FORMULATION

Using the above strain-displacement relations and constitutive equations, the large deflection finite element formulation for a shell element can be developed for both static and dynamic analyses. Although the formulation is general and valid for any shell element, the most commonly employed elements are four-noded, eight-noded and nine-noded isoparametric quadrilateral elements. The procedures used in Chapter 10 can be followed as far as establishing non-linear finite element equilibrium equations is concerned. For the sake of brevity, we only consider formulation for a transient dynamic analysis, from which the formulation for static analysis is apparent.

In transient dynamic case, the principle of virtual work after including the inertia forces can be written as

$$\begin{aligned}&\int_A \{\delta\bar{\varepsilon}\}^T \{\bar{\sigma}\}\, dA + \int_V \rho\{\delta u\}^T \{\ddot{u}\}\, dV \\ &= \int_A \{\delta\bar{u}\}^T \{b\}\, dA + \int_A \{\delta\bar{u}\}^T \{f\}\, dA + \int_l \{\delta\bar{u}^s\}^T \{\bar{t}\}\, dl,\end{aligned} \tag{11.16}$$

in which $\{\ddot{u}\} = [\ddot{u}, \ddot{v}, \ddot{w}]^T$ is the acceleration vector, $\{b\}$ is a vector of body forces per unit area acting at the shell mid-surface, $\{f\}$ contains distributed loading components per unit area acting at the shell mid-surface, $\{\bar{t}\}$ is a vector of prescribed stress resultants acting along the edge(s) of the shell, A is the area of the shell mid-surface and l denotes the length of the edges on which $\{\bar{t}\}$ is prescribed. The load terms are given as

$$\{b\} = \begin{bmatrix} b_x & b_y & b_z & 0 & 0 \end{bmatrix}^T,$$

$$\{f\} = \begin{bmatrix} f_x & f_y & f_z & 0 & 0 \end{bmatrix}^T$$

and
$$\{\bar{t}\} = \begin{bmatrix} \bar{N}_x & \bar{N}_y & \bar{N}_z & \bar{M}_x & \bar{M}_y \end{bmatrix}^T. \tag{11.17}$$

In Eq. (11.17) the five components of the load vectors correspond to the loads associated with the three deflections and the two rotations respectively. $\bar{N}_x, \bar{N}_y, \bar{N}_z$ are prescribed stress resultants (forces per unit length) in x, y and z directions respectively and $\bar{M}_x, \bar{M}_y$ are prescribed moment resultants (moments per unit length) about the y- and x-axes respectively.

Using the relation between $\{u\}$ and $\{\bar{u}\}$, Eq. (11.16) can be rewritten as

$$\begin{aligned}&\int_A \{\delta\bar{\varepsilon}\}^T \{\bar{\sigma}\}\, dA + \int_A \{\delta\bar{u}\}^T [P]\{\ddot{\bar{u}}\}\, dA \\ &= \int_A \{\delta\bar{u}\}^T \{b\}\, dA + \int_A \{\delta\bar{u}\}^T \{f\}\, dA + \int_l \{\delta\bar{u}^s\}^T \{\bar{t}\}\, dl,\end{aligned} \tag{11.18}$$

where
$$\{\ddot{\bar{u}}\} = \begin{bmatrix} \ddot{u}_0 & \ddot{v}_0 & \ddot{w}_0 & \ddot{\theta}_x & \ddot{\theta}_y \end{bmatrix}^T$$

and
$$[P] = \begin{bmatrix} p_1 & 0 & 0 & p_2 & 0 \\ 0 & p_1 & 0 & 0 & p_2 \\ 0 & 0 & p_1 & 0 & 0 \\ p_2 & 0 & 0 & p_3 & 0 \\ 0 & p_2 & 0 & 0 & p_3 \end{bmatrix}, \tag{11.19}$$

in which the different terms are as follows:

$$\left(p_1, p_2, p_3\right) = \int_{-h/2}^{h/2} \left(1, z, z^2\right) \rho dz. \tag{11.20}$$

The variations of the displacements and accelerations within each element e are related to its nodal displacements $\{U^e\}$ and nodal accelerations $\left\{\ddot{U}^e\right\}$ by the following equations:

$$\{\bar{u}\} = [N]\{U^e\}$$

and

$$\{\ddot{\bar{u}}\} = [N]\{\ddot{U}^e\}, \tag{11.21}$$

where

$$[N] = [[N_1] \ [N_2] \ \cdots \ [N_n]] \tag{11.22}$$

and

$$\{U^e\} = \begin{bmatrix} u_1 & v_1 & w_1 & \theta_{x1} & \theta_{y1} & \cdots & u_n & v_n & w_n & \theta_{xn} & \theta_{yn} \end{bmatrix}^T. \tag{11.23}$$

The subscript n in Eqs. (11.22) and (11.23) refers to the number of nodes in the shell element. A typical sub-matrix $[N_r]$ in Eq. (11.22) is given as $[N_r] = N_r\,[I]$, where N_r is the rth shape function of the element and $[I]$ is an identity matrix of size 5×5 corresponding to the number of degrees of freedom.

In large displacement problem, strain at any point within an element consists of both linear and non-linear parts and can be written as

$$\{\bar{\varepsilon}\} = \{\bar{\varepsilon}_L\} + \{\bar{\varepsilon}_{NL}\}. \tag{11.24}$$

The linear part of the strain within each element e are related to the nodal displacements as

$$\{\bar{\varepsilon}_L\} = [B_L]\{U^e\}, \tag{11.25}$$

where $[B_L]$ is the strain-displacement matrix pertaining to the linear part of the strain and can be evaluated to be

$$[B_L] = [[B_{L1}] \ [B_{L2}] \ \cdots \ [B_{Ln}]]^T, \tag{11.26}$$

where$[B_{Lr}]$, $r = 1$ to n are strain-displacement submatrices of size 8×5 and are defined as

$$[B_{Lr}] = \begin{bmatrix} \frac{\partial N_r}{\partial x} & 0 & \frac{\partial N_r}{\partial y} & 0 & 0 & c_R \frac{\partial N_r}{\partial y} & 0 & \frac{-N_r}{R_x} \\ 0 & \frac{\partial N_r}{\partial y} & \frac{\partial N_r}{\partial x} & 0 & 0 & -c_R \frac{\partial N_r}{\partial x} & \frac{-N_r}{R_y} & 0 \\ \frac{N_r}{R_x} & \frac{N_r}{R_y} & 0 & 0 & 0 & 0 & \frac{\partial N_r}{\partial y} & \frac{\partial N_r}{\partial x} \\ 0 & 0 & 0 & \frac{\partial N_r}{\partial x} & 0 & \frac{\partial N_r}{\partial y} & 0 & N_r \\ 0 & 0 & 0 & 0 & \frac{\partial N_r}{\partial y} & \frac{\partial N_r}{\partial x} & N_r & 0 \end{bmatrix}^T .$$

The non-linear part of the strain in Eq. (11.24) can be expressed as

$$\{\bar{\varepsilon}_{NL}\} = \frac{1}{2}[A]\{g\}, \tag{11.27}$$

where $[A]$ is a matrix of slopes calculated as

$$[A] = \begin{bmatrix} \frac{\partial u_0}{\partial x} & \frac{\partial v_0}{\partial x} & \frac{\partial w_0}{\partial x} - \frac{u_0}{R_x} & 0 & 0 & 0 \\ 0 & 0 & 0 & \frac{\partial u_0}{\partial y} & \frac{\partial v_0}{\partial y} & \frac{\partial w_0}{\partial y} - \frac{v_0}{R_y} \\ \frac{\partial u_0}{\partial y} & \frac{\partial v_0}{\partial y} & \frac{\partial w_0}{\partial y} - \frac{v_0}{R_y} & \frac{\partial u_0}{\partial x} & \frac{\partial v_0}{\partial x} & \frac{\partial w_0}{\partial x} - \frac{u_0}{R_x} \\ 0 & 0 & 0 & 0 & 0 & 0 \\ 0 & 0 & 0 & 0 & 0 & 0 \\ 0 & 0 & 0 & 0 & 0 & 0 \\ 0 & 0 & 0 & 0 & 0 & 0 \\ 0 & 0 & 0 & 0 & 0 & 0 \end{bmatrix} \tag{11.28}$$

and $\{g\}$ is defined as

$$\{g\} = \left[\frac{\partial u_0}{\partial x} \quad \frac{\partial v_0}{\partial x} \quad \left(\frac{\partial w_0}{\partial x} - \frac{u_0}{R_x} \right) \quad \frac{\partial u_0}{\partial y} \quad \frac{\partial v_0}{\partial y} \quad \left(\frac{\partial w_0}{\partial y} - \frac{v_0}{R_y} \right) \right]^T . \tag{11.29}$$

$\{\bar{\varepsilon}_{NL}\}$ within each element e can be related to the nodal displacements as

$$\{\bar{\varepsilon}_{NL}\} = \frac{1}{2}[A][G]\{U^e\} = \frac{1}{2}[B_{NL}]\{U^e\}, \tag{11.30}$$

where $[G]$ is a matrix of shape function derivatives defined by the relation

$$\{g\} = [G]\{U^e\} \tag{11.31}$$

and $\frac{1}{2}[B_{NL}]$ is the strain-displacement matrix contributed from the non-linear part of the strain displacement relations.

The total strain can thus be expressed as

$$\{\bar{\varepsilon}\} = \left[[B_L] + \frac{1}{2}[B_{NL}]\right]\{U^e\} = [\bar{B}]\{U^e\}. \tag{11.32}$$

Now, the variation of strain is given by

$$\{\delta\bar{\varepsilon}\} = \{\delta\bar{\varepsilon}_L\} + \{\delta\bar{\varepsilon}_{NL}\}. \tag{11.33}$$

$\{\delta\bar{\varepsilon}_{NL}\}$ is given as

$$\{\delta\bar{\varepsilon}_{NL}\} = \frac{1}{2}[A]\{\delta g\} + \frac{1}{2}[\delta A]\{g\}.$$

It can easily be shown that $[A]\{\delta g\} = [\delta A]\{g\}$ and thus

$$\{\delta\bar{\varepsilon}_{NL}\} = [A]\{\delta g\}. \tag{11.34}$$

Thus, we have

$$\{\delta\bar{\varepsilon}\} = [B_L]\{\delta U^e\} + [A][G]\{\delta U^e\} = [[B_L] + [B_{NL}]]\{\delta U^e\} = [B]\{\delta U^e\}. \tag{11.35}$$

Rewriting Eq. (11.18) as a sum of integration over the areas and edges of all finite elements and substituting for variations of the displacements and strains within each element, we obtain

$$\sum_e \{\delta U^e\}^T \left[\int_{A^e} [B]^T \{\bar{\sigma}\}\, dA + \int_{A^e} [N]^T [P]\{\ddot{u}\}\, dA\right] =$$
$$\sum_e \{\delta U^e\}^T \left[\left(\int_{A^e} [N]^T \{b\}\, dA\right) + \left(\int_{A^e} [N]^T \{f\}\, dA\right) + \left(\int_{l^e} [N^s]^T \{\bar{t}\}\, dl\right)\right]. \tag{11.36}$$

In this equation, the displacement interpolation matrices $[N^s]$ corresponding to the element edges are obtained from the element displacement interpolation matrices $[N]$ by substituting element edge coordinates.

Since $\{\delta U\}$ is arbitrary, it can be removed from both the sides of Eq. (11.36). The resulting equation of equilibrium can be written as

$$\sum_e \left(\int_{A^e} [B]^T \{\bar{\sigma}\} dA + \int_{A^e} [N]^T [P] \{\ddot{u}\} dA \right) =$$
$$\sum_e \left(\int_{A^e} [N]^T \{b\} dA + \int_{A^e} [N]^T \{f\} dA + \int_{l^e} \left[N^s\right]^T \{\bar{t}\} dl \right). \quad (11.37)$$

The foregoing equation can be written at time t_{n+1} in a non-linear sense as

$$\{\Psi_{n+1}\} = \sum_e \left(\int_{A^e} [B]^T \{\bar{\sigma}_{n+1}\} dA + \int_{A^e} [N]^T [P] \{\ddot{u}_{n+1}\} dA \right) - \{F_{n+1}\}, \quad (11.38)$$

where $\{\Psi_{n+1}\}$ is residual or out-of-balance force vector at a particular solution iteration of $(n+1)$th time-step. $\{F_{n+1}\}$ is the applied load vector consisting of terms in the right hand side of Eq. (11.37).

Using Newmark time integration method, the nodal acceleration terms $\{\ddot{U}_{n+1}\}$ can be expressed in terms of nodal displacements $\{U_{n+1}\}$ as given in Eq. (9.29).

Eq. (9.29) along with Eq. (11.21) can be put into Eq. (11.38) to form

$$\{\Psi_{n+1}\} = \sum_e \int_{A^e} [B]^T \{\bar{\sigma}_{n+1}\} dA +$$
$$\frac{1}{\beta(\Delta t)^2} \left(\sum_e \int_{A^e} [N]^T [P][N] dA \right) \{U_{n+1}\} - \{F^M_{n+1}\} - \{F_{n+1}\}. \quad (11.39)$$

In Eq. (11.39), $\{F^M_{n+1}\}$ can be referred to as the load vector due to inertia term at time t_{n+1} and is given as

$$\{F^M_{n+1}\} = [M] \left(\frac{1}{\beta(\Delta t)^2} \{U_n\} + \frac{1}{\beta(\Delta t)} \{\dot{U}_n\} + \frac{1-2\beta}{2\beta} \{\ddot{U}_n\} \right). \quad (11.40)$$

In Eq. (11.40), $[M] = \sum_e \left[M^e\right] = \sum_e \int_{A^e} [N]^T [P][N] dA$ is the global mass matrix.

Putting Eq. (11.9) into Eq. (11.39), an effective static equilibrium equation can be obtained as given below

$$\{\Psi_{n+1}\} = \left[\sum_e \int_{A^e} [B]^T \left[\bar{D}\right]\left[\bar{B}\right] dA + \frac{1}{\beta(\Delta t)^2}\left(\sum_e \int_{A^e} [N]^T [P][N] dA\right)\right]$$
$$\{U_{n+1}\} - \{F^M_{n+1}\} - \{F_{n+1}\}. \tag{11.41}$$

Eq. (11.41) can be written as

$$\{\Psi_{n+1}\} = \left[\hat{K}\right]\{U_{n+1}\} - \left\{\hat{F}_{n+1}\right\}, \tag{11.42}$$

where

$$\left[\hat{K}\right] = [K] + \frac{1}{\beta(\Delta t)^2}[M], \tag{11.43}$$

$$\left\{\hat{F}_{n+1}\right\} = \{F^M_{n+1}\} + \{F_{n+1}\} \tag{11.44}$$

and

$$[K] = \sum_e \left[K^e\right] = \sum_e \int_{A^e} [B]^T \left[\bar{D}\right]\left[\bar{B}\right] dA. \tag{11.45}$$

The solution of above non-linear Eq. (11.42) is achieved iteratively. If Newton-Raphson incremental iterative scheme is adopted, it is necessary to find relation between $\{\delta\Psi_{n+1}\}$ and $\{\delta U_{n+1}\}$ as discussed in Chapter 10.

Taking differentiation of residual force (Eq. (11.39)) with respect to $\{U_{n+1}\}$, we get

$$\{\delta\Psi_{n+1}\} = \sum_e \int_{A^e} [B]^T \{\delta\bar{\sigma}_{n+1}\} dA + \sum_e \int_{A^e} [\delta B_{NL}]^T \{\bar{\sigma}_{n+1}\} dA +$$
$$\frac{1}{\beta(\Delta t)^2}\left(\sum_e \int_{A^e} [N]^T [P][N] dA\right)\{\delta U_{n+1}\}, \tag{11.46}$$

where

$$\{\delta\bar{\sigma}_{n+1}\} = \left[\bar{D}\right]\{\delta\bar{\varepsilon}_{n+1}\} = \left[\bar{D}\right][B]\{\delta U^e_{n+1}\}. \tag{11.47}$$

$\left[\delta B_{NL}\right]^T$ in Eq. (11.46) can be written as

$$\left[\delta B_{NL}\right]^T = \left[G\right]^T \left[\delta A\right]^T. \tag{11.48}$$

It can easily be shown that

$$\left[\delta A\right]^T \left\{\bar{\sigma}_{n+1}\right\} = \left[S_{n+1}\right]\left\{\delta g\right\} = \left[S_{n+1}\right]\left[G\right]\left\{\delta U^e_{n+1}\right\}, \tag{11.49}$$

where [*S*] is a matrix of the stress resultant array and is evaluated as

$$\left[S\right] = \begin{bmatrix} N_x\left[I\right] & N_{xy}\left[I\right] \\ sym & N_y\left[I\right] \end{bmatrix}, \tag{11.50}$$

in which [*I*] is an identity matrix of size 3 × 3.

Using Eqs. (11.47) to (11.49), Eq. (11.46) can be written as

$$\begin{aligned}\left\{\delta\Psi_{n+1}\right\} = {} & \sum_e\left(\left(\int_{A^e}\left[B\right]^T\left[\bar{D}\right]\left[B\right]dA\right)\left\{\delta U^e_{n+1}\right\}\right) + \\ & \sum_e\left(\left(\int_{A^e}\left[G\right]^T\left[S_{n+1}\right]\left[G\right]dA\right)\left\{\delta U^e_{n+1}\right\}\right) + \\ & \frac{1}{\beta\left(\Delta t\right)^2}\left(\sum_e\int_{A^e}\left[N\right]^T\left[P\right]\left[N\right]dA\right)\left\{\delta U_{n+1}\right\}\end{aligned}$$

or

$$\left[\hat{K}^T\right]\left\{\delta U_{n+1}\right\} = \left\{\delta\Psi_{n+1}\right\}, \tag{11.51}$$

where the effective tangent stiffness matrix $\left[\hat{K}^T\right]$ is evaluated as

$$\left[\hat{K}^T\right] = \left[K^T\right] + \frac{1}{\beta\left(\Delta t\right)^2}\left[M\right], \tag{11.52}$$

in which

$$\left[K^T\right] = \sum_e\left[K^{T(e)}\right] = \sum_e\left(\int_{A^e}\left[B\right]^T\left[\bar{D}\right]\left[B\right]dA + \int_{A^e}\left[G\right]^T\left[S_{n+1}\right]\left[G\right]dA\right). \tag{11.53}$$

11.4 EXAMPLE PROBLEMS

The above finite element shell formulation is used to solve some example problems of large deformation of beams, plates and shells. Both non-linear static and non-linear transient dynamic analyses are performed. Example problems of laminated composite plates and cylindrical shells are also investigated. Eight noded isoparametric quadrilateral elements are used to discretize the domain.

11.4.1. Example 1: Large deformation of hinged-hinged and pinned-pinned beams

As an example, a benchmark problem of large deformation of a simply-supported isotropic beam subjected to a uniformly distributed load is considered as solved earlier in Chapter 10. Both hinged-hinged (HH) and pinned-pinned (PP) cases are considered here. SB-FEM input file for this example problem is listed in Appendix 8. Using symmetry, half of the beam is discretized into a uniform mesh of 4 × 1 elements. Two different integration schemes i.e. full 3 × 3 scheme and selective (energy-based) 3 × 2 scheme are employed for this problem.

Table 11.1 Central deflection of a simply supported beam under uniformly distributed load (UDL).

UDL per unit length, q	*Present solution*		*Reddy (2004)*	
	3×3 integration rule	*3×2 selective (energy-based) integration rule*	*2×2 integration rule*	*2×1 selective (energy-based) integration rule*
Hinged-hinged (HH) condition				
1	0.514716	0.520834	0.5108	0.5208
2	1.02894	1.04118	0.9739	1.0417
3	1.54218	1.56053	1.3764	1.5625
4	2.05395	2.07843	1.7265	2.0833
5	2.56379	2.59439	2.0351	2.6042
6	3.07124	3.10795	2.3116	3.1250
7	3.57586	3.61869	2.5630	3.6458
8	4.07723	4.12618	2.7930	4.1667
9	4.57496	4.63001	3.0060	4.6875
10	5.06867	5.12983	3.2051	5.2083

Table 11.1 *(Contd.)*

Table 11.1 *(Contd.)*

		Pinned-pinned (PP) condition		
1	0.366608	0.368480	0.3669	0.3687
2	0.543847	0.545384	0.5424	0.5466
3	0.662668	0.663919	0.6601	0.6663
4	0.754424	0.755450	0.7510	0.7591
5	0.830297	0.831140	0.8263	0.8361
6	0.895601	0.896290	0.8912	0.9027
7	0.953303	0.953860	0.9485	0.9617
8	1.00524	1.00568	1.0002	1.0150
9	1.05264	1.05297	1.0473	1.0638
10	1.09635	1.09659	1.0908	1.1089

It is clear from Table 11.1 that the present element model is able to capture both the behaviors correctly. It can also be observed that the selective integration scheme which is free of shear locking provides a better result. All the subsequent example problems are solved using a selective integration scheme.

11.4.2 Example 2: Clamped 16-ply symmetric laminate subjected to a uniform load

Next a 16-ply $[\pm 45/0_2/\mp 45/90_2]_S$ graphite-epoxy square laminate of size b = 254 mm and thickness h = 2.114 mm is considered. The material properties and loading conditions are same as taken in Chapter 10. Figure 11.2 gives the comparison between present analysis and the result by Noor (1985) for laminate central deflection of the mid-ply (w_{max}). A good agreement can be observed.

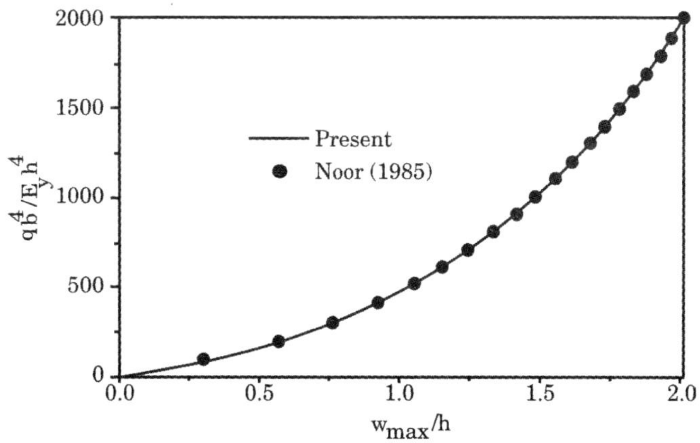

Fig. 11.2 Load-deflection curve for 16-ply $[\pm 45/0_2/\mp 45/90_2]_S$ square laminate (size b = 254 mm and thickness h = 2.114 mm) with clamped edges and subjected to a uniform load

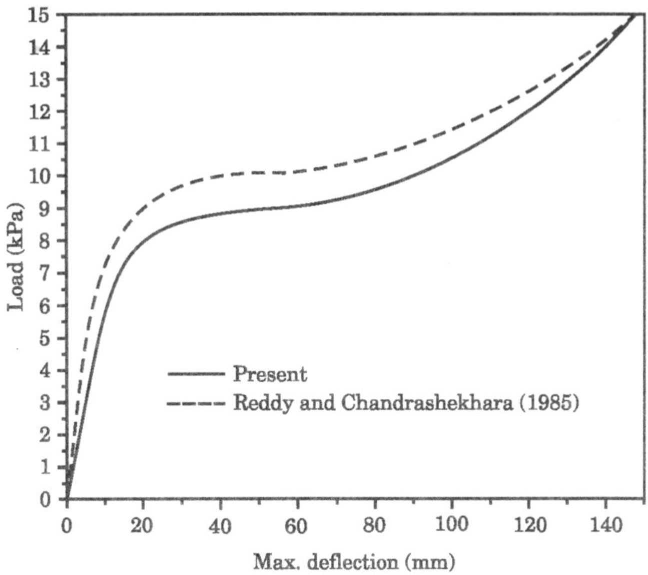

Fig. 11.3 Load-deflection curve for clamped [0/90] asymmetric cross-ply cylindrical shell panel subjected to a uniform load

11.4.3 Example 3: Clamped [0/90] asymmetric cross-ply cylindrical shell panel subjected to a uniform load

In this example, the asymmetric cross-ply cylindrical shell panel investigated by Reddy and Chandrashekhara (1985) is considered. This problem is same as investigated in Chapter 10. The load-deflection curve of the non-linear solution is shown in Figure 11.3 along with the result of Reddy and Chandrashekhara (1985). Fairly good agreement can be seen between the two results.

11.4.4 Example 4: Impact analysis of a graphite/epoxy cylindrical shell

A T300/976 graphite/epoxy cylindrical shell with lay-up $[90_4/0_8/90_4]$ and clamped on their edges is considered. With reference to Figure 11.4, geometric properties are: $a = b = 100$ mm; $R_x = R =$ a; and $R_y = \infty$. The impactor is a steel mass of 200 gm having a half sphere head of 10 mm diameter and an initial velocity of 5 ms^{-1}. The material properties of fiberite T300/976 graphite/epoxy composite are taken as follows (Choi and Chang, 1992):

ply thickness = 0.14224 mm; density, $\rho = 1540$ kg m^{-3};

$E_x = 156$ GPa; $E_y = E_z = 9.09$ GPa; $G_{xy} = G_{xz} = 6.96$ GPa;

$G_{yz} = 3.24$ GPa; $\nu_{xy} = \nu_{xz} = 0.228$; $\nu_{yz} = 0.4$.

The solution methodology for impact analysis of laminated cylindrical shells is same as that of laminated plates discussed in Chapter 9 except that the modified constant of the Hertz contact law for cylindrical shell during loading is given by (Goldsmith, 1960)

$$k = \frac{4}{3}\frac{\left[1/r_I + 1/(2R)\right]^{-1/2}}{\left[\left(1-\nu_I^2\right)/E_I + 1/E_y\right]},$$

in which R is the radius of the cylindrical shell and all other terms have the same meaning as in Chapter 9.

The results of contact force and shell centre displacement are presented in Figure 11.5 for both linear and non-linear analyses of shell. An increase in contact period and reduction in peak contact force occur by using a non-linear analysis when compared with the linear one. This indicates that the overall stiffness of the laminate decreases when geometrical non-linearity is incorporated in the solution (Kumar, 2008). The shell element results are also compared with those obtained by 3D non-linear formulation using eight noded brick elements (Kumar, 2010a). A good agreement can be observed between the two solutions.

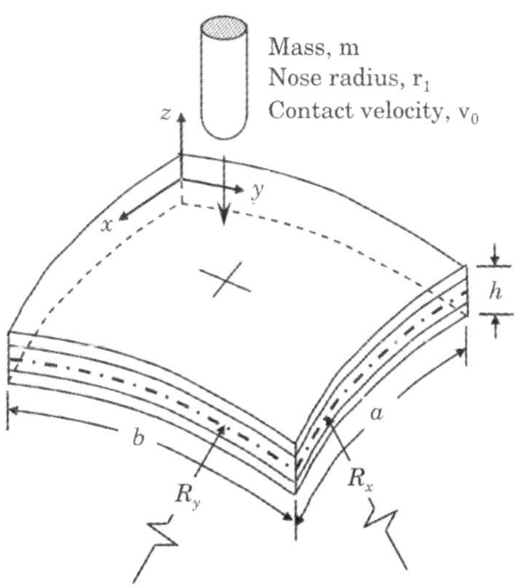

Fig. 11.4 Problem description of impact on a general doubly curved shell

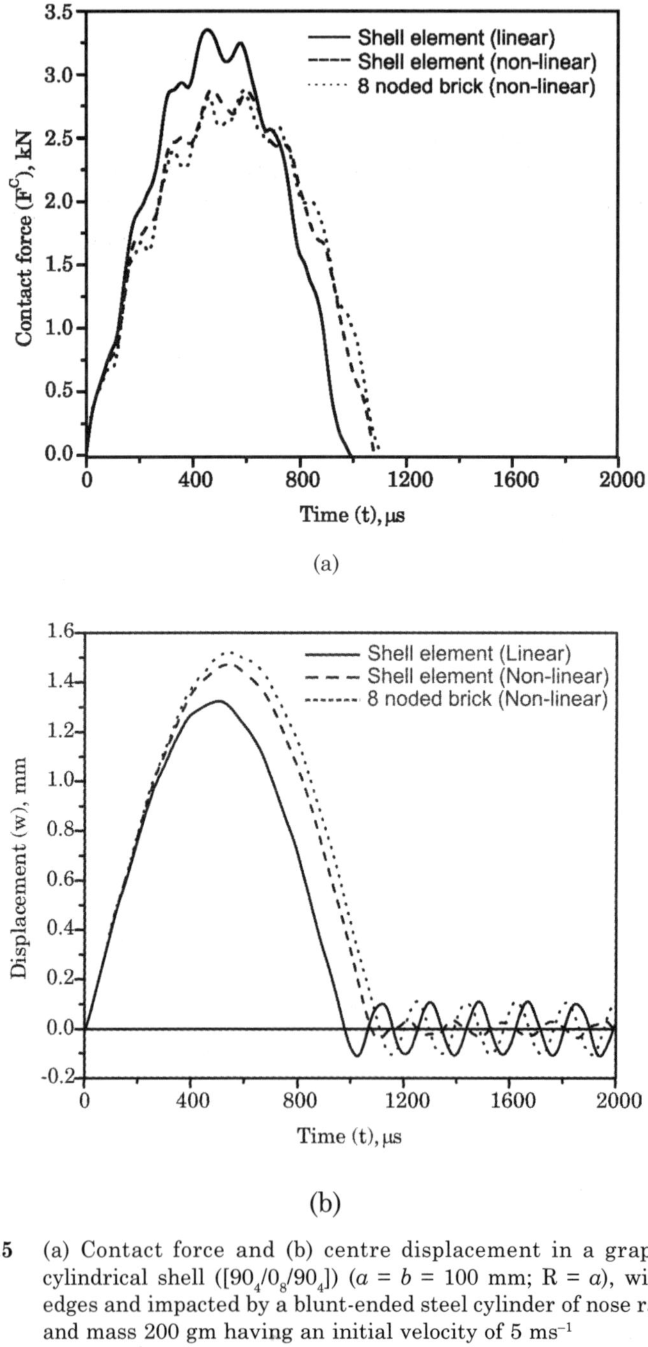

Fig. 11.5 (a) Contact force and (b) centre displacement in a graphite/epoxy cylindrical shell ($[90_4/0_8/90_4]$) ($a = b = 100$ mm; R = a), with clamped edges and impacted by a blunt-ended steel cylinder of nose radius 5 mm and mass 200 gm having an initial velocity of 5 ms^{-1}

12

Heat Transfer Analysis

12.1 INTRODUCTION

In the previous chapters, we considered the finite element formulation and solution of problems in the field of structural analysis. However, the finite-element method which originated in the area of structural mechanics, has been rapidly expanded to a wide range of non-structural problems. One such field is that of heat transfer analysis. In this chapter, we discuss the finite element formulation for the solution of heat transfer problems. Heat transfer occurs when there is a temperature difference within a body or between a body and its surroundings. Heat is transferred in the form of conduction, convection and thermal radiation. Determining temperature distribution in a body is important in many problems in engineering practice.

12.2 HEAT BALANCE EQUATION

The temperature distribution in a body (Figure 12.1) can be calculated using the governing partial differential equation shown in Eq. (12.1) as

$$\rho c\frac{\partial T}{\partial t}+\rho c\left(u_x\frac{\partial T}{\partial x}+u_y\frac{\partial T}{\partial y}+u_z\frac{\partial T}{\partial z}\right)$$

$$=\frac{\partial}{\partial x}\left(K_x\frac{\partial T}{\partial x}\right)+\frac{\partial}{\partial y}\left(K_y\frac{\partial T}{\partial y}\right)+\frac{\partial}{\partial z}\left(K_z\frac{\partial T}{\partial z}\right)+Q_g, \tag{12.1}$$

where $T(x,y,z,t)$ is the temperature; ρ is density; c is specific heat; K_x, K_y and K_z are thermal conductivities in x-, y- and z-directions respectively; u_x, u_y, u_z is the velocity vector for mass transport of heat; t is time; and Q_g is a heat generation term representing the heat released per unit volume in the body.

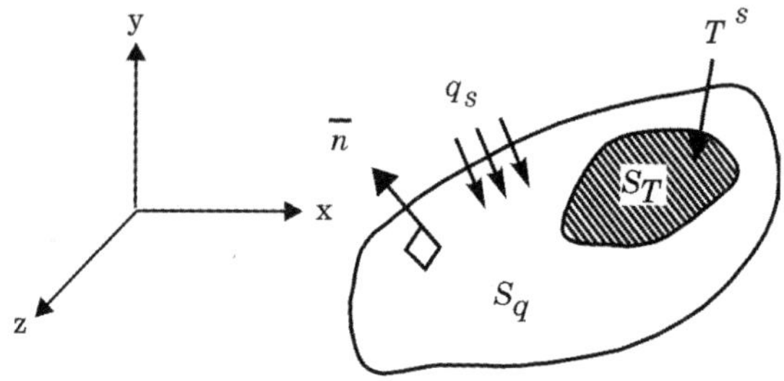

Fig. 12.1 A body subjected to heat transfer (prescribed temperature T^s on surface S_T and prescribed heat flux input q_s on surface S_q)

Eq. (12.1) is used in the modeling of non-stationary processes. Processes involving stationary temperature fields are simulated by the steady state convection-diffusion equation given as

$$\frac{\partial}{\partial x}\left(K_x\frac{\partial T}{\partial x}\right)+\frac{\partial}{\partial y}\left(K_y\frac{\partial T}{\partial y}\right)+\frac{\partial}{\partial z}\left(K_z\frac{\partial T}{\partial z}\right)$$

$$-\rho c\left(u_x\frac{\partial T}{\partial x}+u_y\frac{\partial T}{\partial y}+u_z\frac{\partial T}{\partial z}\right)+Q_g=0\cdot \quad (12.2)$$

Eq. (12.1) may be rewritten in the form

$$\rho c\dot{T}+\rho c\{u\}^T(\{\Delta\}T)-\{\Delta\}^T([D]\{\Delta\}T)-Q_g=0, \quad (12.3)$$

where

$$\{\Delta\}=\begin{bmatrix}\frac{\partial}{\partial x} & \frac{\partial}{\partial y} & \frac{\partial}{\partial z}\end{bmatrix}^T$$

$$[D]=\begin{bmatrix}K_x & 0 & 0\\ 0 & K_y & 0\\ 0 & 0 & K_z\end{bmatrix}$$

$$\{u\} = \begin{bmatrix} u_x & u_y & u_z \end{bmatrix}^T. \tag{12.4}$$

Premultiplying Eq. (12.3) by a virtual change in temperature δT and integrating over the volume yields

$$\int_V \delta T \rho c \dot{T} dV + \int_V \delta T \rho c \{u\}^T (\{\Delta\} T) dV - \int_V \delta T \{\Delta\}^T ([D]\{\Delta\} T) dV - \int_V \delta T Q_g dV = 0. \tag{12.5}$$

Using integration by parts, third term of the above equation can be expressed as

$$\int_V \delta T \{\Delta\}^T ([D]\{\Delta\} T) dV = \int_V \{\Delta\}^T (\delta T [D]\{\Delta\} T) dV - \int_V \{\Delta\}^T (\delta T)([D]\{\Delta\} T) dV. \tag{12.6}$$

By using divergence theorem, first term of the right side of the above equation is

$$\int_V \{\Delta\}^T (\delta T [D]\{\Delta\} T) dV = \int_{S_q} \{n\}^T (\delta T [D]\{\Delta\} T) dS = \int_{S_q} \delta T \{n\}^T ([D]\{\Delta\} T) dS, \tag{12.7}$$

where $\{n\}$ denotes the unit outward normal to the boundary surface. Substituting from Eqs. (12.6) and (12.7) into Eq. (12.5), we obtain

$$\int_V \delta T \rho c \dot{T} dV + \int_V \delta T \rho c \{u\}^T (\{\Delta\} T) dV + \int_V \{\Delta\}^T (\delta T)([D]\{\Delta\} T) dV - \int_V \delta T Q_g dV - \int_{S_q} \delta T q_s dS = 0, \tag{12.8}$$

where q_s is the heat flux across the boundary surface S_q and is given as

$$q_s = \{n\}^T ([D]\{\Delta\} T). \tag{12.9}$$

Conservation of energy at the massless surface of the body at any given instant may be expressed as follows:

$$q_s = q_f + q_c + q_r$$

or

$$q_s = q_f + h\left(T_e - T\right) + \sigma\varepsilon\left(T_e^4 - T^4\right), \tag{12.10}$$

where q_f is the specified heat flux, q_c is the heat flux due to free convection and q_r is the heat flux due to radiation. In Eq. (12.10), h is the heat convection coefficient for the environment, T_e is the environment temperature, σ is the Stefan-Boltzman constant and ε is the emissivity of the material surface.

12.3. FINITE-ELEMENT FORMULATION

The temperature field in Eq. (12.8) is approximated within an element by

$$T = \sum_{\alpha=1}^{n} N_\alpha T_\alpha = [N]\{T^e\}, \tag{12.11}$$

where N_a is the shape function and T_a is the temperature at α th node of the element. n is the number of nodes in the element. $\{T^e\}$ is a vector of element nodal temperatures and shape function matrix $[N]$ is given by

$$[N] = [N_1 \quad N_2 \quad \cdots \quad N_n]. \tag{12.12}$$

The combination $\{\Delta\}T$ inside the element is written as

$$\{\Delta\}T = [B]\{T^e\}, \tag{12.13}$$

where

$$[B] = \{\Delta\}[N]$$

or

$$[B] = \begin{bmatrix} \dfrac{\partial N_1}{\partial x} & \cdots & \dfrac{\partial N_n}{\partial x} \\ \dfrac{\partial N_1}{\partial y} & \cdots & \dfrac{\partial N_n}{\partial y} \\ \dfrac{\partial N_1}{\partial z} & \cdots & \dfrac{\partial N_n}{\partial z} \end{bmatrix}. \tag{12.14}$$

Discretizing Eq. (12.8) for m number of elements and substituting Eqs. (12.11) and (12.13) into it, we obtain

$$\sum_{e=1}^{m}\Big[\int_{V^e}\rho c\{\delta T^e\}^T[N]^T[N]\{\dot{T}^e\}dV + \int_{V^e}\rho c\{\delta T^e\}^T[N]^T\{u\}^T[B]\{T^e\}dV + \int_{V^e}\{\delta T^e\}^T[B]^T[D][B]\{T^e\}dV - \int_{V^e}\{\delta T^e\}^T[N]^T Q_g dV - \int_{S_q^e}\{\delta T^e\}^T[N^s]^T q_s dS\Big] = 0\,, \tag{12.15}$$

where $[N^s]$ is the element surface interpolation function matrix obtained from the volume interpolation matrix $[N]$ by substituting element surface natural coordinates. In the above equation, same interpolation function is used to approximate $\dot{T}$ within the element. Also the equivalences of $[N]\{T^e\} = \{T^e\}^T[N]^T$ and $\{\Delta\}^T(\delta T) = (\{\Delta\}\delta T)^T$ have been used in the equation.

With Boolean connectivity between local nodes in the element and global nodes, the above equation can be rewritten as

$$\{\delta T\}^T\Bigg[\left(\sum_{e=1}^{m}\int_{V^e}\rho c[N]^T[N]dV\right)\{\dot{T}\} + \left(\sum_{e=1}^{m}\int_{V^e}\rho c[N]^T\{u\}^T[B]dV\right)\{T\} + \left(\sum_{e=1}^{m}\int_{V^e}[B]^T[D][B]dV\right)\{T\} - \sum_{e=1}^{m}\left(\int_{V^e}[N]^T Q_g dV + \int_{S_q^e}[N^s]^T q_s dS\right)\Bigg] = 0\,. \tag{12.16}$$

Because of arbitrariness of $\{\delta T\}$, the following system of equation is obtained:

$$\left(\sum_{e=1}^{m}\int_{V^e}\rho c[N]^T[N]dV\right)\{\dot{T}\} + \left[\sum_{e=1}^{m}\left(\int_{V^e}\rho c[N]^T\{u\}^T[B]dV + \int_{V^e}[B]^T[D][B]dV\right)\right]\{T\} - \sum_{e=1}^{m}\left(\int_{V^e}[N]^T Q_g dV + \int_{S_q^e}[N^s]^T q_s dS\right) = 0\,. \tag{12.17}$$

Eq. (12.17) can be expressed in the form

$$[C]\{\dot{T}\}+[K]\{T\}=\{Q\}, \tag{12.18}$$

where $[C]$ is the heat capacity matrix, $[K]$ is the heat conduction matrix, $\{Q\}$ is the thermal load vector, $\{T\}$ is the vector of nodal point temperatures, and $\{\dot{T}\}$ the vector of nodal point temperature-rates.

The matrices $[C]$ and $[K]$, and the vector $\{Q\}$ are as given below:

$$[C]=\sum_{e=1}^{m}[C^e]=\sum_{e=1}^{m}\int_{V^e}\rho c[N]^T[N]dV$$

$$[K]=\sum_{e=1}^{m}[K^e]=\sum_{e=1}^{m}\left(\int_{V^e}[B]^T[D][B]dV+\int_{V^e}\rho c[N]^T\{u\}^T[B]dV\right)$$

$$\{Q\}=\sum_{e=1}^{m}\{Q^e\}=\sum_{e=1}^{m}\left(\int_{V^e}[N]^T Q_g dV+\int_{S_q^e}[N^s]^T q_s dS\right). \tag{12.19}$$

The surface heat flux vector in the second term of the vector $\{Q^e\}$ has several components as described in Eq. (12.10) and is expressed by

$$\int_{S_q^e}[N^s]^T q_s dS=\int_{S_f^e}[N^s]^T q_f dS+\int_{S_c^e}[N^s]^T h(T_e-T)dS+$$
$$\int_{S_r^e}[N^s]^T\sigma\varepsilon\left(T_e^4-T^4\right)dS. \tag{12.20}$$

The boundary conditions such as convection terms in Eq. (12.20) consist of unknown temperatures and thus can be split into two parts, in which one containing the unknown temperatures is added to the matrix $[K^e]$. Boundary conditions such as the radiation term in Eq. (12.20) are applied using previous iteration values for surface temperatures. As a result, matrix $[K]$ and vector $\{Q\}$ are modified to be

$$[K]=\sum_{e=1}^{m}\Big(\int_{V^e}[B]^T[D][B]dV+\int_{V^e}\rho c[N]^T\{u\}^T[B]dV+\int_{S_c^e}h[N^s]^T[N^s]dS+$$
$$\int_{S_r^e}\sigma\varepsilon\left([N^s]\{T^e\}\right)^3[N^s]^T[N^s]dS\Big) \tag{12.21}$$

and

$$\{Q\} = \sum_{e=1}^{m} \left(\int_{V^e} [N]^T Q_g dV + \int_{S_f^e} \left[N^s\right]^T q_f dS + \int_{S_c^e} hT_e \left[N^s\right]^T dS + \int_{S_r^e} \sigma\varepsilon T_e^4 \left[N^s\right]^T dS \right). \tag{12.22}$$

The transient equilibrium equation given by Eq. (12.18) can be integrated step-by-step with respect to time using an approximation of the type

$$T_{n+1} = T_n + \Delta t\left[(1-\beta)\dot{T}_n + \beta\dot{T}_{n+1}\right], \tag{12.23}$$

where Δt is the time step (increment of time in one step), subscript n denotes the values at nth time step (*i.e.* at a time point, say t_n) and the subscript (n+1) denotes the values at (n+1)th time step (*i.e.* at time point t_{n+1} or $t_n + \Delta t$). β is a parameter varying between 0 and 1. For unconditional stability, β should be greater than 0.5 and a value of 0.5 is generally chosen.

Now, we have from the above equation at time t_{n+1}

$$\dot{T}_{n+1} = \frac{T_{n+1}}{\beta\Delta t} - \frac{T_n}{\beta\Delta t} - \left(\frac{1-\beta}{\beta}\right)\dot{T}_n = \frac{T_{n+1}}{\beta\Delta t} + \hat{\dot{T}}_n. \tag{12.24}$$

The hat over $\dot{T}_n$ denotes

$$\hat{\dot{T}}_n = -\frac{T_n}{\beta\Delta t} - \left(\frac{1-\beta}{\beta}\right)\dot{T}. \tag{12.25}$$

Substituting Eq. (12.24) into Eq. (12.18), we obtain

$$\left[[K] + \frac{1}{\beta\Delta t}[C]\right]\{T_{n+1}\} = \{Q_{n+1}\} - [C]\left\{\hat{\dot{T}}_n\right\}. \tag{12.26}$$

12.4. TWO-DIMENSIONAL AXISYMMETRIC CASE

The general finite element heat transfer equation developed above can easily be deduced for special cases such as two-dimensional problems or axisymmetric problems. Assuming steady state and neglecting internal convection terms and radiation terms, we have for two-dimensional axisymmetric element (Figure 12.2) with z-axis as the axis of symmetry,

$$\begin{aligned}\left[K^e\right] &= \int_{V^e} [B]^T [D][B]\,dV + \int_{S_c^e} h\left[N^s\right]^T \left[N^s\right] dS \\ &= 2\pi \iint [B]^T [D][B]\,r\,dr\,dz + 2\pi \int_{l_c^e} h\left[N^s\right]^T \left[N^s\right] r\,dl\end{aligned} \tag{12.27}$$

and

$$\begin{aligned}\left\{Q^e\right\} &= \int_{S_f^e} q_f \left[N^s\right]^T dS + \int_{S_c^e} h T_e \left[N^s\right]^T dS \\ &= 2\pi \int_{l_f^e} q_f \left[N^s\right]^T r\,dl + 2\pi \int_{l_c^e} h\, T_e \left[N^s\right]^T r\,dl\,.\end{aligned} \tag{12.28}$$

In above equations, *dl* is the differential length along the edge(s) of the element on which the respective boundary conditions are prescribed.

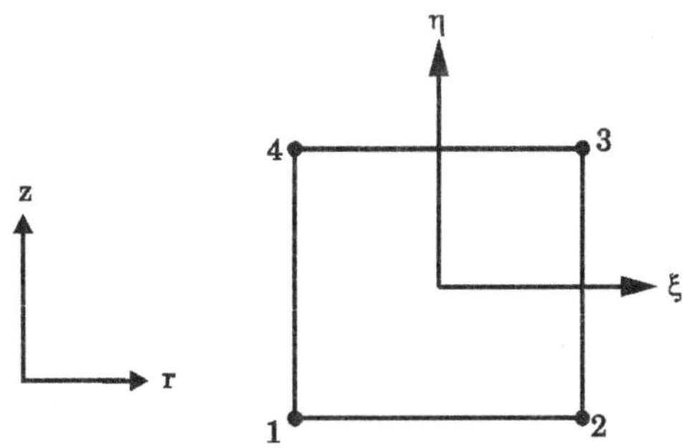

Fig. 12.2 A typical two-dimensional axisymmetric element

Matrix [B] in Eq. (12.27) is given by

$$[B] = \left[\frac{\partial}{\partial r} \quad \frac{\partial}{\partial z}\right]^T [N]. \tag{12.29}$$

Conductivity matrix [*D*] in Eq. (12.27) can be represented for two-dimensional axisymmetric analysis as

$$\left[D\right] = \begin{bmatrix} K_r & 0 \\ 0 & K_z \end{bmatrix}. \tag{12.30}$$

12.5. EXAMPLE PROBLEMS

12.5.1 Example 1: Heat transfer along a fin of rectangular cross-section

A simple example problem of the temperature distribution in a fin of rectangular cross-section, as shown in Figure 12.3, is solved by the finite element method. The fin is rectangular in shape, 80 mm long (L), 20 mm wide (W), and 4 mm thick (T). The thermal conductivity of the fin material is 17 W $m^{-1}K^{-1}$, and heat transfer coefficient and temperature of the surrounding medium are 25 $Wm^{-2}K^{-1}$ and 20 °C respectively. It is assumed that the heat transfer from the right end of the fin is negligible. The problem is solved using four-noded quadrilateral thermal elements. The calculated temperature at the tip of the fin is found to be 33.80 °C, and compare well with the theoretical result (Kakac and Yener, 1985) which is calculated as 33.53 °C.

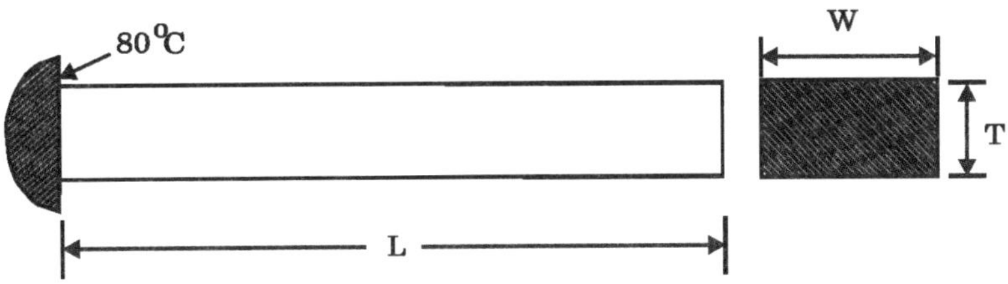

Fig. 12.3 A fin of a rectangular cross-section

12.5.2 Example 2: Temperature profile in a blast furnace hearth

The blast furnace is a shaft-type furnace with a vertical steel stack (lined with refractory bricks) superimposed over a crucible-like hearth. Iron ore, coke and limestone are charged into the top of the furnace, and preheated air is blown into the bottom. After a certain time interval, the raw materials descend to the bottom of the hearth where they become the final product of liquid slag and liquid iron. From time to time, these liquid products are drained out of the furnace through tapping holes. Heat transfer analysis of the hearth using finite element method is one of the important numerical studies of the blast furnace (Kumar, 2005).

In this example, calculation of temperature profile in different refractories of hearth zone (upto the height near the taphole) is carried out for two different designs of operating blast furnaces, which are designated as BF-A and BF-B. An axisymmetric finite element analysis is performed on symmetric half portion of the hearth (Figure 12.4) with appropriate boundary conditions. BF-A has a hearth inner diameter of

9.7 m and produces about 3050 tons/day of hot metal at a working volume of approximately 2000 m^3. BF-B has hearth inner diameter of 9.2 m and produces about 2700 tons/day of hot metal at a working volume of approximately 1580 m^3. BF-A hearth comprises cast iron stave cooling by water at the side-walls and underhearth air cooling while BF-B utilises water spray cooling at the side-walls and underhearth water cooling. Figures 12.5 and 12.6 show schematic sketches and necessary dimensions (in metre) of the symmetric half portion of the two hearths along with different materials used. The refractories are denoted in coded form. The thermal conductivities of these materials for the two hearths are presented in Tables 12.1 and 12.2 respectively.

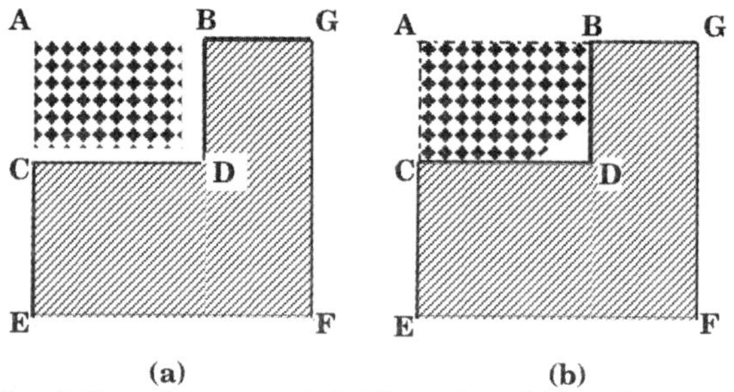

Fig. 12.4 Simulation of symmetric half portion of blast furnace hearth with two different states of coke-bed (a) floating and (b) sitting with an annular zone of variable porosity in the corner

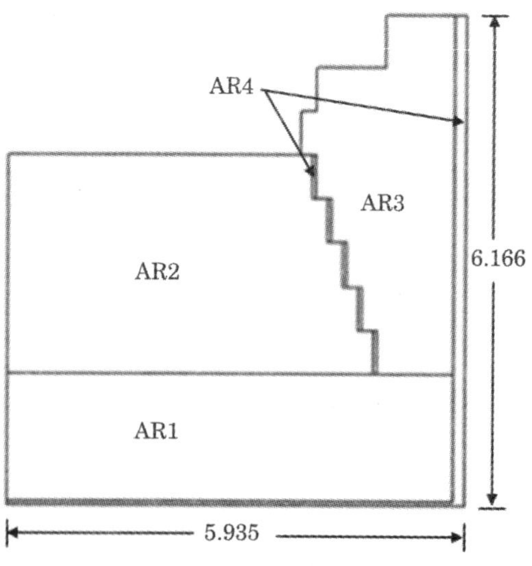

Fig.12.5 Schematic diagram of the symmetric half portion of BF-A hearth (drawn to scale)

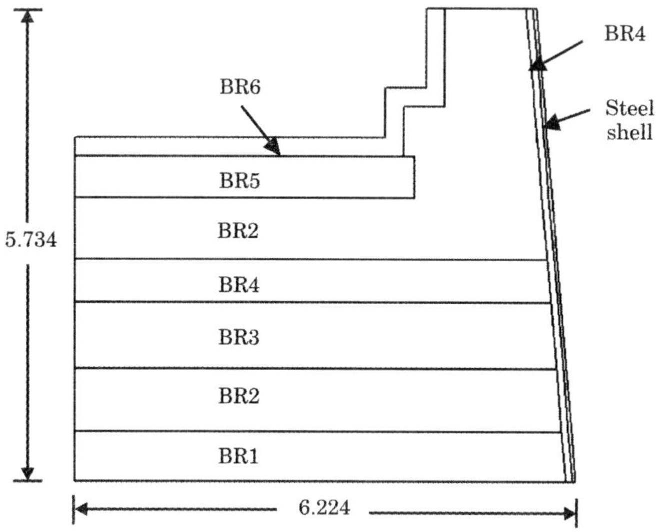

Fig. 12.6 Schematic diagram of the symmetric half portion of BF-B hearth (drawn to scale)

Table 12.1 Thermal conductivities of various materials used in BF-A hearth ($W\ m^{-1}K^{-1}$).

AR1	$k_z = 101\text{-}0.0577\ T$, $k_r = 0.9\ k_z$
AR2	$k = 0.8 + 0.00056\ T$
AR3	$k = 3.55+0.00179\ T$
AR4	$k = 3.15+0.00167\ T$
T = temperature (in °C)	

Table 12.2 Thermal conductivities of various materials used in BF-B hearth ($W\ m^{-1}K^{-1}$).

BR1	116.3
BR2	10.0
BR3	3.0
BR4	9.3
BR5	1.5
BR6	1.1
Steel Shell	52.0

BF-A hearth

With reference to Figure 12.4, hot metal-refractory interfaces CD and BD are set at constant temperature of hot metal taken to be 1450 °C. Convective boundary conditions are specified at the side-wall surface FG and bottom surface EF by taking in plant data for average temperatures (T_w= 30 °C, T_a= 35 °C) and approximate convective heat transfer coefficients (h_w =300 W $m^{-2}K^{-1}$, h_a =10 W $m^{-2}K^{-1}$) for the cooling water and air respectively. The refractory wall surface at the top of the hearth, BG and the symmetry line CE are set to be adiabatic (zero heat flux). The SB-FEM input file listing for this problem is depicted in Appendix 8. The calculated temperature profile (in °C) including 1150 °C isotherm is shown in Figure 12.7.

BF- B hearth

Convective boundary conditions are specified at the side-wall surface FG and bottom surface EF (Figure 12.4) by taking in plant data for average temperature (T_w= 32 °C) and approximate convective heat transfer coefficient (h_w = 300 W $m^{-2}K^{-1}$) for the cooling water. Since the appropriate convective heat transfer coefficients of cooling water for the two different cooling systems could not be calculated, these data are assumed to be same for both the systems. All other boundary conditions are kept same as in the BF-A hearth. The calculated temperature profile (in °C) including 1150 °C isotherm is shown in Figure 12.8. From the result, it can be seen that several isotherms are concentrated towards the upper portion of the hearth. This can be attributed to very low conductivity value of the protective layer material (BR6) used in the region which was taken into consideration in the analysis of this hearth.

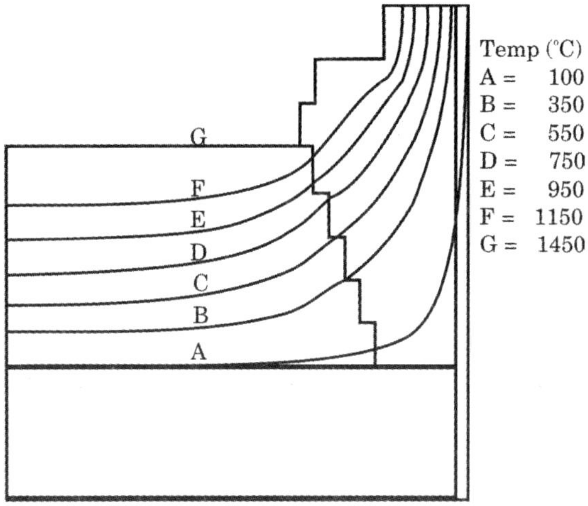

Fig. 12.7 Temperature profile in different refractories of the hearth zone in the initial design of BF-A

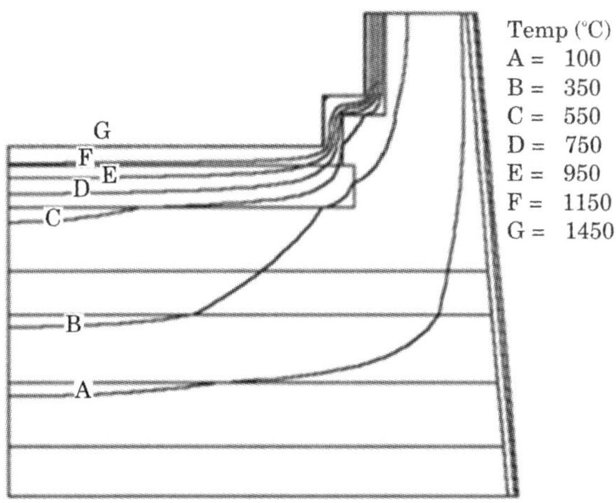

Fig. 12.8 Temperature profile in different refractories of the hearth zone in the initial design of BF-B

13

Metal Working Analysis

13.1. INTRODUCTION

Metal working or metal forming is defined as an operation in which an initially simple part, for example a billet or sheet blank, is plastically deformed between tools (or dies) to obtain the desired final configuration. The metal forming processes can be classified into two broad categories: (a) bulk forming processes and (b) sheet-metal forming processes. In both cases, the surfaces of the deforming material and of the tools are in contact, and friction between them has a major influence on the process. In bulk forming which includes forging, rolling, extrusion and drawing, the input material is in billet, slab or rod form, and a considerable increase in the surface-to-volume ratio occurs in the formed part. The workpiece undergoes large plastic deformation resulting in an appreciable change in shape or cross-section (Figure 13.1). In sheet forming, a sheet blank is plastically deformed into a three-dimensional object without any significant changes in sheet thickness and surface characteristics.

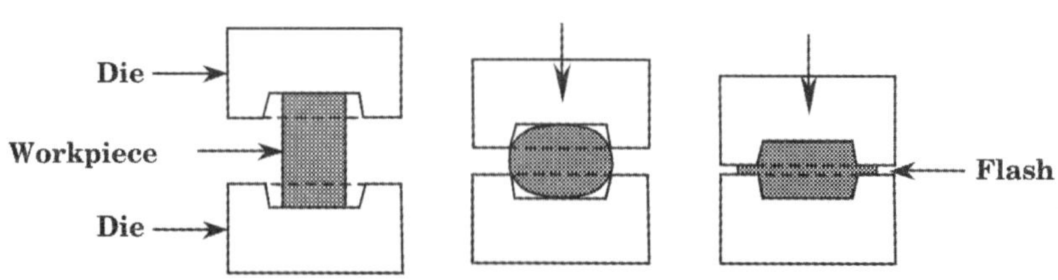

Fig. 13.1 An example of metal forming using closed-die forging

In the metal forming processes, the controlled change of the workpiece geometry in the deformation zone is influenced by various parameters *e.g.* plastic properties of workpiece materials, tribological conditions in the contact zone between workpiece and tool, tool geometry, tool materials, process temperature and velocity. The optimum layout of a metal forming process therefore requires a deep and broad knowledge of all of the above-mentioned parameters. In particular, it requires a good understanding of the interaction between these parameters, including the influence of the machine tool involved with regard to process-time relations.

As soon as the first steps were taken in the development of metal forming technology from shop floor experience level to increasingly scientifically determined high technology level, the modeling of metal forming processes was born in order to quantitatively describe the essential characteristics of a process. In the past a number of approximate methods of analysis have been developed and applied to various forming processes (Kobayashi *et al.*, 1989). The methods most well known are the slab method, the slip-line field method, the visioplasticity method; upper-(and lower-) bound techniques and Hill's general method.

The slab method considers the stresses on a plane perpendicular to the metal flow direction. A slab, straight or curved, of infinitesimal thickness is selected in this plane at any arbitrary point in the deformed metal. The forces on the slab are balanced and this will result in a differential equation of static equilibrium. By analytical or numerical integration of the differential equation and with the introduction of the boundary conditions, it is possible to determine the forming forces and to obtain other information.

The slip-line field method yields solutions for the cases of plane plastic flow for perfectly plastic materials (constant yield stress). For plain strain plastic deformation, slip lines indicate the direction of the maximum shear stresses and according to the von Mises flow rule, also the directions of maximum shear strain rates. The normal stresses are everywhere equal to the hydrostatic pressure. Along slip lines, therefore, the normal strain rates are equal to zero. In addition to information about the stresses, the method can also be used to calculate velocity fields through plasticity equations. A detailed discussion for slip line method and its applications can be found in Ewing (1967) and Johnson *et al.* (1970, 1982).

The visioplasticity method originated by Thomsen *et al.* (1954) combines experiment and analysis. In metal forming processes, it is possible to determine the velocity fields approximately by a parting of the workpiece and then inscribing a network of lines (grid patterns) which deforms during operation on the parted surface. From the two consecutive deformed grid patterns, the instantaneous velocity components can be determined. From the calculated velocity vectors, strain rates are calculated and stress distributions are obtained from plasticity equations. Although difficulties can arise when calculating the stress field, this method serves as a useful input for upper-bound applications.

The upper- and lower-bound methods employ the extremum principles of the von Mises plasticity theory to obtain upper and lower bounds respectively on the external forces need to carry out a metal working operation (Prager and Hodge, 1941). Upper bound theorem states that the rate of work done by the actual surface tractions with prescribed velocities is less than or equal to the rate of work done by the surface tractions corresponding to any other kinematically admissible velocity field. In this method, an admissible velocity field, which satisfies the incompressibility of metals and the prescribed velocity boundary conditions, is guessed based on this theorem. The total forming energy and also the forming load are obtained from the computed deformation, shear and friction energies based on the above velocity field. A lower bound method, on the other hand, considers that the rate of work done by the actual surface tractions with prescribed velocities is greater than or equal to the rate of work done by the surface tractions corresponding to any other statically admissible stress field.

Hill (1963) presented a general method of analysis for metal-working processes when the plastic flow is unconstrained. The method is based on a criterion of approximation derived from the interpretation of the virtual work-rate principle. He illustrated this method with analyses of compression with barreling, spread in bar drawing, and thickness change in tube sinking.

These methods have been useful in predicting forming load, overall geometry changes of deforming workpieces and qualitative modes of metal flow, and in determining approximate optimum process conditions. However, accurate determination of the effects of various process parameters on the detailed metal flow became possible only when the finite-element method was developed for the metal working analyses. Since then, the finite-element method has assumed steadily increased importance in simulation of metal-forming processes.

13.2. FEM FORMULATION IN METAL FORMING

The success of the finite-element method in the field of structural analysis stimulated the work of extending the application of this method to the areas of metal working. Extensive references on the earlier development of the finite-element method to metal-forming applications can be found in the papers of Kobayashi (1982, 1985).

Complete finite-element analysis of metal forming problems involves large strain, and both material and geometric nonlinearities have to be included. There are two broad approaches to the simulation of metal forming. The first approach is known as the flow formulation, which usually uses rigid-plastic or rigid-viscoplastic material models. Since in metal forming the plastic (non-linear) strains are much larger than the elastic strains, the latter are neglected. This allows the use of a simpler rigid-plastic model. The constitutive equations for this approach are identical to those for a non-Newtonian fluid. Nodal velocities are the unknowns in the flow formulation, and are related to the strain rates by standard kinematic expressions. The constitutive equation now relates the stresses and the strain rates. The yield stress can be introduced as a function of strain rate, strain and temperature and any type of approximating function can be used. As no elastic strains are accounted for, the large steps can be used in modeling metal-forming procedures which makes the method very economical, especially for modeling hot deformation. However, because of same reason, predictions regarding residual stresses cannot be made with the rigid-plastic finite-element method. The method can be implemented using an Eulerian description in which the nodes are fixed in space and the material flows through the mesh. Then it can be effectively used for steady state processes, like modeling the contact region in rolling.

The second one is a solid-state incremental approach which usually uses elastic-plastic or elastic-viscoplastic material models. The displacements are usually the primary nodal variables in this formulation, being related to the strains by standard expressions. The method assumes that the material deformation includes a small, recoverable elastic part and a much larger, non-recoverable plastic part. A large number of constitutive models exist. The elastic strains in combinations with different models for the plastic behaviour are possible. Furthermore, it is possible to implement constraints for treating incompressible deformation. The method is more general than the flow formulation. It can give details regarding deformation loads, stresses and strains, and residual stresses. However, because of the large change in the material flow behavior between elastic and plastic deformation and the need to check the status of each element, the deformation steps must be small, and this makes the method uneconomical. The method can be implemented using a Lagrangian description of the motion in which the nodes, and thereby the elements, are following the material flow. However in Lagrangian reference frame, the large deformations will cause distorted elements that eventually will lead to a failure in the analysis.

The Arbitrary Lagrangian Euler formulation (ALE) is used as a combination of Lagrangian and Eulerian frames for metal deformation analysis. In this the nodes on the boundary follows the material flow and the interior nodes can be chosen to lay arbitrarily along the flow line in each time increment. This formulation will reduce the distortion of the elements. This method has the advantages of both Lagrangian and Euler formulations.

13.3. RIGID-PLASTIC FINITE ELEMENT FORMULATION

Rigid-plastic finite-element models have become the most efficient tools in the simulation of metal forming processes. They combine the good accuracy of the finite-element technique with reasonable computing time and computer memory requirement. The main advantage of this method is that the power of deformation is calculated as a product of the invariants of the stress tensor and the strain-rate tensor, which makes the solution insensitive to rotations. The viscoplastic effect can be introduced either in the Levy-Mises flow rule using a yield stress sensitive to strain rate or using the Norton-Hoff law (Pietrzyk, 2000).

13.3.1. Rate form of virtual work

To formulate the rigid-plastic finite element method, we begin with the rate form of the virtual work equations. For a plastically deforming body of volume V under traction $\{t\} = [t_1 \quad t_2 \quad t_3]^T$ prescribed on a part of the surface S_F and velocity $\{u\} = [u_1 \quad u_2 \quad u_3]^T$ prescribed on the remainder of the surface, S_U ; the rate form of virtual work equations after neglecting body forces, can be written as

$$\int_V \sigma_{ij}\dot{\varepsilon}_{ij}dV - \int_{S_F} t_i u_i dS = 0. \tag{13.1}$$

In Eq. (13.1), σ_{ij} is the Cauchy stress tensor and $\dot{\varepsilon}_{ij}$ is the strain rate (infinitesimal strain increment) given by following expressions:

$$\dot{\varepsilon}_{ij} = \frac{1}{2}\left(\frac{\partial u_i}{\partial x_j} + \frac{\partial u_j}{\partial x_i}\right). \tag{13.2}$$

13.3.2. Flow rule

In the flow theory of elasticity, increments of strain are related to stresses through a flow rule of the form

$$\dot{\varepsilon}_{ij} = \gamma \frac{\partial G}{\partial \sigma_{ij}}, \tag{13.3}$$

where $G = G(\sigma_{ij}, \varepsilon_{ij})$ is a plastic potential function and the scalar γ is a function of the yield criterion F; $\gamma = \gamma\left[F(\sigma_{ij}, \varepsilon_{ij})\right]$.

If $F < 0$ the material is rigid, while for $F > 0$ the material has yielded and has a non-zero strain rate. Assuming "associated flow", that is, plastic potential and the yield function coincide, allows us to write $G = F$, which implies

$$\dot{\varepsilon}_{ij} = \gamma \frac{\partial F}{\partial \sigma_{ij}}. \tag{13.4}$$

For an isotropic, work-hardening material, the von Mises yield criterion F may be defined as

$$F = \sqrt{J_2} - k, \tag{13.5}$$

where J_2 is the second invariant of the deviatoric stress tensor σ'_{ij}, as given below (See Appendix - 2) and k is a parameter dependent upon the work–hardening characteristics of the material,

$$J_2 = \tfrac{1}{2}\sigma'_{ij}\sigma'_{ij}. \tag{13.6}$$

By comparison with a uniaxial tension test (in which, $\sigma_{11} = Y$ and all other stress components are zero), it can be shown that

$$k = \frac{Y}{\sqrt{3}}, \tag{13.7}$$

where Y is the yield stress in uniaxial tension. Now, evaluating the derivative in Eq. (13.4), we get

$$\frac{\partial F}{\partial \sigma_{ij}} = \frac{1}{2\sqrt{J_2}} \frac{\partial}{\partial \sigma_{ij}} [J_2] = \frac{1}{2\sqrt{J_2}} \frac{\partial}{\partial \sigma_{ij}} \left(\tfrac{1}{2}\sigma'_{ij}\sigma'_{ij}\right) = \frac{\sigma'_{ij}}{2\sqrt{J_2}}. \tag{13.8}$$

Using Eqs. (13.4) and (13.8), the flow rule may be written as

$$\dot{\varepsilon}_{ij} = \frac{\gamma \sigma'_{ij}}{2\sqrt{J_2}}. \tag{13.9}$$

Now, squaring both sides of Eq. (13.9) gives

$$\dot{\varepsilon}_{ij}\dot{\varepsilon}_{ij} = \frac{\gamma^2 \sigma'_{ij}\sigma'_{ij}}{4J_2} = \frac{1}{2}\gamma^2. \tag{13.10}$$

For a von Mises material model, the effective stress $\bar{\sigma}$ can be shown to be (Wagoner and Chenot, 1997)

$$\bar{\sigma} = \sqrt{\tfrac{3}{2}\sigma'_{ij}\sigma'_{ij}} \tag{13.11}$$

while the effective strain rate $\dot{\bar{\varepsilon}}$ may be written as

$$\dot{\bar{\varepsilon}} = \sqrt{\tfrac{2}{3}\dot{\varepsilon}_{ij}\dot{\varepsilon}_{ij}}. \tag{13.12}$$

Using the above definition for effective strain rate yields the following relation for the scalar γ:

$$\gamma = \sqrt{2\dot{\varepsilon}_{ij}\dot{\varepsilon}_{ij}} = \sqrt{3}\dot{\bar{\varepsilon}}. \tag{13.13}$$

Using Eqs. (13.6), (13.9) and (13.13), we obtain

$$\dot{\varepsilon}_{ij} = \frac{3}{2}\frac{\dot{\bar{\varepsilon}}}{\bar{\sigma}}\sigma'_{ij} \tag{13.14}$$

or

$$\sigma'_{ij} = \frac{2}{3}\frac{\bar{\sigma}}{\dot{\bar{\varepsilon}}}\dot{\varepsilon}_{ij}, \tag{13.15}$$

which is the Levy-Mises equation.

In general terms, the flow stress $\bar{\sigma}$ is dependent on temperature, strain and strain-rates and the general constitutional model for the rigid-viscoplastic materials can be written as

$$\bar{\sigma} = \bar{\sigma}\left(\bar{\varepsilon}, \dot{\bar{\varepsilon}}, T\right). \tag{13.16}$$

13.3.3. Variational equation

Using contracted (vectorial) notation for stresses and strain rates, the rate-of-virtual work expression (Eq. (13.1)) may be written under the condition of plastic incompressibility (neglecting the rate of work done by hydrostatic stress components) as

$$I = \int_V \{\dot{\varepsilon}\}^T \{\sigma'\}\, dV - \int_{S_F} \{u\}^T \{t\}\, dS, \tag{13.17}$$

where $\{\sigma'\}$ is a vector of six deviatoric stress components. The above equation in itself, doesn't satisfy incompressibility requirement i.e. volumetric strain-rate, $\dot{\varepsilon}_v = \dot{\varepsilon}_{ii} = 0$. In order to enforce incompressibility constraint, the above functional is modified by adding the term $\int_V \lambda \dot{\varepsilon}_v dV$ so that

$$I = \int_V \{\dot{\varepsilon}\}^T \{\sigma'\}\, dV + \int_V \lambda \dot{\varepsilon}_v dV - \int_{S_F} \{u\}^T \{t\}\, dS, \tag{13.18}$$

where λ is a Lagrangian multiplier. In the third term, the traction t_i will be the friction force per unit area at the die-workpiece interface. To obtain the equilibrium equations, the functional (13.18) is made stationary with respect to the kinematically admissible velocity fields which satisfy compatibility and the velocity boundary conditions. Thus, we have

$$\delta I = \int_V \{\delta\dot{\varepsilon}\}^T \{\sigma'\}\, dV + \int_V \lambda \delta\dot{\varepsilon}_v dV + \int_V \dot{\varepsilon}_v \delta\lambda dV - \int_{S_F} \{\delta u\}^T \{t\}\, dS = 0. \tag{13.19}$$

In Eq. (13.19), $\{\delta u\}$ and $\delta\lambda$ are arbitrary variations, and $\{\delta\dot{\varepsilon}\}$ and $\delta\dot{\varepsilon}_v$ are the variations in strain-rate derived from $\{\delta u\}$.

As an alternative to the Lagrangian multiplier approach, the incompressibility condition may be enforced by means of a penalty function. Modifying the functional in Eq. (13.17) by adding the term $\frac{\zeta}{2}\int_V \dot{\varepsilon}_v^2 dV$ and taking the variation of the functional, the equilibrium equation may be expressed as

$$\delta I = \int_V \{\delta\dot{\varepsilon}\}^T \{\sigma'\} dV + \int_V \zeta\, \dot{\varepsilon}_v \delta\dot{\varepsilon}_v dV - \int_{S_F} \{\delta u\}^T \{t\} dS = 0, \qquad (13.20)$$

where ζ, a penalty constant, is a very large positive constant.

Equation (13.19) or (13.20) is the basic equation for the finite-element formulation.

13.3.4. Finite element formulation

Eq. (13.20) is rewritten as a sum of integration over the volume and areas of all finite elements, *i.e.*

$$\sum_e \left(\int_{V^e} \{\delta\dot{\varepsilon}\}^T \{\sigma'\} dV + \int_{V^e} \zeta\dot{\varepsilon}_v \delta\dot{\varepsilon}_v dV - \int_{S_F^e} \{\delta u\}^T \{t\} dS \right) = 0. \qquad (13.21)$$

In FEM, a continuous velocity field over each element can be defined uniquely in terms of velocities of associated nodal points by using the shape function. Equation (13.21) can thus be expressed in terms of nodal point velocities $\{U^e\}$ and their variations $\{\delta U^e\}$ for each element.

We have the following relation inside each element:

$$\{u\} = [N]\{U^e\},$$

[*N*] being the shape function matrix.

Eq. (13.2) can be written as

$$\{\dot{\varepsilon}\} = [\Delta]\{u\} = [B]\{U^e\},$$

in which [Δ] is the derivative operator matrix given by Eq. (2.23), and [*B*] = [Δ] [*N*] is called the strain-rate matrix.

Eq. (13.15) can be written in matrix form as

$$\{\sigma'\} = \frac{\bar{\sigma}}{\dot{\bar{\varepsilon}}}[D]\{\dot{\varepsilon}\}. \qquad (13.22)$$

The diagonal matrix $[D]$ has $\frac{2}{3}$ and $\frac{1}{3}$ as its components; corresponding to normal strain rate and engineering shear strain rate, respectively.

The volumetric strain rate $\dot{\varepsilon}_v$ is given by

$$\dot{\varepsilon}_v = \dot{\varepsilon}_1 + \dot{\varepsilon}_2 + \dot{\varepsilon}_3 = [C]\{U^e\}, \tag{13.23}$$

where $[C]$ is a column matrix so that $C_J = B_{1J} + B_{2J} + B_{3J}$, where B_{IJ} is an element in the Ith row and Jth column of the strain-rate matrix $[B]$.

The effective strain rate $\dot{\bar{\varepsilon}}$ (Eq. 13.12) is defined in terms of strain-rate components as

$$\left(\dot{\bar{\varepsilon}}\right)^2 = \{\dot{\varepsilon}\}^T [D]\{\dot{\varepsilon}\} = \{U^e\}^T [B]^T [D][B]\{U^e\} = \{U^e\}^T [P]\{U^e\}. \tag{13.24}$$

In Eq. (13.24), $[P] = [B]^T [D][B]$ is used for simplicity.

Putting values in Eq. (13.21), we have

$$\sum_e \{\delta U^e\}^T \left[\left(\int_{V^e} \frac{\bar{\sigma}}{\dot{\bar{\varepsilon}}}[B]^T [D][B]dV\right)\{U^e\} + \varsigma\left(\int_{V^e} [C]^T [C]dV\right)\{U^e\}\right] - \sum_e \{\delta U^e\}^T \left(\int_{S_F^e} [N^s]^T \{t\}\, dS\right) = 0$$

or

$$\{\delta U\}^T \left[\sum_e \left(\int_{V^e} \frac{\bar{\sigma}}{\dot{\bar{\varepsilon}}}[B]^T [D][B]dV + \int_{V^e} \varsigma[C]^T [C]dV\right)\right]\{U\} - \{\delta U\}^T \left[\sum_e \left(\int_{S_F^e} [N^s]^T \{t\}\, dS\right)\right] = 0. \tag{13.25}$$

Since $\{\delta U\}$ is arbitrary, Eq. (13.25) can be written as

$$\left[\sum_e \left(\int_{V^e} \frac{\bar{\sigma}}{\dot{\bar{\varepsilon}}}[B]^T [D][B]dV + \int_{V^e} \varsigma[C]^T [C]dV\right)\right]\{U\} - \sum_e \left(\int_{S_F^e} [N^s]^T \{t\}\, dS\right) = 0. \tag{13.26}$$

Now the terms in Eq. (13.26) can be defined as follows:

Element stiffness matrix,

$$\left[K^e\right] = \int_{V^e} \frac{\bar{\sigma}}{\dot{\bar{\varepsilon}}} [B]^T [D][B] dV \tag{13.27}$$

Element penalty function matrix,

$$\left[K_\lambda^e\right] = \int_{V^e} \varsigma [C]^T [C] dV \tag{13.28}$$

Element surface force vector,

$$\left\{F_s^e\right\} = \int_{S_F^e} \left[N^s\right]^T \{t\} dS. \tag{13.29}$$

Putting the terms defined above into the Eq. (13.26), we get the finite element equation of equilibrium as

$$[K]\{U\} = \{F\}, \tag{13.30}$$

where

global stiffness matrix is,

$$[K] = \sum_e \left(\left[K^e\right] + \left[K_\lambda^e\right]\right) \tag{13.31}$$

and global force vector is,

$$\{F\} = \{F_s\} = \sum_e \left\{F_s^e\right\}. \tag{13.32}$$

In metal-forming, the equation of equilibrium given by Eq. (13.26) is nonlinear and the solution must be obtained iteratively. The equation can be written in a non-linear sense as

$$\{\Psi\} = \sum_e \left(\int_{V^e} \frac{\bar{\sigma}}{\dot{\bar{\varepsilon}}} [P]\left\{U^e\right\} dV + \int_{V^e} \varsigma [C]^T [C]\left\{U^e\right\} dV \right) - \sum_e \left(\int_{S_F^e} \left[N^s\right]^T \{t\} dS \right), \tag{13.33}$$

where $\{\Psi\}$ is residual of the nodal point force vector at a particular solution iteration.

Newton-Raphson method is generally adopted to solve Eq. (13.33). The method consists of linearization and application of suitable convergence criteria to obtain the final solution. Linearization is achieved by Taylor expansion near an assumed solution point $\{U\}=\{U_0\}$ (initial guess), namely

$$\left\{\Psi_{U_0+\delta U}\right\}=\left\{\Psi_{U_0}\right\}+\left\{\delta\Psi_{U_0}\right\}\approx\{0\}. \tag{13.34}$$

In Eq. (13.34), $\{\delta\Psi\}$ can be referred to as vector of residual (out-of-balance) force increments due to $\{\delta U\}$, being the first order correction of the velocity. It is, therefore, necessary to find relation between $\{\delta\Psi\}$ and $\{\delta U\}$ in the Newton-Raphson scheme.

Taking variation of residual force vector (Eq. (13.33)) with respect to $\{U\}$, we obtain

$$\begin{aligned}\{\delta\Psi\}=\sum_e\left(\int_{V^e}\frac{\bar{\sigma}}{\dot{\bar{\varepsilon}}}[P]\{\delta U^e\}dV+\int_{V^e}[P]\{U^e\}\left(\frac{1}{\dot{\bar{\varepsilon}}}\frac{\partial\bar{\sigma}}{\partial\dot{\bar{\varepsilon}}}-\frac{\bar{\sigma}}{\dot{\bar{\varepsilon}}^2}\right)\frac{1}{\dot{\bar{\varepsilon}}}\{U^e\}^T[P]\{\delta U^e\}dV\right)+\\+\sum_e\int_{V^e}\varsigma[C]^T[C]\{\delta U^e\}dV\end{aligned} \tag{13.35}$$

or

$$\begin{aligned}\{\delta\Psi\}=\Bigg[\sum_e\left(\int_{V^e}\frac{\bar{\sigma}}{\dot{\bar{\varepsilon}}}[P]dV+\int_{V^e}\left(\frac{1}{\dot{\bar{\varepsilon}}}\frac{\partial\bar{\sigma}}{\partial\dot{\bar{\varepsilon}}}-\frac{\bar{\sigma}}{\dot{\bar{\varepsilon}}^2}\right)\frac{1}{\dot{\bar{\varepsilon}}}[P]\{U^e\}\{U^e\}^T[P]dV\right)+\\+\sum_e\int_{V^e}\varsigma[C]^T[C]dV\Bigg]\{\delta U\}.\end{aligned} \tag{13.36}$$

In Eq. (13.35), in the second term,

$$\delta\left(\frac{\bar{\sigma}}{\dot{\bar{\varepsilon}}}\right)=\left(\frac{1}{\dot{\bar{\varepsilon}}}\frac{\partial\bar{\sigma}}{\partial\dot{\bar{\varepsilon}}}-\frac{\bar{\sigma}}{\dot{\bar{\varepsilon}}^2}\right)\delta\dot{\bar{\varepsilon}}=\left(\frac{1}{\dot{\bar{\varepsilon}}}\frac{\partial\bar{\sigma}}{\partial\dot{\bar{\varepsilon}}}-\frac{\bar{\sigma}}{\dot{\bar{\varepsilon}}^2}\right)\frac{1}{\dot{\bar{\varepsilon}}}\delta\left(\tfrac{1}{2}\dot{\bar{\varepsilon}}^2\right)$$

is used.

Eq. (13.36) may be written as

$$\left[K^T\right]\{\delta U\} = \{\delta \Psi\}, \tag{13.37}$$

$[K^T]$ being called the tangent stiffness matrix.

Eq. (13.34) can thus be written as

$$\left[K^T\right]\{\delta U\} = -\left\{\Psi_{U_0}\right\}. \tag{13.38}$$

Once the solution of Eq. (13.38) for the velocity correction term $\{\delta U\}$ is obtained, the assumed velocity $\{U_0\}$ is updated to $\{U_0\} + \alpha\{\delta U\}$, where α is a constant between 0 and 1 called the deceleration coefficient. Iteration is continued until the velocity correction terms become negligibly small. For convergence of Newton-Raphson iterations, the initial guess velocity should be close to the actual solution. When a deformation process is relatively simple, the initial guess velocity can be calculated, for instance, by the upper-bound method (Kobayasi *et al.*, 1989). However, it may not be easy to obtain a good initial guess velocity in case of complex deformation process. In that case, direct iteration method may be appropriate.

Direct iteration method is another technique for solving a non linear equation (Eq. 13.26). In this method, the constitutive equation (Eq. 13.15) is assumed to be linear during each iteration, and thus the term $\bar{\sigma}/\dot{\bar{\varepsilon}}$ in Eq. (13.26) remains constant during each iteration. This method does not require any initial guess velocity. For metal forming applications, the direct iteration method converges fast towards the solution during the earlier stages of iteration. However, as the solution point is approached, the convergence becomes very slow. It is reported that the best computational efficiency can be obtained by using (1) the direct iteration method for generating the initial guess and for cases where the Newton Raphson method does not converge, and (2) the Newton Raphson method for all other cases (Kobayasi *et al.*, 1989).

Two convergence criteria may be used. One measures the error norm of the velocities, $\left\|\{\delta U\}\right\| / \left\|\{U\}\right\|$, where the Euclidean vector is defined as $\left\|\{U\}\right\| = \left(\{U\}^T\{U\}\right)^{1/2}$, and requires such an error norm to decrease from iteration to iteration. The other criterion requires the norm of the residual equations, $\left\|\{\Psi\}\right\|$ to decrease.

In general terms, the first criterion is most useful in the early stages of iteration, when the velocity field is still far from the solution. The second test is most useful when slightly ill-conditioned systems reach the final stage of iterations. The final solution is considered to be attained when the error norm reaches a specified small value, say 5×10^{-5}.

13.4. FRICTION MODELING

Friction conditions between the workpiece and dies greatly influence the metal flow, strain patterns and performances of many metal forming operations. On the other hand, friction itself depends on the forming operation characteristics. There are several parameters that can affect frictional stresses, namely contact pressure; sliding velocity between workpiece and dies; the elastic and plastic properties of the two materials (workpiece and dies); surface roughness; lubrication at the contact surfaces; local temperature at contact points and in lubricants; and so on. With all of these variables influencing friction, it seems necessary that frictional stresses are measured under conditions close to those in forming operations of interest. However, for purposes of simulation, it is necessary to adopt simple laws that allow measurement of a limited number of parameters.

The friction shear stress, f_s, is most commonly expressed by Coulomb's law as

$$f_s = \mu p, \tag{13.39}$$

where p is a compressive normal stress (pressure) to the interface and μ is called the friction coefficient, or Coulomb coefficient.

While Coulomb's law is appropriate for simple friction between non-deforming bodies (which are by definition lightly loaded), it drastically overestimates friction at high contact pressures. A simple thought experiment shows that the frictional stress at the interface cannot exceed the shear strength of the material because the material can slide at this shear stress even if the interface is firmly bonded. This extreme condition is called sticking friction. Therefore, according to this concept, the friction shear stress, f_s, may be represented mathematically as

$$f_s = mk, \tag{13.40}$$

where m is a constant known as friction factor and has a value of $0 \leq m \leq 1$, and k is the shear strength of the deforming material (workpiece).

Studies indicate that Eq. (13.40) adequately characterizes the friction condition in bulk forming processes while Eq. (13.39) is commonly used for representation of friction in sheet-metal forming. A reason for this is that the compressive normal stress at the interface in sheet-metal forming is much smaller in magnitude, in comparison with that in bulk forming processes. For various forming conditions, the values of m vary as given below (Kobayasi *et al.*, 1989):

- m=0.05-0.15 in cold forming of sheets, aluminum alloys, and cooper, using conventional phosphate-soap lubricants or oils.
- m=0.2-0.4 in hot forming of sheets, cooper, and aluminum alloys with graphite-based (graphite-water or graphite-oil) lubricants.
- m=0.1-0.3 in hot forming of titanium and high-temperature alloys with glass lubricants.
- m=0.7-1.0 when no lubricant is used, *e.g.*, in hot rolling of plates or slabs and in non-lubricated extrusion of aluminum alloys.

For problems such as ring compression, forging and rolling, the unknown direction of the relative velocity between the die and the workpiece at the interface makes it difficult to handle the friction condition in a straightforward manner. The frictional stress along the contact surface changes its direction at a point (or a region) where the velocity of the deforming material relative to the dies becomes zero. The location of this point (called neutral point) depends mainly on the magnitude of the frictional stress itself and is not known a priori. In order to deal with these situations, a velocity-dependent frictional stress suggested by Chen and Kobayashi (1978) is used as an approximation to the condition of constant frictional stress. This frictional stress function is expressed as

$$\{f_s\} = -mk\left\{\frac{2}{\pi}\tan^{-1}\left(\frac{|V_r|}{a}\right)\right\}\frac{\{V_r\}}{|V_r|}, \tag{13.41}$$

where $\{V_r\}$ is the relative velocity between the die and the workpiece, mk is the magnitude of the frictional stress, and a is a constant some orders of magnitude smaller than the die velocity. The approximation expression (13.41) has been used for the smooth transition of the frictional stress in the range near the neutral point (Chen and Kobayashi, 1978).

13.5. THERMAL FORMULATION

When plastic deformation takes place at high temperatures in the warm and hot forming regimes, material properties can vary considerably with temperature. Heat is generated during a metal forming process, and if dies are at a considerably lower temperature than the workpiece, the heat loss by conduction to the dies and by radiation and convection to the environment can result in severe temperature gradients within the workpiece. Thus, the consideration of temperature effects in the analysis of metal-forming problems is very important. Furthermore, at elevated temperatures, plastic deformations can induce phase transformations and alterations in grain structures that, in turn, can modify the flow stress of the workpiece material as well as other mechanical properties. Considering hot strip rolling process, the heat transfer mechanism is very complex and consists of the following main components:

- Heat generation due to plastic deformation inside the deformation body.
- Heat generated due to friction along the roll-strip interface.
- Heat loss between stands due to convection and radiation from the strip to the environment and during descaling or cooling by water.
- Heat loss due to heat flow from the strip to the rolls.

13.5.1. Finite element equation

During the forming process, the temperature distribution in the workpiece can be calculated using the governing partial differential equation shown in Eq. (12.1). The finite element equation of thermal equilibrium is given by Eq. (12.18) or Eq. (12.26). In the present case, heat generation term Q_g represents the heat released per unit volume because of the plastic work, and can be calculated as

$$Q_g = \kappa\bar{\sigma}\dot{\bar{\varepsilon}}, \tag{13.42}$$

where the heat generation efficiency κ, represents the fraction of mechanical energy transformed into heat and is usually assumed to be 0.9. The fraction of the remainder of the plastic deformation energy (1 – κ) is expended to cause changes in dislocation density, grain boundaries, and phases. This energy is usually recoverable by annealing.

Conservation of energy at the surface of the workpiece at any given instant may be expressed as follows:

$$q_s = q_f + q_c + q_{\text{lub}} + q_r$$

or $$q_s = \frac{1}{2}|F_r||V_R| + h(T_e - T) + h_{\text{lub}}(T_D - T) + \sigma\varepsilon(T_e^4 - T^4), \quad (13.43)$$

where q_f is the surface heat generation rate due to friction, q_c is the heat flux due to free convection, q_{lub} is the heat flux due to convection through lubricant and q_r is the heat flux due to radiation. In Eq. (13.43), F_r is the friction force at die-workpiece interface, V_R is the relative velocity of workpiece with respect to the die, h is the heat convection coefficient for the environment, T_e is the environment temperature, h_{lub} is the heat convection coefficient for the lubricant, T_D is the time-dependent die temperature, σ is the Stefan-Boltzman constant and ε is the emissivity of the material surface. The factor ½ appears in the first term on the right hand side of Eq. (13.43) because it is assumed that the heat generated due to friction at the die-workpiece interface is evenly distributed between workpiece and die.

The thermal load vector $\{Q\}$ in Eq. (12.19) has several components as described in Eq. (13.43) and is expressed by

$$\{Q\} = \sum_{e=1}^{m}\{Q^e\} = \sum_{e=1}^{m}\Big(\int_{V^e}[N]^T(\kappa\overline{\sigma}\dot{\overline{\varepsilon}})dV + \int_{S_c^e}\left[N^s\right]^T q_f dS + \int_{S_f^e}\left[N^s\right]^T h(T_e - T)dS +$$
$$+ \int_{S_c^e}\left[N^s\right]^T h_{\text{lub}}(T_D - T)dS + \int_{S_f^e}\left[N^s\right]^T \sigma\varepsilon(T_e^4 - T^4)dS\Big). \quad (13.44)$$

In Eq. (13.44), S_c^e is the portion of the element surface in contact with a die, and S_f^e is the portion of the element surface which is free and not in contact with a die.

Following the step used in Chapter 12, the boundary conditions such as convection terms in Eq. (13.44) can be split into two parts, in which one containing the unknown temperatures is added to the matrix $[K]$. Boundary conditions such as the radiation term in Eq. (13.44) are applied using previous iteration values for surface temperatures. As a result, matrix $[K]$ and vector $\{Q\}$ are modified to be

$$[K] = \sum_{e=1}^{m}\Big(\int_{V^e}[B]^T[D][B]dV + \int_{V^e}\rho c[N]^T\{u\}^T[B]dV + \int_{S_c^e}h_{\text{lub}}\left[N^s\right]^T\left[N^s\right]dS +$$
$$\int_{S_f^e}\left[h + \sigma\varepsilon\left(\left[N^s\right]\{T^e\}\right)^3\right]\left[N^s\right]^T\left[N^s\right]dS\Big) \quad (13.45)$$

and

$$\{Q\} = \sum_{e=1}^{m}\left(\int_{V^e} [N]^T \left(\kappa\bar{\sigma}\dot{\bar{\varepsilon}}\right) dV + \int_{S_c^e} \left[N^s\right]^T q_f dS + \int_{S_c^e} h_{\text{lub}} T_D \left[N^s\right]^T dS + \int_{S_f^e} \left(hT_e + \sigma\varepsilon T_e^4\right)\left[N^s\right]^T dS \right). \tag{13.46}$$

13.6. COMBINED MECHANICAL AND THERMAL ANALYSIS

One approach for solving the combined problem is to alternate between the deformation and thermal analyses iteratively until both solutions converge. The deformation analysis gives the nodal velocities $\{U\}$ (from which strain and strain rate may be calculated) and the thermal solution of the workpiece yields the temperature distribution $\{T\}$ inside the workpiece. There is no internal heat generation in the die and therefore deformation calculations are not necessary for the die. The thermal analysis of the die yields the temperature distribution $\{T_D\}$ in the die. Assuming a converged solution has been obtained for all the above quantities $\{U_n\}$, $\{T_n\}$ and $\{T_{D(n)}\}$ at some time t_n (the subscripts referring to the time step), the next step is to obtain converged solutions for the velocities and temperatures at time step $t_{n+1} = t_n + \Delta t$. The steps necessary to do this are as follows:

(a) The finite element mesh of the workpiece is updated based on velocities $\{U_n\}$ and the time step Δt, to give the mesh configuration at time t_{n+1}. The effective strains are then calculated for the new configuration.

(b) The converged velocities and temperatures for time step t_n are used as initial guesses for the solution at time step t_{n+1}, that is,

$$\left\{U_{n+1}^{(0)}\right\} = \left\{U_n\right\}$$

$$\left\{T_{n+1}^{(0)}\right\} = \left\{T_n\right\}$$

$$\left\{T_{D(n+1)}^{(0)}\right\} = \left\{T_{D(n)}\right\}$$

Here, the superscript is the iteration count for the iterative cycle between deformation and thermal analyses, in a particular time step.

(c) For iteration count $i = 0, 1, 2, \ldots$ until convergence,

(i) Based on $\left\{T_{n+1}^{(i)}\right\}$, perform deformation analysis on workpiece to converge to $\left\{U_{n+1}^{(i+1)}\right\}$.

(ii) Based on $\left\{U_{n+1}^{(i+1)}\right\}$ and $\left\{T_{D(n+1)}^{(i)}\right\}$, perform thermal analysis of the workpiece to converge to $\left\{T_{n+1}^{(i+1)}\right\}$.

(iii) Based on $\left\{T_{n+1}^{(i+1)}\right\}$, perform thermal analysis of die to converge to $\left\{T_{D(n+1)}^{(i+1)}\right\}$.

(iv) Check for convergence, that is check to see if $\left\{U_{n+1}^{(i+1)}\right\}$, $\left\{T_{n+1}^{(i+1)}\right\}$ and $\left\{T_{D(n+1)}^{(i+1)}\right\}$ differ appreciably from $\left\{U_{n+1}^{(i)}\right\}$, $\left\{T_{n+1}^{(i)}\right\}$ and $\left\{T_{D(n+1)}^{(i)}\right\}$, respectively. If yes, convergence is not satisfied. In this case, update i by adding one and return to step (c). If convergence is satisfied, then update n by adding one and return to step (a).

13.7 EXAMPLE PROBLEMS

13.7.1. Example 1: Upsetting with flat dies

Open die forging processes has a series of upsetting and drawing processes. Upsetting is the one of the simple operations routinely and commonly used.

A simple compression of a circular cylinder is simulated using the present formulation. The computational conditions used for the simulation are as follows and taken to be same as taken by Kobayashi *et al.* (1989):

Initial billet dimension	: Diameter = 25.4 mm, Height = 25.4 mm
Die velocity	: –25.4 mm/sec
Friction factor	: 0.5
Flow stress	: 69 $\dot{\bar{\varepsilon}}^{0.1}$ MPa

Upsetting is carried out in between two flat dies. The incremental step size is 2% of the initial height. An axisymmetric analysis is performed and the element type chosen is four noded quadrilateral elements. Total number of nodes and elements are taken to be 25 and 16 respectively. It may be mentioned that these numbers are not enough to obtain accurate results and have been taken only for demonstration purposes. In the mesh of the billet (Figure 13.2), elements are numbered from left to right in each row with rows starting from top of the mesh. Deformed mesh of the billet at die displacements of 0.4H_0 (H_0 = original height of cylinder) is compared in Figure 13.2 with the result available in Kobayashi *et al.* (1989). The present results of effective strain rate and the total effective strain calculated at the centroids of the elements at die displacement of 0.4Ho, are also compared as depicted in Table 13.1. A good agreement is observed between the two results.

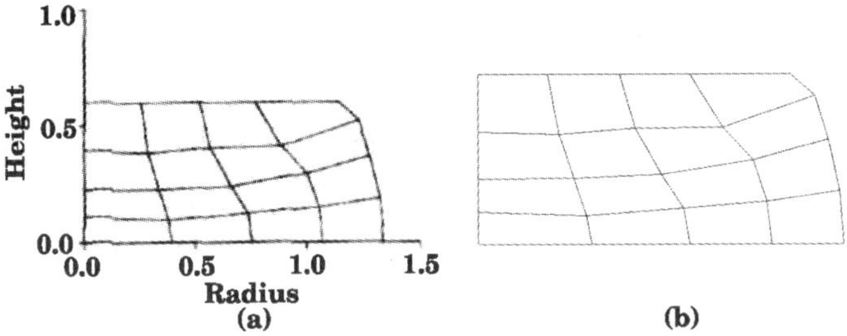

Fig. 13.2 FEM grid distortion after 40% reduction in height; (a) Kobayashi *et al.* (1989) and (b) SB-FEM result

Table 13.1. Comparison of results for upsetting with flat dies.

Element number	Effective strain rate		Total effective strain	
	Present	Kobayashi *et al.* (1989)	Present	Kobayashi *et al.* (1989)
1	1.11775	1.0693530	0.187169	0.1662418
2	0.996432	0.9206089	0.220211	0.1901851
3	1.41993	1.3531200	0.34968	0.3206123
4	1.69186	1.6726366	0.578763	0.5611510
5	1.8042	1.8394359	0.449507	0.4400065
6	1.5006	1.5083918	0.457069	0.4483966
7	2.02892	2.0198449	0.577624	0.5676523
8	1.43195	1.4386273	0.498418	0.4973681
9	2.05395	2.0823323	0.698132	0.7078742
10	2.10121	2.1647993	0.680723	0.6981489
11	1.98973	2.0083485	0.609882	0.6199824
12	1.05892	1.0691494	0.410613	0.4176614
13	2.17026	2.1942566	0.851245	0.8754001
14	2.38695	2.4754310	0.775419	0.8086128
15	1.7116	1.7497068	0.58448	0.6054953
16	1.0519	1.0600883	0.409731	0.4199022

13.7.2. Example 2: Spike forging

In spike forging a cylindrical billet is forged in an impression die containing a central cavity. The deformation characteristics of the spike forging are such that the portion of the material near the outside diameter flows radially, while the portion near the center of the top surface is extruded forming a spike. It has been reported that the material flow, which is characterized by spike height variation, depends on the interface friction as well as the geometries of dies and billets used.

The problem is an axisymmetric rigid-plastic problem and a schematic drawing is shown in Figure 13.3. The billet material used in this analysis has the yield stress (in MPa) given by the following equation:

$$\sigma_y = A\bar{\varepsilon}^{0.21}\dot{\bar{\varepsilon}}^{0.13}, \tag{13.47}$$

where A is a constant equal to $11.3 \exp\left(\frac{3140}{T}\right)$, T being the forging temperature taken to be 1030°C. The undeformed circular billet has the dimensions of 57.2 mm in height and 50.8 mm in diameter. The friction factor used in this analysis is m = 0.5. The velocity of the upper die used for the analysis is 25.4 mm/sec and the bottom die is stationary. The non steady state deformation of spike forging is analyzed in a step by step manner by treating it quasi-linearly during each incremental deformation. The incremental displacement of upper die for each step is chosen to be 0.02 times of the original workpiece height. The number of four noded axisymmetric quadrilateral elements is taken to be either 64 or 256.

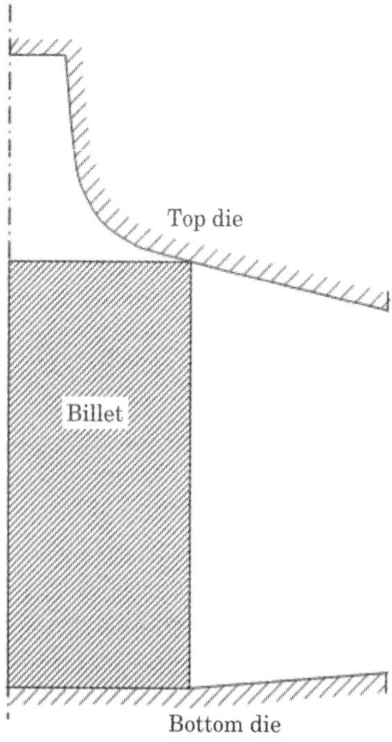

Fig. 13.3 Schematic drawing of spike forging dies and billet

The deformation, metal flow, effective strain rates and effective strains at different die displacements of $0.4H_0$, $0.6H_0$ and $0.72H_0$ are calculated, H_0 being the original height of the billet. The deformed FEM grid at die displacements of $0.6H_0$ is given in Figure 13.4 for the two different meshes. It is to be noted that only segments of dies required to define constraints and friction loading, are modeled and meshed here. Effective strain rate distributions in the grid at die displacements of $0.6H_0$ and $0.72H_0$ are plotted in Figure 13.5 while the effective strain distributions after die displacements of $0.6H_0$ and $0.72H_0$ are plotted in Figure 13.6.

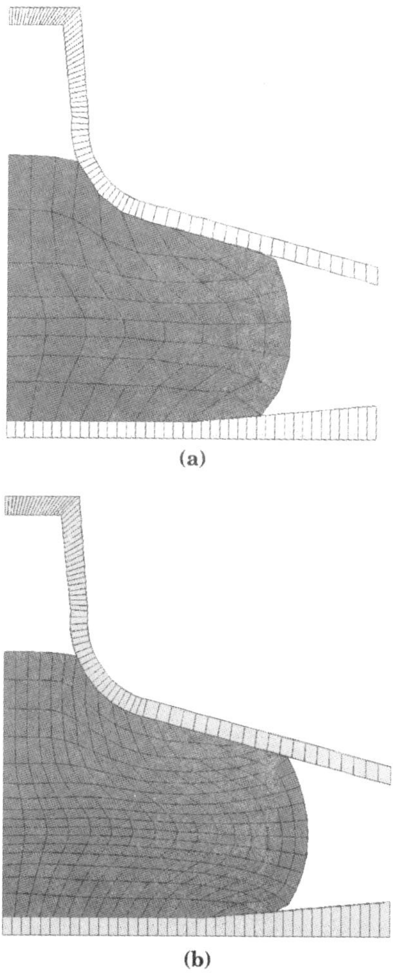

Fig. 13.4 Meshes of the workpiece and dies at a die displacement of $0.6H_0$ during spike forging of a cylindrical steel billet in an impression die ((a) and (b) : two different meshes)

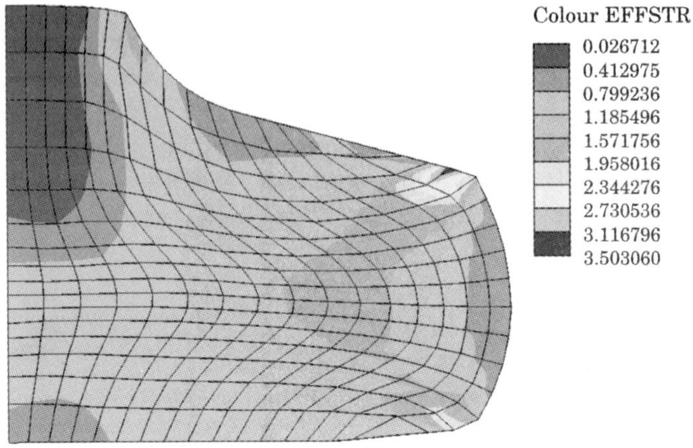

(a)

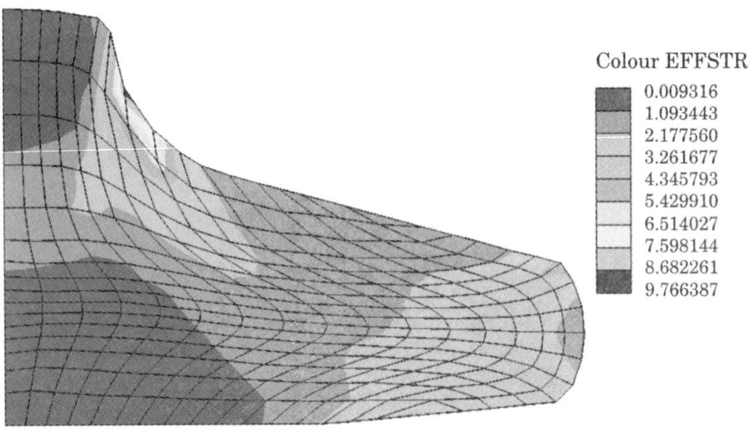

(b)

Fig. 13.5 Effective strain rate distributions at die displacements of (a) $0.6H_0$ and (b) $0.72\ H_0$ during spike forging of a cylindrical steel billet

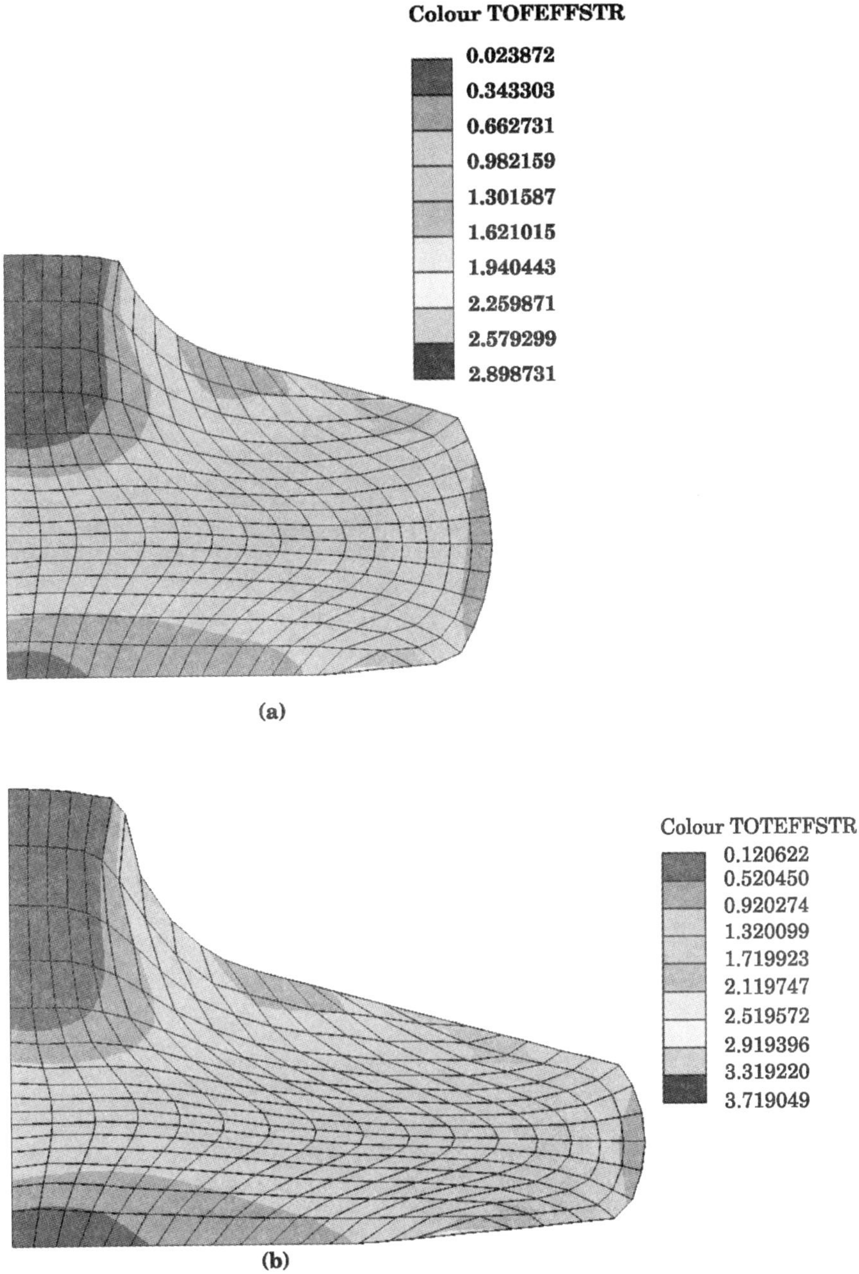

Fig. 13.6 Effective strain distributions after die displacements of (a) $0.6H_0$ and (b) $0.72\ H_0$ during spike forging of a cylindrical steel billet

The metal working finite element analysis results in severe distortions of mesh and it is essential to frequently refine the mesh or modify some elements during the solution phase. Appropriate technique has been incorporated in the SB-FEM code which performs remeshing and maps the data (nodal velocities, effective strains and temperatures) from the old mesh to the new mesh. Figure 13.7 depicts the effective strain rate distribution mapped to the new mesh generated by remeshing after the die displacement of $0.6H_0$.

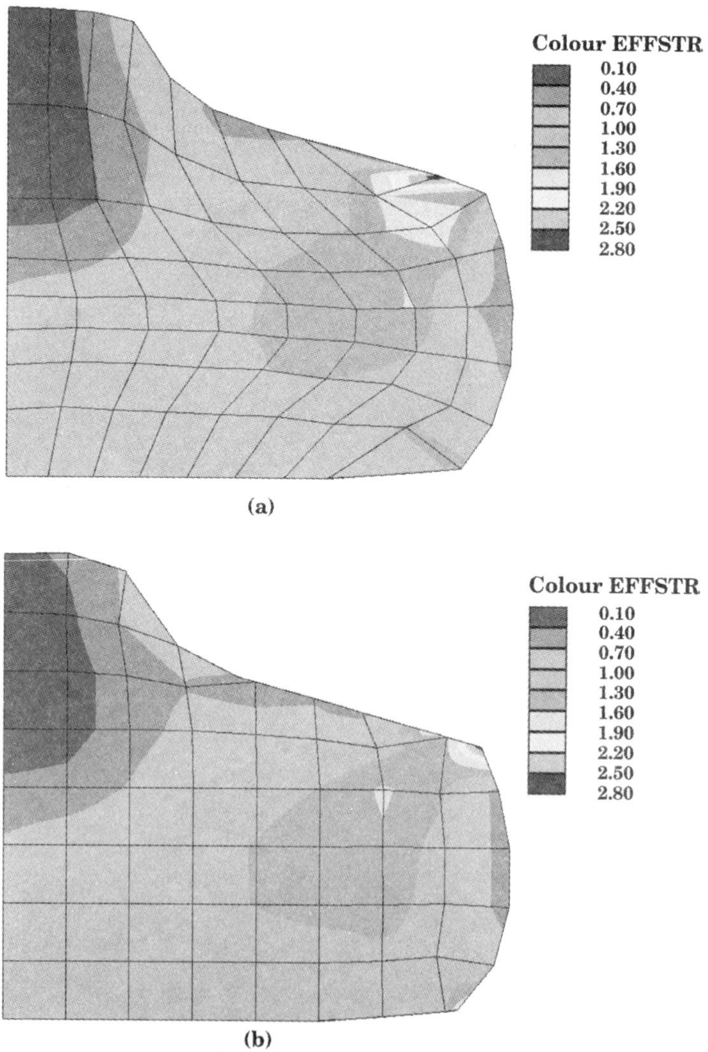

Fig. 13.7 Effective strain rate mapped to the new mesh after remeshing at die displacement of 0.6 H_0 during spike forging of a cylindrical steel billet ((a) : original mesh and (b) : new mesh)

In another investigation, a remeshing technique is employed in which the workpiece is remeshed at regular intervals to remove mesh distortion, and the old results for nodal velocities and effective strains are mapped to the new mesh. The final mesh after die displacement of $0.72H_0$ is depicted in Figure 13.8.

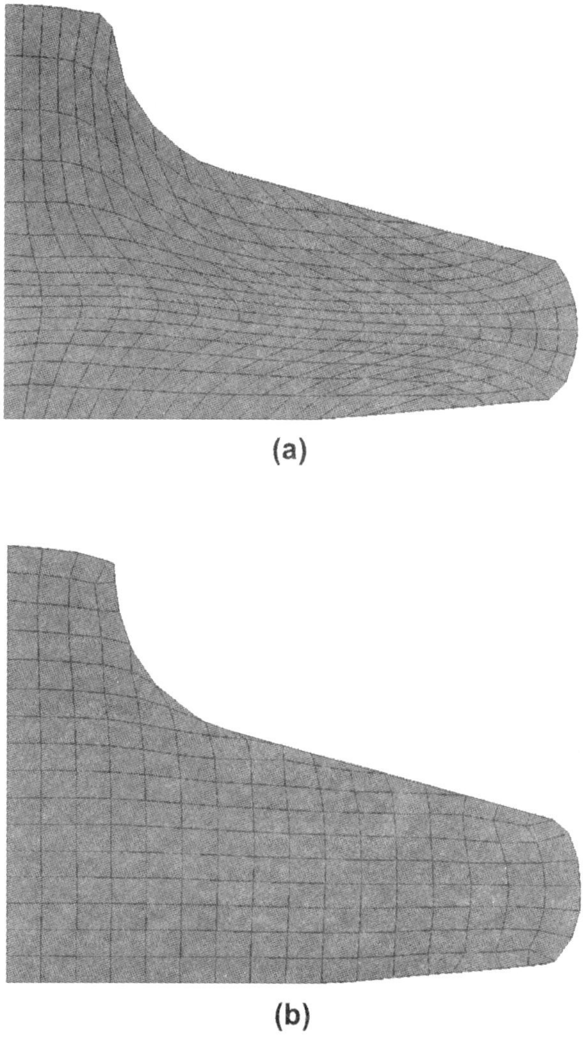

Fig. 13.8 Deformed meshes of the workpiece at a die displacement of $0.72H_0$ during spike forging: (a) without remeshing and (b) when remeshing technique is employed at regular intervals

13.7.3. Example 3: Plane strain rolling of a steel strip

The coupled deformation-thermal formulation is applied to solve an industrial problem of plane strain hot rolling of a steel strip. The flow stress (in MPa) of S45C (AISI 1045) stainless steel is given by the following equation:

$$\sigma_y = 85.6\bar{\varepsilon}^{0.21}\dot{\bar{\varepsilon}}^{0.13}. \tag{13.48}$$

The computational conditions are taken as follows:

Entry thickness of the strip : 8 mm,
Thickness reduction : 37.5 %,
Width of the strip : 20 mm,
Roll diameter : 100 mm,
Peripheral speed of roll : 100 mm/s,
Entry temperature : 1030 °C and
Atmospheric temperature : 20 °C.

Density, thermal conductivity and specific heat of the material are taken to be 7800 Kg/m³, 32 W/mK and 3600 kJ/m³K. Frictional factor is taken to be 0.5. Effective strain rate distribution in the strip at the simulation step when the strip has traveled full into the roll gap is plotted in Figure 13.9. The effective strain developed in the strip after the strip has traveled full into the roll gap is depicted in Figure 13.10.

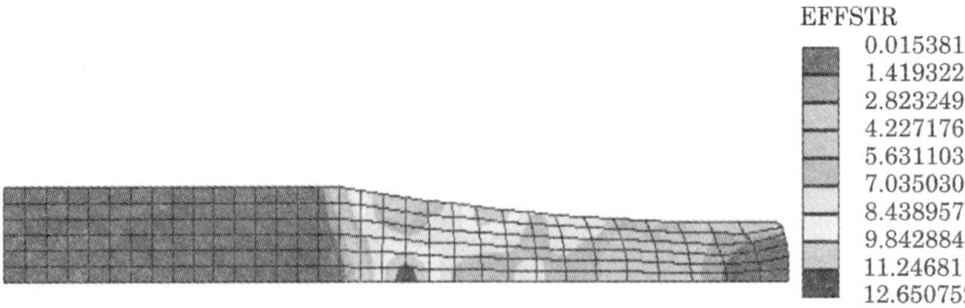

Fig. 13.9 Effective strain rate in the strip when the strip has traveled full into the roll gap

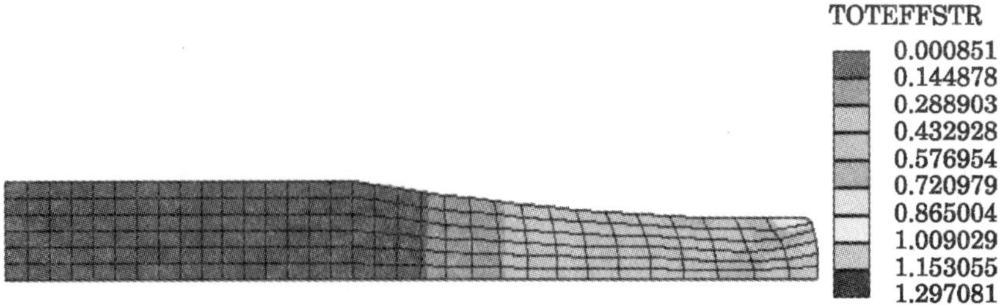

Fig. 13.10 Effective strain in the strip after the strip has traveled full into the roll gap

Temperature distribution

The temperature distribution of the strip during rolling is also calculated. The heat transfer coefficient between the roll and the strip and the roll temperature during contact are taken to be 19.561 kW/ m^2K and 650 °C respectively, although a detailed analysis of these two thermal parameters of the roll would be more appropriate. The temperature distribution (in °C) in the strip after the strip has traveled completely into the roll gap is plotted in Figure 13.11. In the figure, the work roll can be seen to cause considerable chilling of the surface; while at location much away from the roll, the temperature rises due to the plastic work.

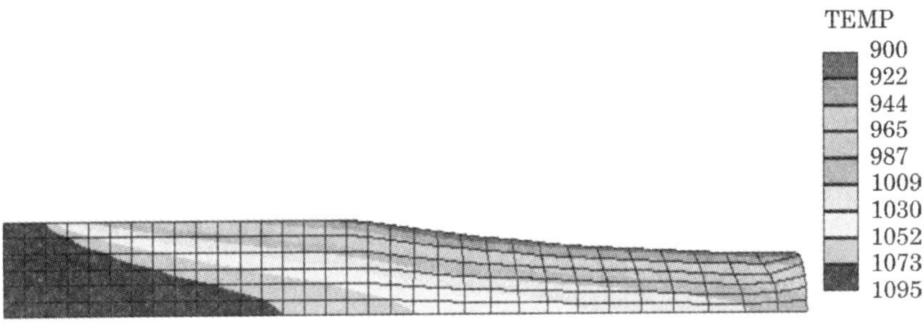

Fig. 13.11 Temperature distribution (in °C) in the strip after the strip has traveled full into the roll gap

13.7.4. Example 4: Analysis of an oval rod rolling process of low carbon steel

Shape rolling processes generally involve large strain rate inhomegeneities and are difficult to simulate. Since the process usually takes place in hot operation, three dimensional mechanical behavior coupled with heat transfer in and around the billet must be taken into account for accurate evaluation of the roll-pass system. In order to simplify the problem from three-dimensional to two-dimensional, a deformation mode method based on a generalized plane strain approach can be used (Xin *et al.*, 1991 and Glowacki *et al.*, 1992). In the deformation mode method, the three dimensional velocity is expressed by the combination of the finite element field on the cross-section of the billet and the deformation modes expressing the velocity component in the rolling direction. Further, the generalized plane strain approach assumes, in the first approximation, that the longitudinal strain is uniformly distributed across the sample. It leads to the zero shear strain rates connected with this direction. In consequence, the non-steady state finite-element solution is applied in the plane perpendicular to the direction of rolling and the plane is allowed to move with the actual sample's velocity.

As an example, an oval rod rolling process of low carbon steel on a two-high rolling mill is considered and the metal flow and heat transfer are simulated using the generalized plane strain approach. For simplicity, the thermal analysis of the rolls is ignored, and the heat transfer between the workpiece and the rolls is simplified (Kumar, 2010e). The following parameters are taken:

Roll diameter	: 200 mm,
Sample size	: 20×20×200 mm,
Initial temperature	: 1000 °C and
Rolling time	: 0.05 s (approximately).

Sample is rolled in the oval groove shown in Figure 13.12 with two different gaps between the rolls. The first has a gap (t) of 4.2 mm which leads to the final height of 13.2 mm with elongation coefficient 1.44. The second has a gap (t) of 1.5 mm which results in the final height of 10.5 mm and elongation coefficient 2.06. The flow stress of the steel is given by the following equation:

$$\sigma_y = k_1 \exp\left(m_1 T + m_4 \bar{\varepsilon}\right)\left(\dot{\bar{\varepsilon}}\right)^{m_2}\left(\bar{\varepsilon}\right)^{m_3}. \tag{13.49}$$

The coefficients in the strain hardening curve are: $k_1 = 4150.0$ MPa; m_1 = -0.00311/K; m_2 = 0.1202; m_3 = 0.2889; m_4 = -0.5006. Because of symmetry, only one quarter of the workpiece is considered for analysis.

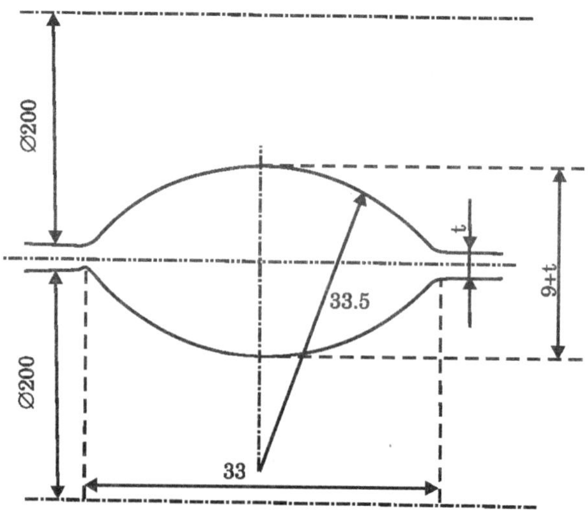

Fig. 13.12 Dimensions of the oval groove

The distributions of the effective strain rate on the sample's cross-section at the end of rolling and the effective strain after rolling are presented in Figures 13.13 and 13.14 respectively. Significant strain inhomogeneity can be observed in the sample. The temperature distribution in the exit plane of the rod is presented in Figure 13.15 which depicts a considerable temperature gradient towards the contact surface of the rolls.

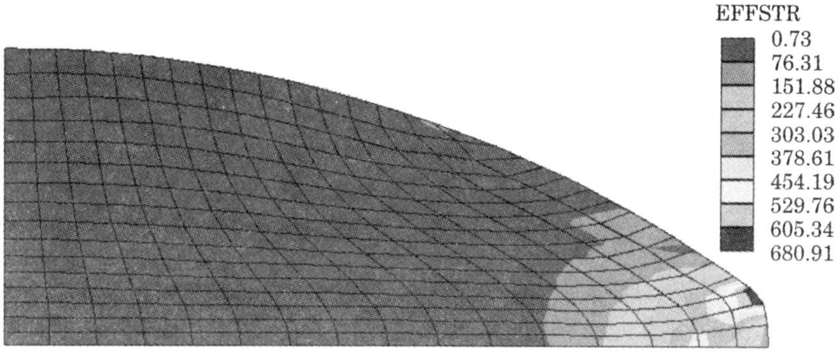

Fig. 13.13 Effective strain rate in the cross-section of the workpiece in the oval groove at the end of rolling

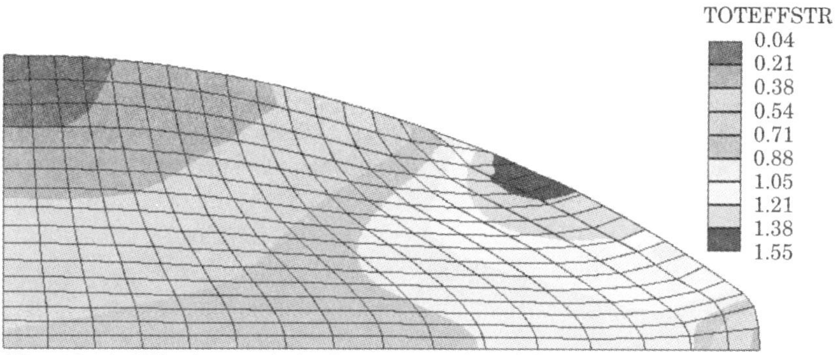

Fig. 13.14 Effective strain in the cross-section of the workpiece in the oval groove after rolling

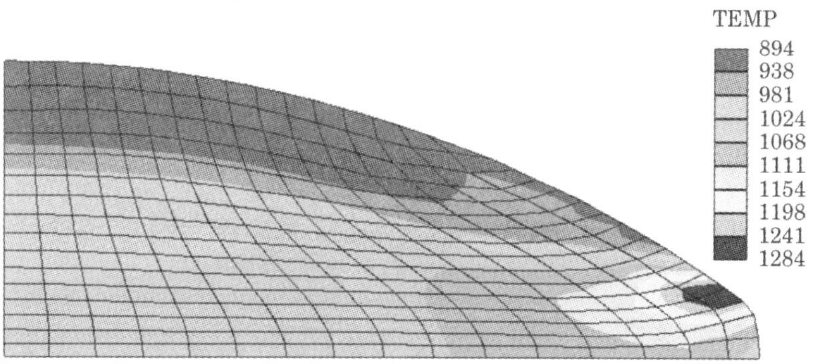

Fig. 13.15 Temperature distribution in the cross-section of the workpiece in the oval groove after rolling

Bibliography

1. Ahmad, S., Irons, B.M. and Zienkiewicz, O.C., 1970. Analysis of thick and thin shell structures by curved finite elements. *Int. J. Numer. Methods Eng.*, 2: 419–451.

2. Argyris, J.H. and Kelsey, S., 1960. Energy Theorems and Structural Analysis, Butterworths, London (Part 1 reprinted from Aircraft Engrg. 26 Oct-Nov, 1954 and 27 April-May, 1955).

3. Argyris, J.H., Balmer, H., Klieber, M. and Hindenlang, U., 1980. Natural description of large inelastic deformations for shells of arbitrary shape - application of Trump element. *Comp. Meth. in Appl. Mech. and Eng.*, 22: 361–389.

4. Backlund, J. and Wennerstrom, H., 1974. Finite element analysis of elasto-plastic shells. *Int. J. Numer. Methods Eng.*, 8: 415–424.

5. Barlow, J., 1976. Optimal stress location in finite element method. *Int. J. Numer. Methods Eng.*, 10: 243–51.

6. Bathe, K.J. and Wilson, E.L., 1978. Numerical Method in Finite Element Analysis, Prentice Hall of India Limited, New Delhi.

7. Bathe, K.J., Dvorkin, E. and Ho, L.W., 1983. Our discrete-Kirchhoff and isoparametric element for non-linear analysis - an assessment. *Comput. Struct.*, 16(1–4): 89–98.

8. Brusa, L. and Nigro, L., 1980. A one-step method for direct integration of structural dynamic equations. *Int. J. Numer. Methods Eng.*, 15: 685–699.

9. Chen, C.C. and Kobayashi, S., 1978. Rigid-plastic finite element analysis of ring compression. Applications of Numerical Methods to Forming Processes, *ASME, AMD*, 28: 163.

10. Choi, H.Y. and Chang, F.K., 1992. A model for predicting damage in graphite/ epoxy laminated composites from low-velocity point impact. *J. Compos. Mater.*, 26: 2134–2169.

11. Clough, R.W., 1960. The finite element method in plane stress analysis. J. Struct. Div., Proc. 2nd ASCE Conf. Electronic computation, Pittsburgh, pa., pp. 345–378.

12. Clough, R.W. and Johnson, C.P., 1968. A finite element approximation for the analysis of thin shells. *Int. J. Solids Struct.*, 4.

13. Courant, R., 1943. Variational methods for the solution of problems of equilibrium and vibration. *Bull. Am. Math. Soc.*, 49: 1–43.
14. Cowper, G.R., 1973. Gaussian Quadrature formulas for triangles. *Int. J. Numer. Methods Eng.*, 7: 405–408.
15. Dym, C.L. and Shames, I.H., 1973. Solid Mechanics – a Variational Approach, International student *edn.*, McGraw-Hill Kogahusha Ltd.
16. Ewing, D.J.F., 1967. A series method for constructing plastic slip line fields. *J. Mech. Phys. Solids*, 15: 105–114.
17. Forry, M., 1968. Variational Calculus in Science and Engineering, McGraw-Hill, New York.
18. Gallagher, R.H., 1975. Finite Element Analysis-Fundamentals, Prentice Hall Inc., Englewood Cliffs, N.J.
19. Gallagher, R.H., 1976. Problems and progress in thin shell finite element analysis. Finite Element for Thin Shells and Curved Members, Ashwell, D.G. and Gallagher, R.H. (*eds.*), pp. 1–14.
20. Glowacki, M., Kedzierski, Z., Kusiak, H., Madej, W. and Pietrzyk, M., 1992. Simulation of metal flow, heat transfer and structure evolution during hot rolling in square-oval-square series. *J. Mater. Process. Technol.*, 34: 509–516.
21. Goldsmith, W., 1960. Impact: The Theory and Physical Behavior of Colliding Solids, Edward Arnold Ltd., London.
22. Hammer, P.C., Marlowe, O.P. and Stroud, A.H., 1956. Numerical integration over simplexes and cones. *Math. Tables Aids Comp.*, 10: 130–137.
23. Harris, C.O., 1959. Introduction to Stress Analysis, The Macmillan Co., New York.
24. Hill, R., 1963. A general method of analysis of metal working processes. *J. Mech. Phys. Solids*, 11: 305.
25. Hinton, E. and Campbell, J., 1974. Local and global smoothing of discontinuous finite element functions using a least square method. *Int. J. Numer. Methods Eng.*, 8: 461–80.
26. Hinton, E., Scott, F.C. and Ricketts, R.E., 1975. Local least squares stress smoothing for parabolic isoparametric elements. *Int. J. Numer. Methods Eng.*, 9: 235–56.
27. Hinton, E. and Owen, D.R.J., 1977. Finite Element Programming, Academic press, London.
28. Hinton, E. and Owen, D.R.J., 1984. Finite Element Software for Plates and Shells, Pineridge press, U.K.
29. Ho-Le, K., 1988. Finite element mesh generation methods: a review and classification. *Computer-Aided Design*, 20(1).
30. Hrenikoff, A., 1941. Solution of problems in elasticity by the framework method. *J. Appl. Mech., Trans. ASME*, 8: 169–175.
31. Idelsohn, S., 1981. On the use of deep, shallow or flat shell finite elements for the analysis of thin shell structures. *Comp. Meth. in Appl. Mech. and Eng.*, 26: 321–330.
32. Irons, B.M., 1970. A frontal solution program. *Int. J. Numer. Methods Eng.*, 2: 5–32.

33. Johnson, W., Sowerby, R. and Haddow, J.B., 1970. Plane strain slip-line fields, American Elsevier, New York.
34. Johnson, W., Sowerby, R. and Venter, R.D., 1982. Plane strain slip-line fields for metal deformation processes, Pergamon press, Oxford.
35. Kakac, S. and Yener, Y., 1985. Heat Conduction, Second *edn.*, Hemisphere publishing corporation, Washington.
36. Kobayashi, S., 1982. A review on the finite element method and metal forming process modelling. *J. Appl. Metal Working*, Vol. 2, p. 163.
37. Kobayashi, S., 1985. Metal forming and the finite element method-past and future. Proceeding of the 25th Int. Conf. Mach Tool Des. Res., Birmingham, p. 17.
38. Kobayasi, S., Oh, S.I. and Altan, T., 1989. Metal Forming and Finite Element Method, Oxford University Press, New York.
39. Krishnamoorthy, C.S., 1994. Finite Element analysis: Theory and Programming, Second *edn.*, Tata McGraw-Hill, New Delhi.
40. Kumar, S., 1998. Finite element analysis of impact-induced deformations, stresses and damages in composite laminates. Ph.D. thesis, Indian Institute of Technology, Kharagpur.
41. Kumar, S., 2005. Heat transfer analysis and estimation of refractory wear in an iron blast furnace hearth using finite element method, *ISIJ Int.*, 45(8): 1122–1128.
42. Kumar, S., Rao, B.N. and Pradhan, B., 2007. Effect of impactor parameters and laminate characteristics on impact response and damage in curved composite laminates. *J. Reinf. Plast. Comp.*, 26(13): 1273–1290.
43. Kumar, S., 2008. Analysis of impact response and damage in laminated composite shell involving large deformation and material degradation. *J. Mech. Mater. Struct.*, 3(9): 1741–1756.
44. Kumar, S., 2010a. Three-dimensional non-linear finite element analysis of impact response and damage in laminated composite shells. *Indian Journal of Aerospace Sciences and Technologies*, 62(2): 109–121.
45. Kumar, S., 2010b. Analysis of impact response and damage in laminated composite cylindrical shells undergoing large deformations, *Struct. Eng. Mech.*, 35(3): 349–364.
46. Kumar, S., 2010c. Object-oriented finite element programming for engineering analysis in C^{++}, *J. Software*, 5(7): 689–696.
47. Kumar, S., 2010d. Object-oriented finite element analysis of metal working processes, *J. Software Engg. Applications*, 3(6): 572–579.
48. Kumar, S., 2010e. Finite element modeling of thermomechanical behavior and microstructural evolution in steel during hot deformation process. Project Report No. SR/ FTP/ETA-31/2005, DST, New Delhi.
49. Kumar, S., 2011a. URL: http://www.softbodh.org.
50. Kumar, S., 2011b. Use of unsymmetric finite element method in impact analysis of composite laminates, *Finite Elements Anal. Des.*, 47(4): 373–377.
51. Liyanapathirana, D.S., Deeks, A.J. and Randolph, M.F., 2000. Application of stress smoothing methods to a finite element pile driving analysis. *Eng. Comput.*,

17(3): 218–234.

52. Noor, A.K., 1985. Hybrid analytical technique for nonlinear analysis of structures. *AIAA J.*, 23: 938–946.

53. Noor, A.K., 1991. Bibliography of books and monographs on finite element technology. *Applied Mech. Rev.*, 44(6).

54. Pagano, N.J., 1970. Exact solutions for rectangular bidirectional composites and sandwich plates. *J. Compos. Mater.*, 4: 20–34.

55. Pandya, B.N. and Kant, T., 1987. A consistent refined theory for flexure of a symmetric laminate. *Mech. Res. Commun.*, 14(2): 107–113.

56. Pawsey, S.F. and Clough, R.W., 1971. Improved numerical integration of thick shell finite elements. *Int. J. Numer. Methods Eng.*, 3: 575–586.

57. Pietrzyk, M., 2000. Finite-element simulation of large plastic deformation. *J. Mater. Process. Technol.*, 106(1–3): 223–229.

58. Pradhan, B. and Kumar, S., 2000. Finite element analysis of low-velocity impact damage in composite laminates. *J. Reinf. Plast. Comp.*, 19(4): 322–339.

59. Prager, W. and Hodge, P.G., 1941. Theory of Perfectly Plastic Solids, Wiley, New York.

60. Rao, S.S., 2004. The Finite Element Method in Engineering, Fourth *edn.*, Elsevier publisher.

61. Reddy, J.N. and Chandrashekhara, K., 1985. Nonlinear analysis of laminated shells including transverse shear strains. *AIAA J.*, 23: 440–441.

62. Reddy, J.N., 2004. An Introduction to Non-Linear Finite Element Analysis, Oxford University Press, Oxford.

63. Roark, R.J. and Young, W.C., 1989. Formulas for Stress and Strain, 6th *edn.*, McGraw-Hill, New York.

64. Sanders, Jr. J.L. 1959. An improved first approximation theory for thin shells. NASA Report R-24.

65. Shephard, M.S., 1988. Approaches to the automatic generation and control of finite element meshes. *Appl. Mech. Rev.*, 41(4).

66. Sokolnikoff, I.S., 1956. Mathematical Theory of Elasticity, McGraw Hill, New York.

67. Stavsky, Y., 1961. Bending and stretching of laminated aelotropic plates. *J. Eng. Mechanics, ASCE*, 87(EM6): 31–56.

68. Stein, M., 1986. Nonlinear theory for plates and shells including the effects of transverse shearing. *AIAA J.*, 24: 1538–1544.

69. Taylor, R.L., Beresford, P.J. and Wilson, E.L., 1976. A non-conforming element for stress analysis. *Int. J. Numer. Methods Eng.*, 10: 1211–1219.

70. Thomsen, E.G., Yang, C.T. and Bierbower, J.B., 1954. An experimental investigation of the mechanical of plastic deformation of metals. *Univ. California Pub. Engg.*, 5.

71. Timoshenko, S.P. and Goodier, J.N., 1934. Theory of Elasticity, Third *edn.*, McGraw Hill, New York.

72. Timoshenko, S., 1956. Strength of Materials, Part-II, Advanced Theory and Problems, Third *edn.*

73. Timoshenko, S. and Woinowsky-Krieger, S., 1959. Theory of Plates and Shells, Second edn., McGraw Hill, New York.

74. Turner, M.J., Clough, R.W., Martin, H.C. and Topp, L.J., 1956. Stiffness and deflection analysis of complex structures. *J. Aero. Sci.*, 23: 805–824.

75. Wagoner, R.H. and Chenot, J.L. 1997. Fundamentals of Metal Forming, John Wiley & Sons, Inc., New York.

76. Washizu, K., 1965. Variational Methods in Elasticity and Plasticity, Second *edn.*, Pergamon Press.

77. Wilson, E.L., Taylor, R.L., Doherty, W.P. and Ghaboussi, J., 1973. Incompatible displacement models. *In*: Fenves, S.J., Perrone, N., Robinson, A.R., and Schnobrich, W.C., (eds.), *Numerical and Computer Methods in Structural Mechanics*, Academic Press, pp. 43–57.

78. Wood, W.L., Bossak, M. and Zienkiewicz, O.C., 1980. An alpha modification of Newmark's method. *Int. J. Numer. Methods Eng.*, 15: 1562–1566.

79. Wood, W.L., 1984. A further look at Newmark, Houbolt, etc., time-stepping formulae. *Int. J. Numer. Methods Eng.*, 20: 1009–1017.

80. Wu, H.T. and Chang, F.K., 1989. Transient dynamic analysis of laminated composite plates subjected to transverse impact. *Comput. Struct.*, 31: 453–466.

81. Xin, P., Aizawa, T. and Kihara, J., 1991. Roll pass evaluation for hot shape rolling processes. *J. Mater. Process. Technol.*, 27: 163–178.

82. Yang, S.H. and Sun, C.T., 1982. Indentation law for composite laminates, *ASTM STP*, 787: 425–449.

83. Zienkiewicz, O.C., Taylor, R.L. and Too, J.M., 1971. Reduced integration technique in general analysis of plates and shells, *Int. J. Numer. Methods Eng.*, 3: 275–290.

84. Zienkiewicz, O.C., 1977. The Finite Element Method, Third *edn.*, McGraw-Hill Limited, London.

85. Zienkiewicz, O.C. and Taylor, R.L., 1989. The Finite Element Method, Vol. 1, Basic formulation and linear problems, McGraw-Hill (U.K.) Limited.

Appendix-1

Some Mathematical Theorems

A1.1. THE GREEN'S THEOREM

Green's theorem is basically integration by parts in two or three-dimensions and is stated as follows:

Let V be a finite domain with boundary S, where f and g are two scalar functions defined on V. Then for any coordinate x_k (k = 1,2, or 3) corresponding to a position vector $\bar{x}$, we have

$$\int_V f\frac{\partial g}{\partial x_k}dV = \int_S fgn_k dS - \int_V \frac{\partial f}{\partial x_k}gdV, \tag{A1.1}$$

where n_k (k = 1, 2, and 3) is the kth component of the outward normal to S.

The proof can be derived as follows.

For simplicity, we assume that a straight line parallel to any of the $\hat{x}_k$ axis intersects the boundary S of the domain in, at most, two points as shown in Figure A1.1 for $k = 3$. A curve C is now defined in $\hat{x}_3$ plane which decompose the boundary S into two parts: S_U (upper) and S_L (lower). The curve is defined so that for any point $p(x_1, x_2)$ on the curve, a straight line containing p and parallel to $\hat{x}_3$ axis intersects S_U at point p_U (x_1, x_2, x_{U3}) and S_L at point p_L (x_1, x_2, x_{L3}).

The x_3 coordinates at point p_U and p_L can be written as

$$x_{U3} = h_U\left(x_1, x_2\right) \text{ and } x_{L3} = h_L\left(x_1, x_2\right). \tag{A1.2}$$

With this notation, the left-hand side of Eq. (A1.1) can be rewritten as

$$\int_V f \frac{\partial g}{\partial x_3} dV = \int_C dx_1 dx_2 \int_{h_L(x_1,x_2)}^{h_U(x_1,x_2)} f \frac{\partial g}{\partial x_3} dx_3. \tag{A1.3}$$

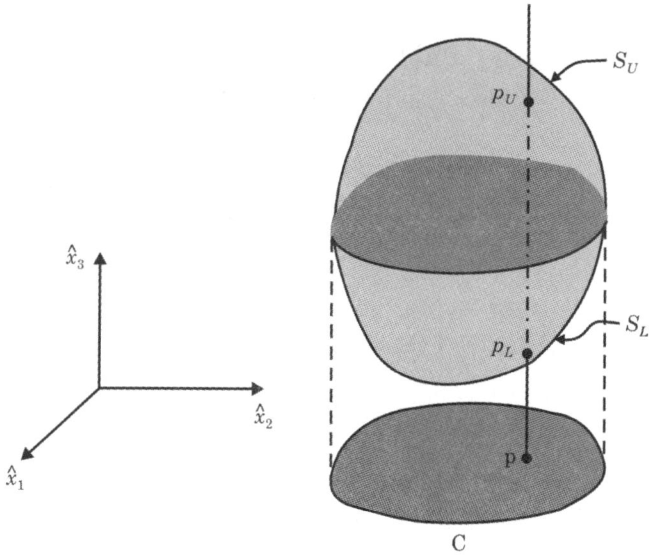

Fig. A1.1 Proof of the Green theorem

In the above equation, the one-dimensional integral can be transformed using integration by parts with respect to x_3 according to

$$\begin{aligned}\int_{h_L(x_1,x_2)}^{h_U(x_1,x_2)} f \frac{\partial g}{\partial x_3} dx_3 &= [fg]_{h_L(x_1,x_2)}^{h_U(x_1,x_2)} - \int_{h_L(x_1,x_2)}^{h_U(x_1,x_2)} \frac{\partial f}{\partial x_3} g dx_3 \\ &= f_U g_U - f_L g_L - \int_{h_L}^{h_U} \frac{\partial f}{\partial x_3} g dx_3 .\end{aligned} \tag{A1.4}$$

In the foregoing equation, the terms f_U and f_L are used to denote the following expressions:

$$f_U = f\left[x_1, x_2, h_U(x_1, x_2)\right] \text{ and } f_L = f\left[x_1, x_2, h_L(x_1, x_2)\right].$$

The notations g_U and g_L also represent the similar expressions.

Now, using Eq. (A1.4), Eq. (A1.3) can be transformed into

$$\int_V f \frac{\partial g}{\partial x_3} dV = \int_C (f_U g_U - f_L g_L) dx_1 dx_2 - \int_C dx_1 dx_2 \int_{h_L}^{h_U} \frac{\partial f}{\partial x_3} g dx_3. \tag{A1.5}$$

If the normal to S is oriented outward (Figure A1.2), we can write on S_U

$$dx_1 dx_2 = \cos\gamma_U dS = n_{U3} dS, \tag{A1.6}$$

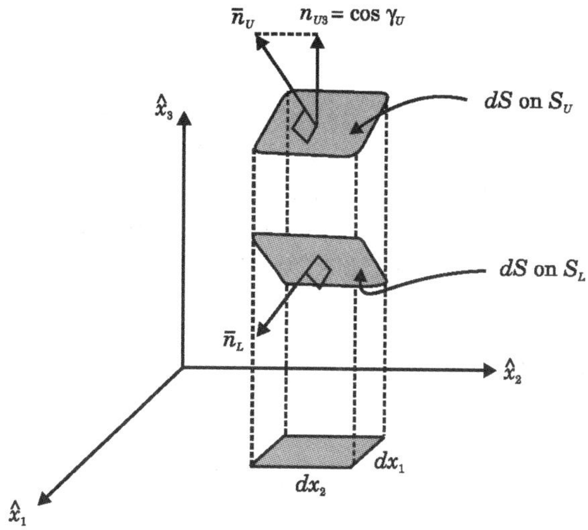

Fig. A1.2 Projection of the differential surface on the $\hat{x}_3$ plane

which means that the projection of the differential surface dS onto the $\hat{x}_3$ plane involves the cosine of the normal $\bar{n}_U$ with the $\hat{x}_3$ axis.

Similarly, the corresponding relation for the lower surface may be written as

$$dx_1 dx_2 = -\cos\gamma_L dS = -n_{L3} dS. \tag{A1.7}$$

Substituting from Eqs. (A1.6) and (A1.7), first integral appearing in the right-hand side of Eq. (A1.5) becomes

$$\int_C (f_U g_U - f_L g_L) dx_1 dx_2 = \int_{S_U} f_U g_U n_{U3} dS + \int_{S_L} f_L g_L n_{L3} dS = \int_S f g n_3 dS. \tag{A1.8}$$

The second integral in the right-hand side of Eq. (A1.5) can be rewritten as a 3-D integral as

$$\int_C dx_1 dx_2 \int_{h_L}^{h_U} \frac{\partial f}{\partial x_3} g dx_3 = \int_V \frac{\partial f}{\partial x_3} g dV. \tag{A1.9}$$

Putting Eqs.(A1.8) and (A1.9) into Eq. (A1.5), we see that the resulting equation is equivalent to Eq. (A1.1), where k is chosen equal to 3.

The above relation was established based on the assumption that a straight line, parallel to any of the $\hat{x}_k$ axis, intersects the boundary in at most two points. To complete the proof, one has to observe that if the straight line intersects the boundary in more than two points, the domain can be decomposed into several sub-domains for which the above assumption is valid and Eq. (A1.1) can be written for each subdomain. Now, it is only to be noted that for those parts of the boundary inside the domain and therefore belonging to two adjacent subdomains, the surface integrals vanish as they correspond to opposite normals.

A1.2. DIVERGENCE THEOREM

The divergence theorem may be stated as follows:

For any vector field $\bar{u}$ defined in V with boundary S, the following equation holds:

$$\int_V \nabla.\bar{u}dV = \int_S \bar{u}.\bar{n}dS. \tag{A1.10}$$

The proof is easily obtained by using Green's theorem. If $f = 1$ and $g = u_k$ are put into Eq. (A1.1), we get

$$\int_V \frac{\partial u_k}{\partial x_k} dV = \int_S u_k n_k dS. \tag{A1.11}$$

In the above equation, no summation is made on the k index, but if we sum over it, we obtain

$$\int_V \sum_k \frac{\partial u_k}{\partial x_k} dV = \int_S \sum_k u_k n_k dS.. \tag{A1.12}$$

Eq. (A1.12) is the component form of Eq. (A1.10), presented with the vector notation. This special derivation of Green's theorem is called divergence theorem.

Writing in matrix form, we have

$$\int_V \{\Delta\}^T \{u\} dV = \int_S \{n\}^T \{u\} dS, \tag{A1.13}$$

where

$$\{\Delta\} = \left[\frac{\partial}{\partial x_1} \quad \frac{\partial}{\partial x_2} \quad \frac{\partial}{\partial x_3}\right]^T. \tag{A1.14}$$

A1.3. ORTHOGONAL CURVILINEAR COORDINATE SYSTEM

In finite element method, in several instances, the element volume or element area (in Cartesian space) with respect to which the integration is made must be transformed to the local curvilinear coordinates (Figure A1.3) defined for the element. This is done using a usual procedure as discussed below.

Suppose that the Cartesian coordinates (x, y, z) are expressed in terms of the orthogonal curvilinear coordinates (ξ, η, ζ) by the equations,

$$x = x(\xi,\eta,\zeta)$$

$$y = y(\xi,\eta,\zeta)$$

$$z = z(\xi,\eta,\zeta), \tag{A1.15}$$

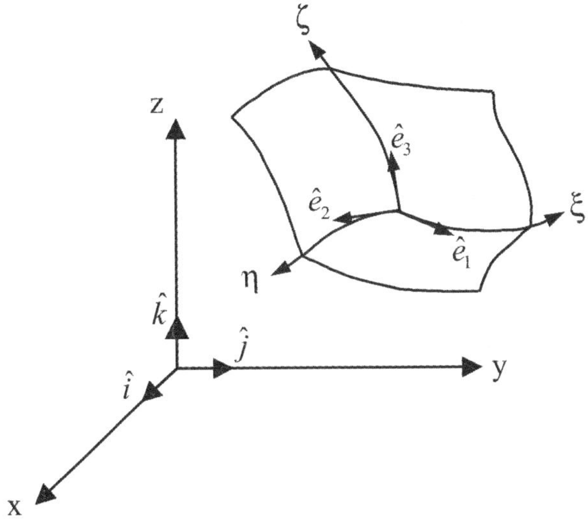

Fig. A1.3 Curvilinear coordinate system

where it is assumed that the correspondence is unique and that the inverse mapping exists.

By definition, a vector tangent to the ξ curve is given by

$$\bar{r}(\xi) = \frac{\partial x}{\partial \xi}\hat{i} + \frac{\partial y}{\partial \xi}\hat{j} + \frac{\partial z}{\partial \xi}\hat{k}. \tag{A1.16}$$

The proof is simple and as follows:

Let $P(x,y,z)$ and $Q(x+\delta x, y+\delta y, z+\delta z)$ be two neighboring points on the ξ curve. Let the arc PQ be $\delta\xi$ and the chord PQ be δc, so that $\lim_{Q\to P}\left(\frac{\delta\xi}{\delta c}\right) = 1$. The direction cosines of chord PQ are : $\frac{\delta x}{\delta c}, \frac{\delta y}{\delta c}, \frac{\delta z}{\delta c}$ *i.e.* $\frac{\delta x}{\delta \xi}\cdot\frac{\delta\xi}{\delta c}, \frac{\delta y}{\delta \xi}\cdot\frac{\delta\xi}{\delta c}, \frac{\delta z}{\delta \xi}\cdot\frac{\delta\xi}{\delta c}$.

When $\delta\xi \to 0, Q \to P$ and PQ tends to a tangent line PT. So, the direction cosines of the tangent to the curve PT are $\frac{\partial x}{\partial \xi}, \frac{\partial y}{\partial \xi}, \frac{\partial z}{\partial \xi}$.

So, an elementary (differential) vector $\overline{d\xi}$ at a point along the ξ curve is given by

$$\overline{d\xi} = \bar{r}(\xi)\, d\xi, \tag{A1.17}$$

where $d\xi = \left|\overline{d\xi}\right|$ is the magnitude of vector $\overline{d\xi}$.

Writing in matrix form elementary vectors for all the three curves, we have

$$\overline{d\xi} = \begin{Bmatrix} \dfrac{\partial x}{\partial \xi} \\ \dfrac{\partial y}{\partial \xi} \\ \dfrac{\partial z}{\partial \xi} \end{Bmatrix} d\xi, \quad \overline{d\eta} = \begin{Bmatrix} \dfrac{\partial x}{\partial \eta} \\ \dfrac{\partial y}{\partial \eta} \\ \dfrac{\partial z}{\partial \eta} \end{Bmatrix} d\eta, \quad \overline{d\zeta} = \begin{Bmatrix} \dfrac{\partial x}{\partial \zeta} \\ \dfrac{\partial y}{\partial \zeta} \\ \dfrac{\partial z}{\partial \zeta} \end{Bmatrix} d\zeta. \tag{A1.18}$$

A1.3.1 Area and volume integrals

If we consider only a two dimensional plane, we have

$$\overline{d\xi} = \begin{Bmatrix} \dfrac{\partial x}{\partial \xi} \\ \dfrac{\partial y}{\partial \xi} \end{Bmatrix} d\xi, \ \overline{d\eta} = \begin{Bmatrix} \dfrac{\partial x}{\partial \eta} \\ \dfrac{\partial y}{\partial \eta} \end{Bmatrix} d\eta. \tag{A1.19}$$

An elementary surface area in the two dimensional plane can be calculated now. As the length of the vector resulting from a cross product of $\overline{d\xi} \times \overline{d\eta}$ is equal to the area of the elementary parallelogram, we can write

$$d(\text{Area}) = \left|\overline{d\xi} \times \overline{d\eta}\right| = \det \begin{bmatrix} \dfrac{\partial x}{\partial \xi} & \dfrac{\partial x}{\partial \eta} \\ \dfrac{\partial y}{\partial \xi} & \dfrac{\partial y}{\partial \eta} \end{bmatrix} .d\xi d\eta = \det\left[J\right] d\xi d\eta. \tag{A1.20}$$

Similarly, an elementary volume is given by the 'triple scalar' or box product as

$$d(\text{Vol}) = \overline{d\xi} \bullet \left(\overline{d\eta} \times \overline{d\zeta}\right) = \det \begin{bmatrix} \dfrac{\partial x}{\partial \xi} & \dfrac{\partial x}{\partial \eta} & \dfrac{\partial x}{\partial \zeta} \\ \dfrac{\partial y}{\partial \xi} & \dfrac{\partial y}{\partial \eta} & \dfrac{\partial y}{\partial \zeta} \\ \dfrac{\partial z}{\partial \xi} & \dfrac{\partial z}{\partial \eta} & \dfrac{\partial z}{\partial \zeta} \end{bmatrix} \cdot d\xi d\eta d\zeta \tag{A1.21}$$

$$= \det\left[J\right] d\xi d\eta d\zeta$$

This follows simply from the geometry. The bracketed vector product, by definition, forms a vector whose length is equal to the parallelogram area with sides tangent to two of the co-ordinates. The second scalar product (multiplication by a length and cosine of the angle between that length and the normal to the parallelogram) establishes the volume of an elementary parallelepiped.

In Eqs. (A1.20) and (A1.21), [*J*] is the Jacobian matrix.

A1.3.2 Surface integrals

In finite element method, surface integrals frequently occur over the surface of an element. In order to evaluate these integrals, the element surface in Cartesian space must be mapped to the curvilinear coordinate system.

In three-dimensional space, an elementary (differential) surface dS will generally lie on a surface where one of the curvilinear coordinates (say ζ) is constant. In that case, the elementary surface area will be equal to the length of the vector resulting from a cross product of $\overline{d\xi} \times \overline{d\eta}$. Thus $dS = \left|\overline{d\xi} \times \overline{d\eta}\right|$, where

$$\overline{d\xi} \times \overline{d\eta} = \begin{Bmatrix} \dfrac{\partial x}{\partial \xi} \\ \dfrac{\partial y}{\partial \xi} \\ \dfrac{\partial z}{\partial \xi} \end{Bmatrix} \times \begin{Bmatrix} \dfrac{\partial x}{\partial \eta} \\ \dfrac{\partial y}{\partial \eta} \\ \dfrac{\partial z}{\partial \eta} \end{Bmatrix} d\xi d\eta = \det \begin{bmatrix} \hat{i} & \hat{j} & \hat{k} \\ \dfrac{\partial x}{\partial \xi} & \dfrac{\partial y}{\partial \xi} & \dfrac{\partial z}{\partial \xi} \\ \dfrac{\partial x}{\partial \eta} & \dfrac{\partial y}{\partial \eta} & \dfrac{\partial z}{\partial \eta} \end{bmatrix} d\xi d\eta. \quad \text{(A1.22)}$$

For two dimensions an elementary line length dl will have a magnitude equal to $\left|\overline{d\xi}\right|$ on constant η surface (edge), where

$$\overline{d\xi} = \begin{Bmatrix} \dfrac{\partial x}{\partial \xi} \\ \dfrac{\partial y}{\partial \xi} \end{Bmatrix} d\xi. \quad \text{(A1.23)}$$

Appendix-2

Basic Equations of Structural Mechanics

A2.1. STRESS AND STRAIN

A2.1.1. Stress

(a) The stresses shown in Figure A2.1 are positive. σ_x, σ_y, σ_z are normal stresses and $\tau_{xy} = \tau_{yx}$, $\tau_{yz} = \tau_{zy}$, $\tau_{xz} = \tau_{zx}$ are shear stresses.

(b) In stress τ_{xy}, x is the plane on which the shear stress acts and y is the axis to which the stress is parallel.

(c) Normal stresses are positive, if they point in the direction of the outward normal. So, tensile stresses are positive and compressive stresses are negative.

(d) A positive shearing stress points in the positive direction of the coordinate axis of the second subscript if it acts on a surface with an outward normal in the positive direction. Conversely, if the outward normal of the surface is in negative direction, then the positive shearing stress points in the negative direction of the coordinate axis of the second subscript.

(e) In suffix (index or indicial) notation, stresses are described by means of a symmetric stress tensor σ_{ij}, in which the suffixes i and j take the values 1, 2 and 3. The components σ_{11}, σ_{22} and σ_{33} are called the normal stresses while the components $\sigma_{12} = \sigma_{21}$, $\sigma_{23} = \sigma_{32}$ and $\sigma_{13} = \sigma_{31}$ are called shear stresses.

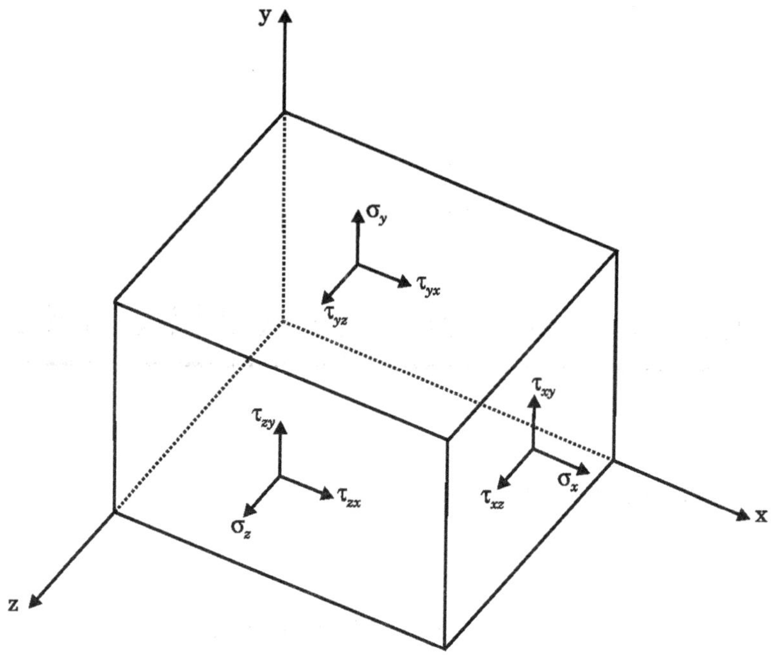

Fig. A2.1 State of stress acting on the faces of a cube (rear faces are not shown)

Stresses and equilibrium

In the Cartesian coordinate system, the equilibrium equations in the presence of body forces $\{b\} = [b_1, b_2, b_3]^T$ (which is distributed force per unit volume acting within the boundary of the body), is given as

$$\frac{\partial \sigma_{ij}}{\partial x_j} + b_i = 0, \text{ or } \sigma_{ij,j} + b_i = 0. \tag{A2.1}$$

In the suffix notation used in Eq. (A2.1), a recurring letter suffix indicates that the sum must be formed of all terms obtainable by assigning to the suffix the values 1, 2 and 3; and a comma followed by, say, '$_i$' stands for partial differentiation with respect to the coordinate x_i. For example, for $i = 1$ the equilibrium equation in x_1 direction may be written in expanded form (referred to as unabridged notation) as

$$\frac{\partial \sigma_{11}}{\partial x_1} + \frac{\partial \sigma_{12}}{\partial x_2} + \frac{\partial \sigma_{13}}{\partial x_3} + b_1 = 0. \tag{A2.2}$$

Traction (stress) vector

The Euler–Cauchy stress principle states that upon any surface (real or imaginary) that passes though an internal material point and divides the body, the action of one part of the body on the other is equivalent (equipollent) to the system of distributed forces and couples on the surface dividing the body, and it is represented by a vector field $\{t\} = [t_1, t_2, t_3]^T$, called the traction or stress vector. The traction vector $\{t\}$ is assumed to depend continuously on the surface's unit outward normal vector $\{n\} = [n_1, n_2, n_3]^T$. Depending on the orientation of the surface plane at the point under consideration, the traction vector may not necessarily be perpendicular to that plane, *i.e.* parallel to $\{n\}$. According to Cauchy's stress theorem, the traction vector $\{t\}$ associated with a plane with normal unit vector $\{n\}$ can be expressed as a function of the stress vectors on the planes perpendicular to the coordinate axes, *i.e.* in terms of the components of the stress tensor σ_{ij}. The relation is given in suffix notation as

$$t_i = \sigma_{ij} n_j. \tag{A2.3}$$

As an example, for $i = 1$ the equation in x_1 direction may be written in unabridged notation as

$$t_1 = \sigma_{11} n_1 + \sigma_{12} n_2 + \sigma_{13} n_3. \tag{A2.4}$$

Equilibrium with tractions

Let us consider an external traction force vector (distributed force per unit surface area) $\{t\} = [t_1, t_2, t_3]^T$ acting on the boundary surface S of a body. The equilibrium of the stresses along the boundary surface is then given by Eq. (A2.3).

Principal stresses and stress invariants

There is a particular orientation of coordinate axes for which the stress tensor becomes diagonal; that is no shear stresses appear. The physical picture is that at every point in a stressed body, there are three orthogonal planes for which the corresponding traction vector are normal to these planes. In order to find these directions, we write the mathematical condition that the traction vector operating on an area

dS (with unit normal vector $\hat{n}$) surrounding the point, is parallel to $\hat{n}$, that is

$$\{t\} = \lambda\{n\} \tag{A2.5}$$

or

$$[\sigma]\{n\} = \lambda\{n\}$$

or

$$\begin{bmatrix} \sigma_{11} & \sigma_{12} & \sigma_{13} \\ \sigma_{21} & \sigma_{22} & \sigma_{23} \\ \sigma_{31} & \sigma_{32} & \sigma_{33} \end{bmatrix} \begin{Bmatrix} n_1 \\ n_2 \\ n_3 \end{Bmatrix} = \lambda \begin{Bmatrix} n_1 \\ n_2 \\ n_3 \end{Bmatrix}$$

or

$$\begin{bmatrix} \sigma_{11} - \lambda & \sigma_{12} & \sigma_{13} \\ \sigma_{21} & \sigma_{22} - \lambda & \sigma_{23} \\ \sigma_{31} & \sigma_{32} & \sigma_{33} - \lambda \end{bmatrix} \begin{Bmatrix} n_1 \\ n_2 \\ n_3 \end{Bmatrix} = 0, \tag{A2.6}$$

where λ is a constant of proportionality.

The above equation is an example of standard eigenvalue and eigenvector problem. The problem has a non-trivial solution only when

$$\begin{vmatrix} \sigma_{11} - \lambda & \sigma_{12} & \sigma_{13} \\ \sigma_{21} & \sigma_{22} - \lambda & \sigma_{23} \\ \sigma_{31} & \sigma_{32} & \sigma_{33} - \lambda \end{vmatrix} = 0. \tag{A2.7}$$

Expanding the determinant yields a cubic equation (known as characteristics equation) given as

$$\lambda^3 - I_1\lambda^2 - I_2\lambda - I_3 = 0, \tag{A2.8}$$

where I_1, I_2, and I_3 are given as

$$I_1 = \sigma_{11} + \sigma_{22} + \sigma_{33},$$

$$I_2 = -\left(\sigma_{11}\sigma_{22} + \sigma_{22}\sigma_{33} + \sigma_{33}\sigma_{11}\right) + \sigma_{23}^2 + \sigma_{31}^2 + \sigma_{12}^2$$

and

$$I_3 = \det[\sigma] = \begin{vmatrix} \sigma_{11} & \sigma_{12} & \sigma_{13} \\ \sigma_{21} & \sigma_{22} & \sigma_{23} \\ \sigma_{31} & \sigma_{32} & \sigma_{33} \end{vmatrix}. \tag{A2.9}$$

The cubic equation has real roots only *i.e.* not imaginary due to the symmetry of the stress tensor. Therefore, Eq. (A2.8) will yield three values of λ, which represents the magnitude of stress vector operating on three orthogonal planes where no shear stress components work. That is, the three roots, say λ_1, λ_2 and λ_3 represent the three principal stresses (σ_1, σ_2, σ_3), without regard to the orientation of the special directions $\{n^{(1)}\}$, $\{n^{(2)}\}$, and $\{n^{(3)}\}$, along which they are oriented. Thus, the characteristics equation (Eq. (A2.8)) is independent of the orientation of the coordinate system and therefore, coefficients I_1, I_2 and I_3 are called invariants (first, second and third invariants respectively).

For each eigenvalue, there is a non-trivial solution for $\{n\}$ in the Eq. (A2.6). These solutions are the principal directions or eigenvectors $\{n^{(1)}\}$, $\{n^{(2)}\}$ and $\{n^{(3)}\}$ defining the plane where the principal stresses act. For example, for principal stress σ_1 the eigenvector equation becomes

$$\begin{bmatrix} \sigma_{11}-\sigma_1 & \sigma_{12} & \sigma_{13} \\ \sigma_{21} & \sigma_{22}-\sigma_1 & \sigma_{23} \\ \sigma_{31} & \sigma_{32} & \sigma_{33}-\sigma_1 \end{bmatrix} \begin{Bmatrix} n_1^{(1)} \\ n_2^{(1)} \\ n_3^{(1)} \end{Bmatrix} = \begin{Bmatrix} 0 \\ 0 \\ 0 \end{Bmatrix}. \qquad \text{(A2.10)}$$

Because the above equation is not sufficient to find a unique $\{n^{(1)}\}$ (because the equation is equally valid for any length of $\{n^{(1)}\}$), we can complete the equation by specifying that $\left(n_1^{(1)}\right)^2 + \left(n_2^{(1)}\right)^2 + \left(n_3^{(1)}\right)^2 = 1$.

The principal directions are parallel to the three principal stress vector and normal to the three principal planes that transmit the principal stresses. The principal stresses and principal directions characterize the stress at a point and are independent of the orientation.

Deviatoric stress tensor and its invariants

Some materials are virtually insensitive to pressure in comparison with their responses to stresses that cause shape distortion. Examples include nearly all liquids and most isotropic solids undergoing plastic deformation. In cases where the hydrostatic pressure is not of significant interest, a different stress tensor can be defined that is unique for all states of stress that differ only by the applied pressure. To do this, the stress tensor can be broken down into two components: stresses that tend to change the volume of the stressed body and stresses that tend to distort the body. The former stresses are referred to as hydrostatic stress tensor or volumetric stress tensor or mean normal stress tensor, and the latter as deviatoric stress tensor. Since the hydrostatic stresses maintain the original proportions of the volume, it makes sense that they would be equal. Thus the hydrostatic stress tensor is given as $\sigma_m \delta_{ij}$, in which σ_m is the mean of the principal stresses given as

$$\sigma_m = \frac{\sigma_1 + \sigma_2 + \sigma_3}{3} = \frac{\sigma_{11} + \sigma_{22} + \sigma_{33}}{3} = \frac{\sigma_{ii}}{3} \tag{A2.11}$$

and δ_{ij} is the Kronecker delta defined as $\delta_{ij} = \begin{cases} 1, \text{if } i = j \\ 0, \text{if } i \neq j \end{cases}$.

The deviatoric stress tensor can be obtained by subtracting the hydrostatic stress tensor from the original stress tensor as

$$\sigma'_{ij} = \sigma_{ij} - \frac{\delta_{ij}\sigma_{ii}}{3}. \tag{A2.12}$$

As σ'_{ij} is a second order tensor, it also has a set of invariants, which can be obtained using the same procedure used to calculate the invariants of the stress tensor σ_{ij}. It can be shown that the principal directions of the deviatoric stress tensor σ'_{ij} are the same as the principal directions of the stress tensor σ_{ij}. The characteristic equation is given as

$$\left|\sigma'_{ij} - \lambda\delta_{ij}\right| = \lambda^3 - J_1\lambda^2 - J_2\lambda - J_3 = 0, \tag{A2.13}$$

where J_1, J_2 and J_3 are the first, second, and third deviatoric stress invariants, respectively. Their values are the same regardless of the orientation of the coordinate system chosen. These deviatoric stress invariants are given as

$$J_1 = \sigma'_{kk} = 0,$$

$$J_2 = \tfrac{1}{2}\sigma'_{ij}\sigma'_{ij}$$

and

$$J_3 = \det\left(\sigma'_{ij}\right) = \det\left[\sigma'\right]. \tag{A2.14}$$

Because $\sigma'_{kk} = 0$, the stress deviator tensor is in a state of pure shear.

A2.1.2. Strain

(a) Strains are represented in a vector form that corresponds to the six independent components of stress,

$$\{\varepsilon\} = \begin{bmatrix} \varepsilon_x & \varepsilon_y & \varepsilon_z & \gamma_{xy} & \gamma_{yz} & \gamma_{xz} \end{bmatrix}^T. \quad \text{(A2.15)}$$

(b) Tensile strains are positive and compressive strains are negative.

(c) In (n, t, z) axes, shearing strain γ_{nt} measures the amount by which the right angle between the n- and t- directions decreases as the material deforms.

(d) Shearing strains that decrease the angle between the two lines at the origin of coordinates are positive and those that increase the angle are negative.

(e) In indicial notation, the state of strain at a point is defined by means of a symmetric strain tensor ε_{ij} in which ε_{11}, ε_{22} and ε_{33} are normal strains and $\varepsilon_{12} = \varepsilon_{21}, \varepsilon_{23} = \varepsilon_{32}, \varepsilon_{13} = \varepsilon_{31}$ are shear strains.

Strain- displacement relation

If displacement of a point in a general continuum with coordinates $x_k, k = 1, 2, 3$ at time t is described by a vector $\{u\} = [u_1, u_2, u_3]^T$, the strain tensor is defined by means of the strain-displacement relationship given as

$$\varepsilon_{ij} = \frac{1}{2}\left(u_{i,j} + u_{j,i}\right). \quad \text{(A2.16)}$$

It is to be noted that these linear relations hold for small deformations only.

A2.2. STRESS-STRAIN RELATIONS

Stresses are related to strains by means of the constitutive relations which for linear elasticity have a linear, homogeneous form of a generalized Hooke's law as

$$\sigma_{ij} = C_{ijkl}\varepsilon_{kl}. \quad \text{(A2.17)}$$

Since both the components of the stress and strain tensors are symmetric, it follows that $C_{ijkl} = C_{ijlk} = C_{jikl} = C_{jilk}$ so that a maximum of only thirty-six of the possible eighty-one parameters could be independent.

The symmetry in the stress and strain tensors can be further exploited to simplify the generalized expression. Due to the symmetry that $\sigma_{ij} = \sigma_{ji}$, only six independent constants can appear in the stress tensor. These six independent constants of stress and strain can be "contracted" to a single index notation by the following standard convention:

$$\begin{aligned} &\sigma_{ii} = \sigma_i, \varepsilon_{ii} = \varepsilon_i, (i = 1,2,3) \\ &\tau_{12} = \sigma_{12} = \sigma_4, \gamma_{12} = 2\varepsilon_{12} = \varepsilon_4, \\ &\tau_{23} = \sigma_{23} = \sigma_5, \gamma_{23} = 2\varepsilon_{23} = \varepsilon_5, \\ &\tau_{13} = \sigma_{13} = \sigma_6, \gamma_{13} = 2\varepsilon_{13} = \varepsilon_6 . \end{aligned} \quad \text{(A2.18)}$$

This contraction of stress and strain tensors to vectors (with six components) means that the fourth order elastic constant tensor may be contracted to two-index notation. Following this convention, the generalized Hooke's law relationship can be compactly written in matrix notation as

$$\{\sigma\} = [C]\{\varepsilon\}, \quad \text{(A2.19)}$$

where [*C*] is a six-by-six matrix of the contracted elements, C_{pq}, called the stiffness co-efficients.

The number of constants required to describe the behaviour of the material may be further reduced by considering the strain energy density function. If there exists a strain energy function *W*,

$$W = \frac{1}{2}\sigma_i\varepsilon_i = \frac{1}{2}C_{ij}\varepsilon_j\varepsilon_i \quad (\text{i, j} = 1, 2, \cdots, 6), \quad \text{(A2.20)}$$

with the property

$$\frac{\partial W}{\partial \varepsilon_i} = \sigma_i,$$

then we can always suppose that the quadratic form (A2.20) is symmetric which yields that the stiffness matrix [*C*] of equation (A2.19) must be symmetric: $C_{ij} = C_{ji}$ $(i, j = 1, 2, ..., 6)$. Thus the number of independent elastic constants is reduced from thirty-six to twenty-one. It is now generally accepted that the number of independent elastic constants in the generalized Hooke's law is 21 in the most general form of anisotropy.

Materials symmetry induced by the nature of the microstructure can dramatically reduce the number of independent parameters required to describe the behavior of materials. For example, if a material exhibits symmetry with respect to a plane, the number of independent elastic constants becomes 13. If there is symmetry with respect to three mutually perpendicular planes, the number becomes 9. This type of symmetry is called orthotropic symmetry and is usually the lowest symmetry exhibited by practical composite materials. Nine independent material descriptors are required to completely specify the mechanical response characteristic of this symmetry class. The structure of the elastic constants matrix associated with this symmetry becomes

$$\begin{Bmatrix} \sigma_1 \\ \sigma_2 \\ \sigma_3 \\ \tau_{12} \\ \tau_{23} \\ \tau_{13} \end{Bmatrix} = \begin{bmatrix} C_{11} & C_{12} & C_{13} & 0 & 0 & 0 \\ C_{12} & C_{22} & C_{23} & 0 & 0 & 0 \\ C_{13} & C_{23} & C_{33} & 0 & 0 & 0 \\ 0 & 0 & 0 & C_{44} & 0 & 0 \\ 0 & 0 & 0 & 0 & C_{55} & 0 \\ 0 & 0 & 0 & 0 & 0 & C_{66} \end{bmatrix} \begin{Bmatrix} \varepsilon_1 \\ \varepsilon_2 \\ \varepsilon_3 \\ \gamma_{12} \\ \gamma_{23} \\ \gamma_{13} \end{Bmatrix}. \tag{A2.21}$$

The usual expression of Hooke's law in terms of the elastic constants, $\{\sigma\} = [C]\{\varepsilon\}$ suggests that the strain is the specified quantity so that the stress $\{\sigma\}$ is the dependent variable. In most practical situations, the converse is true: stress is specified and strain is to be determined. It is therefore, useful to express the strain in terms of the stress, i.e.

$$\{\varepsilon\} = [S]\{\sigma\}, \tag{A2.22}$$

where $[S]$ is the compliance matrix comprised of 36 elements S_{ij} $(i, j = 1, 2,....6)$. The quantity $[S]$ is the inverse of $[C]$, in the matrix sense $[S] = [C]^{-1}$.

The compliance constants are related to the conventional engineering constants through the relationships

$$S_{11} = \frac{1}{E_1}, S_{22} = \frac{1}{E_2}, S_{33} = \frac{1}{E_3}$$

$$S_{12} = -\frac{\nu_{12}}{E_1} = S_{21} = -\frac{\nu_{21}}{E_2}$$

$$S_{13} = -\frac{\nu_{13}}{E_1} = S_{31} = -\frac{\nu_{31}}{E_3}$$

$$S_{23} = -\frac{\nu_{23}}{E_2} = S_{32} = -\frac{\nu_{32}}{E_3}$$

$$S_{44} = \frac{1}{G_{12}}, \quad S_{55} = \frac{1}{G_{23}}$$

$$S_{66} = \frac{1}{G_{13}} \tag{A2.23}$$

with $\nu_{ij} = \nu_{ji}(E_i / E_j)$ with $i, j = 1, 2, 3$.
In matrix notation, we can write

$$[S] = \begin{bmatrix} \frac{1}{E_1} & -\frac{\upsilon_{12}}{E_1} & -\frac{\upsilon_{13}}{E_1} & 0 & 0 & 0 \\ -\frac{\upsilon_{21}}{E_2} & \frac{1}{E_2} & -\frac{\upsilon_{23}}{E_2} & 0 & 0 & 0 \\ -\frac{\upsilon_{31}}{E_3} & -\frac{\upsilon_{32}}{E_3} & \frac{1}{E_3} & 0 & 0 & 0 \\ 0 & 0 & 0 & \frac{1}{G_{12}} & 0 & 0 \\ 0 & 0 & 0 & 0 & \frac{1}{G_{23}} & 0 \\ 0 & 0 & 0 & 0 & 0 & \frac{1}{G_{13}} \end{bmatrix}$$

In equation (A2.23), E_i is the Young's modulus in the x_i direction, ν_{ij} is Poisson's ratio relating ε_i to $\frac{\sigma_j}{E_i}$, and G_{ij} is the shear modulus in the $i-j$ plane.

The elastic constants, C_{ij} can be obtained through the matrix relationships $[C] = [S]^{-1}$. The results are summarized below:

$$C_{11} = E_1\left(1 - \frac{E_3}{E_2}\nu_{23}^2\right)\Delta$$

$$C_{12} = C_{21} = \left(E_2\nu_{12} + E_3\nu_{13}\nu_{23}\right)\Delta$$

$$\text{a}\, C_{13} = C_{31} = E_3\left(\nu_{12}\nu_{23} + \nu_{13}\right)\Delta$$

$$C_{22} = E_2\left(1 - \frac{E_3}{E_1}\nu_{13}^2\right)\Delta$$

$$C_{23} = C_{32} = \frac{E_3}{E_1}\left(E_1\nu_{23} + E_2\nu_{12}\nu_{13}\right)\Delta$$

$$C_{33} = E_3\left(1 - \frac{E_2}{E_1}\nu_{12}^2\right)\Delta$$

$$C_{44} = G_{12}$$

$$C_{55} = G_{23}$$

$$C_{66} = G_{13} \tag{A2.24}$$

with $$\Delta^{-1} = 1 - 2\frac{E_3}{E_1}\nu_{12}\nu_{23}\nu_{13} - \nu_{13}^2\frac{E_3}{E_1} - \nu_{23}^2\frac{E_3}{E_2} - \nu_{12}^2\frac{E_2}{E_1}$$

Transversely isotropic symmetry is the next higher symmetry class. For this material class, the properties in a plane perpendicular to the unique axis are independent of the direction within the plane *i.e.* a rotational symmetry of property exists within the plane. The general structure of the orthotropic material matrix is preserved for this symmetry class; however, this special condition reduces the number of independent descriptors to five. The stress-strain relations can be simplified based on the invariance of elastic constants for an arbitrary rotation of the $x_2 - x_3$ axes about x_1-axis, and we have

$$
\begin{aligned}
C_{12} &= C_{13} \\
C_{22} &= C_{33} \\
C_{44} &= C_{66} \\
C_{55} &= \frac{C_{22} - C_{23}}{2}.
\end{aligned}
\tag{A2.25}
$$

The symmetry as given by the foregoing equation results in following relations among conventional engineering constants:

$$
\begin{aligned}
E_2 &= E_3 \\
\nu_{12} &= \nu_{13} \\
G_{12} &= G_{13} \\
G_{23} &= \frac{E_2}{2(1+\nu_{23})}.
\end{aligned}
\tag{A2.26}
$$

If the material is elastically isotropic, *i.e.* if the properties are identical in all directions, then the number of independent elastic constants reduces to 2. For isotropic materials, we have the following [S] and [C] matrices:

$$
[S] = \begin{bmatrix}
\frac{1}{E} & -\frac{\upsilon}{E} & -\frac{\upsilon}{E} & 0 & 0 & 0 \\
-\frac{\upsilon}{E} & \frac{1}{E} & -\frac{\upsilon}{E} & 0 & 0 & 0 \\
-\frac{\upsilon}{E} & -\frac{\upsilon}{E} & \frac{1}{E} & 0 & 0 & 0 \\
0 & 0 & 0 & \frac{1}{G} & 0 & 0 \\
0 & 0 & 0 & 0 & \frac{1}{G} & 0 \\
0 & 0 & 0 & 0 & 0 & \frac{1}{G}
\end{bmatrix}
\tag{A2.27}
$$

with $G = \dfrac{E}{2(1+\nu)}$ and

$$[C]=\frac{E}{(1+\upsilon)(1-2\upsilon)}\begin{bmatrix} 1-\upsilon & \upsilon & \upsilon & 0 & 0 & 0 \\ \upsilon & 1-\upsilon & \upsilon & 0 & 0 & 0 \\ \upsilon & \upsilon & 1-\upsilon & 0 & 0 & 0 \\ 0 & 0 & 0 & 0.5-\upsilon & 0 & 0 \\ 0 & 0 & 0 & 0 & 0.5-\upsilon & 0 \\ 0 & 0 & 0 & 0 & 0 & 0.5-\upsilon \end{bmatrix} \tag{A2.28}$$

A2.2.1. Two-dimensional case (Plane stress)

The assumption of plane stress is applicable for bodies whose dimension is very small in one of the coordinate directions. Thus the analysis of a thin member, subjected to a uniform in-plane loading over its thickness can be made using the assumption of plane stress (Figure A2.2). Here the notation (x,y,z) is used instead of (1,2,3). In plane stress distribution, it is assumed that

$$\sigma_z = \tau_{yz} = \tau_{xz} = 0,$$

where z represents the direction perpendicular to the plane of the member; and the remaining stress components do not vary through the thickness of the member (*i.e.* in the z direction).

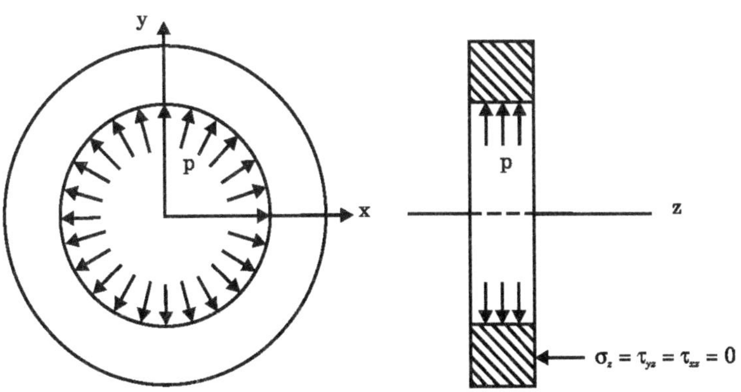

Fig. A2.2. Example of a plane stress problem. A thin member under in-plane loading.

Considering only the plane stress components, $[S]$ and $[C]$ matrices can be obtained as

$$[S] = \begin{bmatrix} \frac{1}{E} & -\frac{\upsilon}{E} & 0 \\ -\frac{\upsilon}{E} & \frac{1}{E} & 0 \\ 0 & 0 & \frac{2(1+\upsilon)}{E} \end{bmatrix} \tag{A2.29}$$

and

$$[C] = \frac{E}{1-\upsilon^2} \begin{bmatrix} 1 & \upsilon & 0 \\ \upsilon & 1 & 0 \\ 0 & 0 & \frac{(1-\upsilon)}{2} \end{bmatrix}. \tag{A2.30}$$

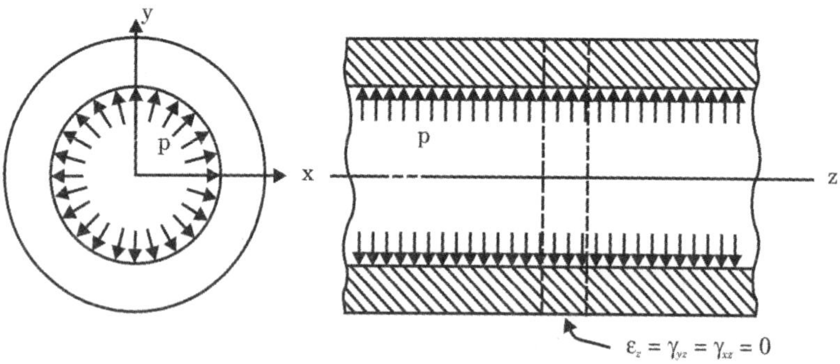

Fig. A2.3. Example of a plane strain problem. A long body of uniform cross section and subjected to transverse loading along its length.

A2.2.2. Two-dimensional case (Plane strain)

The assumption of plane strain is applicable for bodies that are long and whose geometry and loading do not vary significantly in the longitudinal direction. Thus, the analysis of dams. cylinders, and retaining walls can be made using the assumption of plane strain. A long body of uniform cross section subjected to transverse loading along its length is depicted in Figure A2.3. The figure also shows a small thickness in the loaded area that can be considered to be in a plane strain condition. In plane strain distribution, it is assumed that

$$\varepsilon_z = \gamma_{yz} = \gamma_{xz} = 0.$$

In this case, a normal stress σ_z exists in addition to the three plane stress components. Therefore, [S] matrix can be written as

$$[S] = \begin{bmatrix} \frac{1}{E} & -\frac{\upsilon}{E} & -\frac{\upsilon}{E} & 0 \\ -\frac{\upsilon}{E} & \frac{1}{E} & -\frac{\upsilon}{E} & 0 \\ 0 & 0 & 0 & \frac{1}{G} \end{bmatrix}. \tag{A2.31}$$

In addition, we have

$$\varepsilon_z = 0 = -\frac{\upsilon\sigma_x}{E} - \frac{\upsilon\sigma_y}{E} + \frac{\sigma_z}{E}$$

On eliminating σ_z and solving for the three remaining stresses we obtain the matrix [C] as

$$[C] = \frac{E}{(1+\upsilon)(1-2\upsilon)} \begin{bmatrix} 1-\upsilon & \upsilon & 0 \\ \upsilon & 1-\upsilon & 0 \\ 0 & 0 & \frac{(1-2\upsilon)}{2} \end{bmatrix}. \tag{A2.32}$$

A2.2.3. Axisymmetric case

An axisymmetric solid (also called body of revolution) is defined as a three-dimensional body that is developed by rotation of a planar section about an axis (Figure A2.4). For convenience, cylindrical coordinate system (r, θ, z) is used in which r, θ and z denote the radial, circumferential and axial directions respectively.

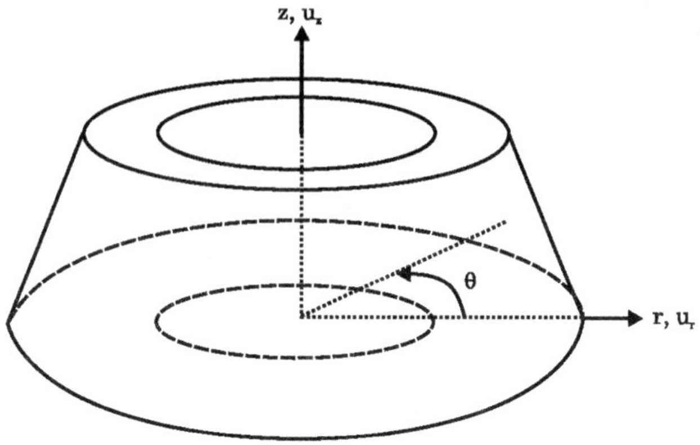

Fig. A2.4 An axisymmetric solid

The problem of stress distribution in the axisymmetric solid under axisymmetric loading is of considerable practical interest. The problems can be mathematically presented as two-dimensional. Because of symmetry about the circumferential direction, the two components of displacements in any plane section of the body along its axis of symmetry define completely the state of strain and therefore the state of stress. The strain and stress vectors in this case can be represented as

$$\{\varepsilon\} = \begin{Bmatrix} \varepsilon_r \\ \varepsilon_z \\ \varepsilon_\theta \\ \gamma_{rz} \end{Bmatrix} = \begin{Bmatrix} \dfrac{\partial u_r}{\partial r} \\ \dfrac{\partial u_z}{\partial z} \\ \dfrac{u_r}{r} \\ \dfrac{\partial u_r}{\partial z} + \dfrac{\partial u_z}{\partial r} \end{Bmatrix}$$

and $$\{\sigma\} = \begin{Bmatrix} \sigma_r \\ \sigma_z \\ \sigma_\theta \\ \tau_{rz} \end{Bmatrix}, \qquad \text{(A2.33)}$$

in which u_r and u_z are the displacement components in radial and axial directions respectively. It is to be noted that the strain components also include a hoop (circumferential) normal strain ε_θ which is induced automatically by the radial displacement. Also the stress components include the hoop normal stress σ_θ due to the presence of ε_θ.

Using Eqs. (2.27) and (2.28), the compliance matrix [*S*] and the elasticity matrix [*C*] can be written as

$$[S] = \frac{1}{E}\begin{bmatrix} 1 & -\upsilon & -\upsilon & 0 \\ -\upsilon & 1 & -\upsilon & 0 \\ -\upsilon & -\upsilon & 1 & 0 \\ 0 & 0 & 0 & 2(1+\upsilon) \end{bmatrix} \tag{A2.34}$$

and

$$[C] = \frac{E}{(1+\upsilon)(1-2\upsilon)}\begin{bmatrix} 1-\upsilon & \upsilon & \upsilon & 0 \\ \upsilon & 1-\upsilon & \upsilon & 0 \\ \upsilon & \upsilon & 1-\upsilon & 0 \\ 0 & 0 & 0 & \frac{(1-2\upsilon)}{2} \end{bmatrix}. \tag{A2.35}$$

A2.2.4. Temperature effects

For isotropic materials, a temperature rise $\Delta T(x, y, z)$ results in initial strains, $\{\varepsilon_0\}$ given as

$$\{\varepsilon_0\} = \begin{bmatrix}\alpha\Delta T & \alpha\Delta T & \alpha\Delta T & 0 & 0 & 0\end{bmatrix}^T, \tag{A2.36}$$

where α is the coefficient of thermal expansion (/°C) of the material. The strains do not cause any stresses when the body is free to deform. The stress-strain relation then becomes

$$\{\sigma\} = [C](\{\varepsilon\} - \{\varepsilon_0\}). \tag{A2.37}$$

In plane stress condition, we have

$$\{\varepsilon_0\} = [\alpha\Delta T \quad \alpha\Delta T \quad 0]^T \tag{A2.38}$$

and in plane strain condition, the constraint that $\varepsilon_z = 0$ results in the following temperature strains:

$$\{\varepsilon_0\} = (1+\upsilon)[\alpha\Delta T \quad \alpha\Delta T \quad 0]^T \tag{A2.39}$$

For axisymmetric analysis, the initial strains are given as

$$\{\varepsilon_0\} = [\alpha\Delta T \quad \alpha\Delta T \quad \alpha\Delta T \quad 0]^T. \tag{A2.40}$$

Orthotropic materials

Orthotropic materials also have thermal properties that vary with direction. If ΔT denotes the temperature change from the stress free state, then the local initial thermal strain is

$$\{\varepsilon_0\} = [\alpha_x\Delta T \quad \alpha_y\Delta T \quad \alpha_z\Delta T \quad 0 \quad 0 \quad 0]^T. \tag{A2.41}$$

α_x, α_y, and α_z are the principal coefficients of thermal expansion in principal material axis directions (x,y,z).

Appendix-3

Constitutive Relations for Composites

A3.1 INTRODUCTION TO FIBRE-REINFORCED COMPOSITES

Fiber reinforced composites, in many respects, represent excellent examples of the principles of material design where the performance of the whole is greater than the sum of its parts. Fibers, light weight, immensely strong and stiff, but easily damaged and in a form of limited engineering applications, and the matrix, comparatively weak and often brittle and not usually attractive for structural load-bearing applications. Together, however, they offer a vast array of materials and great scope for optimization in the true sense of the word. This is not only limited to mechanical characteristics but also extends to thermal properties, creep and fatigue, ballistic performance and chemical resistance.

Fiber reinforced Plastic (FRP) Composites are finding increasing use in various structural applications, particularly in aerospace sector. The potential advantages of these materials over traditional materials include higher specific strength and stiffness, superior corrosion resistance, and improved fatigue properties. Flexibility of manufacture is also a unique characteristic of these materials. Large and complex structures can be fabricated in one piece minimizing tooling costs and obviating the need for joints and fastenings. Another important benefit with composites is the ability to tailor properties or materials, for example by changing fibre orientation or type of reinforcement, thereby allowing the material itself to be designed to suit operating conditions. Many types of plastic resins, both thermosetting and thermoplastic, are used as matrices. Polyesters and epoxies form the basic resin range, but these are augmented by a whole range of specialist materials including polyimides, polybismaleimides (BMI) and multifunctional epoxies amongst the thermosetting polymers and polyphenylene sulphide (PPS), polyether sulfone (PES) and polyether ether ketone (PEEK) amongst the thermoplastic systems. The vast majority of applications using fibre-reinforced polymers employ either glass, carbon/graphite or aramid fibres. However, two other fibres, boron and ultra high molecular weight polyethylene (UHMWPE) are also of interest.

There are several factors which contribute to the engineering performance of a fiber-reinforced composites. Among the most important ones are the orientation, length, shape and composition of the fibers; the mechanical properties of the matrix; and the integrity of the bond between fibers and matrix.

Fiber orientation (how the individual strands are positioned) determines the mechanical strength of the composite and the direction in which that strength will be the greatest. There are several ways of thinking about fiber orientation and how orientation affects the strength of a composite. First, however, it must be realized that unidirectionally oriented fibers provide maximum composite strength and modulus when loads are applied in the direction of the fibers. In contrast, when loads are applied at even very small angles from the fiber direction, composite strength is drastically lowered. Since very few structures are loaded unidirectionally, it is necessary in most cases to mix orientations in a given part or structure. Some of the types of possible orientations for both continuous and discontinuous (short) fibers are shown in Figure A3.1.

A3.1.1 Laminated construction

It is common for individual plies of unidirectional fibers to be combined to form a more complex construction. Such a laminate may contain many individual layers, each at different orientations with respect to one another. The sequence of plies is determined by design considerations so that the directional dependence of strength and stiffness matches that of the loading environment. Clearly an infinite array of laminates is possible, but there are a number of technically important configurations. A symmetric laminate is one where there is a plane of symmetry of laminae in terms of thickness, orientation and properties with respect to the center-line. Another form of construction is the balanced laminate where the number of layers in orientations $\theta_1, \theta_2, \theta_3, ...$, are matched by an equal number in the $-\theta_1, -\theta_2, -\theta_3, ...$, directions. A balanced laminate can be, but not necessarily, symmetrical. One form of balanced construction which regularly occurs in practice is the angle ply. Here there are equal numbers of +θ and –θ layers. The cross-ply is also commonly found and here the layers are oriented at 90° with respect to one another. It is possible to have a laminate where the lamina are so arranged that, to an engineering approximation, the properties do not vary with direction. Such laminates are known as quasi-isotropic and are often used where the applied loadings are such that no one direction predominates. The significance of lamina type is not just limited to simple considerations of stiffness and strength but symmetry and lamina stacking sequence can have a profound effect on the performance of materials. One such typical construction for stacking sequence [0/90/45/-45]s is shown in Figure A3.2.

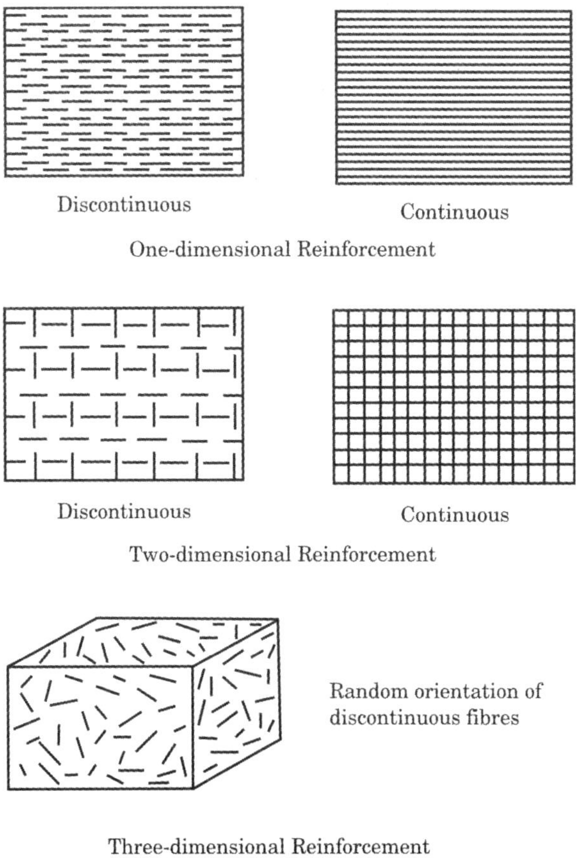

Fig. A3.1 Some possible ways to orient fibers in fiber-reinforced composites.

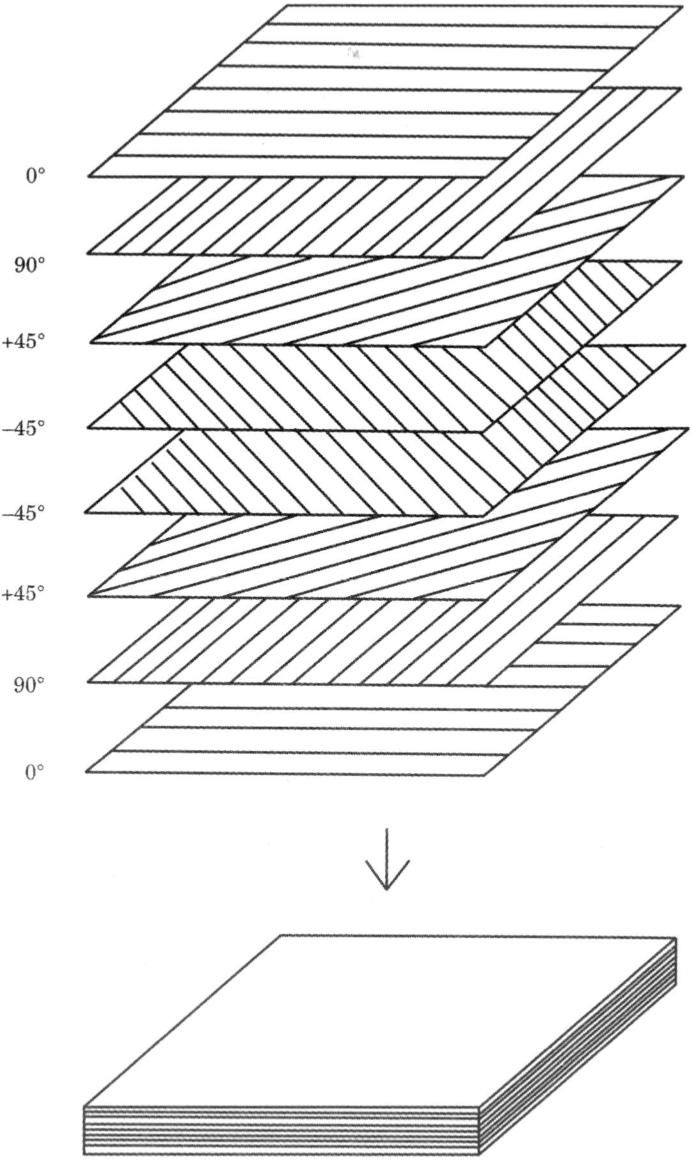

Fig. A3.2 Laminated construction.

A3.2 STRESS-STRAIN RELATION FOR LAMINATED COMPOSITES

A3.2.1 Unidirectional composite lamina

Consider an orthotropic composite lamina shown in Figure A3.3 with the principal material directions designated by x-y-z axes with x-axis representing the fiber direction. The stress-strain relationship can be represented in x-y-z coordinate system as given below

$$\begin{Bmatrix} \sigma_x \\ \sigma_y \\ \sigma_z \\ \tau_{xy} \\ \tau_{yz} \\ \tau_{xz} \end{Bmatrix} = \begin{bmatrix} \bar{C}_{11} & \bar{C}_{12} & \bar{C}_{13} & 0 & 0 & 0 \\ \bar{C}_{12} & \bar{C}_{22} & \bar{C}_{23} & 0 & 0 & 0 \\ \bar{C}_{13} & \bar{C}_{23} & \bar{C}_{33} & 0 & 0 & 0 \\ 0 & 0 & 0 & \bar{C}_{44} & 0 & 0 \\ 0 & 0 & 0 & 0 & \bar{C}_{55} & 0 \\ 0 & 0 & 0 & 0 & 0 & \bar{C}_{66} \end{bmatrix} \begin{Bmatrix} \varepsilon_x \\ \varepsilon_y \\ \varepsilon_z \\ \gamma_{xy} \\ \gamma_{yz} \\ \gamma_{xz} \end{Bmatrix}, \tag{A3.1}$$

where $\left[\bar{C}\right]$ is the elasticity matrix in material coordinate system. A bar over C is used to signify that the matrix is in material coordinate system. The elastic constants $\bar{C}_{ij}$ can be represented from Eq. (A2.24) by interchanging the subscripts 1, 2, and 3 with x, y and z respectively.

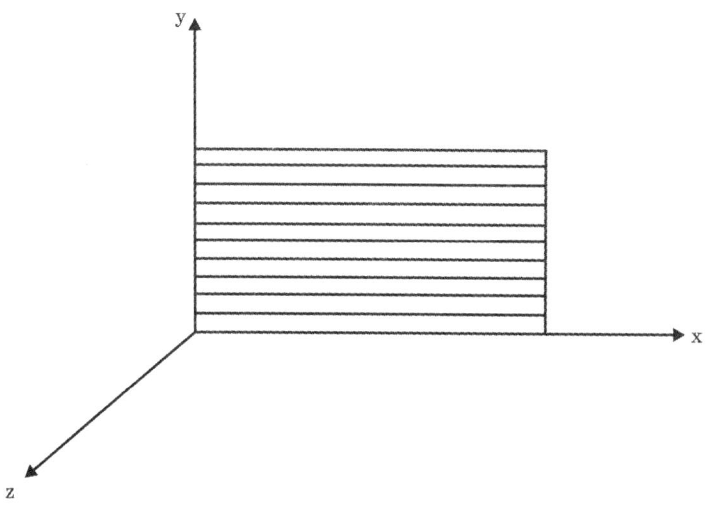

Fig. A3.3 Material coordinate system for a unidirectional lamina.

Unidirectional lamina can also be considered transversely isotropic, with y-z plane (normal to the fibers) as the plane of symmetry. The stress-strain relations can be simplified based on the invariance of elastic constants for an arbitrary rotation of the y-z axes about x-axis; and we have, from Eq. (A2.26), the following relations among conventional engineering constants in material coordinate system:

$$E_y = E_z$$

$$\nu_{xy} = \nu_{xz}$$

$$G_{xy} = G_{xz}$$

$$G_{yz} = \frac{E_y}{2\left(1+\nu_{yz}\right)}. \tag{A3.2}$$

A3.2.2 Orientation transformations

Equation (A3.1) represents the stress-strain relations for a unidirectional lamina in its principal material coordinate system. However a composite laminate is constructed by stacking several unidirectional laminae in a specified sequence of orientation. That is, the principal material directions of each lamina make different angles with a common set of reference axes. For the purpose of analysis and synthesis of laminated structures it is necessary to refer the stress-strain relation to a common reference coordinate system. Therefore, a

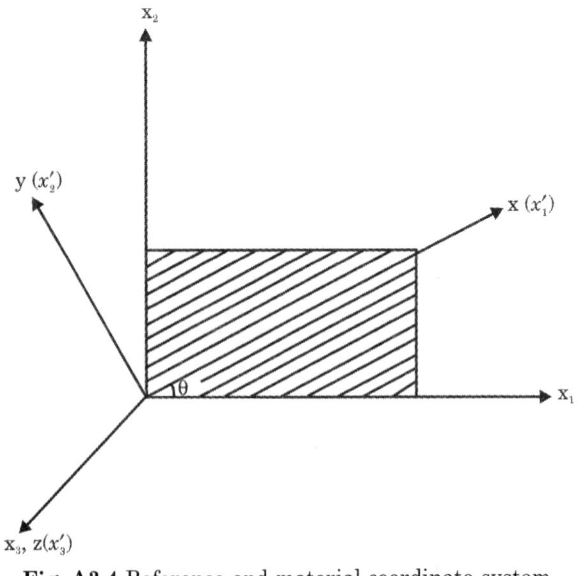

Fig. A3.4 Reference and material coordinate system.

method of transforming stress-strain relations from one coordinate system to another is needed.

As shown in Figure A3.4, the principal material axes (x-y-z axes) of a lamina is oriented at an angle θ with the reference coordinate system (x_1-x_2-x_3 axes).

Now representing x - y - z axes as $x_1' - x_2' - x_3'$, the direction cosines for the transformation of x_1, x_2, x_3 coordinates to the x_1', x_2', x_3' are

$$[\Omega] = \begin{bmatrix} \omega_{11} & \omega_{12} & \omega_{13} \\ \omega_{21} & \omega_{22} & \omega_{23} \\ \omega_{31} & \omega_{32} & \omega_{33} \end{bmatrix} = \begin{bmatrix} \cos\theta & \sin\theta & 0 \\ -\sin\theta & \cos\theta & 0 \\ 0 & 0 & 1 \end{bmatrix}, \quad \text{(A3.3)}$$

where ω_{ij} is the cosine of the angle between x_i' and x_j, i.e.,

$$\omega_{ij} = \cos\left(x_i', x_j\right). \quad \text{(A3.4)}$$

Since, strain and stress are second order tensors, they transform according to the following equations:

$$\varepsilon_{ij}' = \omega_{ik}\omega_{jl}\varepsilon_{kl} \quad \text{(A3.5)}$$

$$\sigma_{ij}' = \omega_{ik}\omega_{jl}\sigma_{kl} \quad \text{(A3.6)}$$

or, in matrix notation as

$$[\varepsilon'] = [\Omega]\ [\varepsilon]\ [\Omega]^T \quad \text{(A3.7)}$$

$$[\sigma'] = [\Omega]\ [\sigma]\ [\Omega]^T. \quad \text{(A3.8)}$$

A more compact form may be obtained from the various orientation relationships by the use of the contracted notations of strains and stresses. In the contracted notation, the strain transformation takes on the form

$$\{\varepsilon'\} = [T_\varepsilon]\ \{\varepsilon\}. \quad \text{(A3.9)}$$

The elements of $[T_\varepsilon]$ can be obtained by comparing the expanded forms of the strain tensor $[\varepsilon]$ and the contracted strain vector $\{\varepsilon\}$ and are given as

$$[T_\varepsilon] =$$

$$\begin{bmatrix} \omega_{11}^2 & \omega_{12}^2 & \omega_{13}^2 & \omega_{12}\omega_{13} & \omega_{11}\omega_{13} & \omega_{11}\omega_{12} \\ \omega_{21}^2 & \omega_{22}^2 & \omega_{23}^2 & \omega_{22}\omega_{23} & \omega_{21}\omega_{23} & \omega_{21}\omega_{22} \\ \omega_{31}^2 & \omega_{32}^2 & \omega_{33}^2 & \omega_{32}\omega_{33} & \omega_{31}\omega_{33} & \omega_{31}\omega_{32} \\ 2\omega_{21}\omega_{31} & 2\omega_{22}\omega_{32} & 2\omega_{23}\omega_{33} & \omega_{22}\omega_{33}+\omega_{32}\omega_{23} & \omega_{21}\omega_{33}+\omega_{31}\omega_{23} & \omega_{21}\omega_{32}+\omega_{31}\omega_{22} \\ 2\omega_{11}\omega_{31} & 2\omega_{12}\omega_{32} & 2\omega_{13}\omega_{33} & \omega_{12}\omega_{33}+\omega_{32}\omega_{13} & \omega_{11}\omega_{33}+\omega_{31}\omega_{13} & \omega_{11}\omega_{32}+\omega_{31}\omega_{12} \\ 2\omega_{11}\omega_{21} & 2\omega_{12}\omega_{22} & 2\omega_{13}\omega_{23} & \omega_{12}\omega_{23}+\omega_{22}\omega_{13} & \omega_{11}\omega_{23}+\omega_{21}\omega_{13} & \omega_{11}\omega_{22}+\omega_{21}\omega_{12} \end{bmatrix}$$

Substituting values from Eq. (A3.3), we get

$$[T_\varepsilon] = \begin{bmatrix} m^2 & n^2 & 0 & 0 & 0 & mn \\ n^2 & m^2 & 0 & 0 & 0 & -mn \\ 0 & 0 & 1 & 0 & 0 & 0 \\ 0 & 0 & 0 & m & -n & 0 \\ 0 & 0 & 0 & n & m & 0 \\ -2mn & 2mn & 0 & 0 & 0 & m^2-n^2 \end{bmatrix}, \tag{A3.10}$$

where

$$\begin{aligned} m &= \cos\theta \\ n &= \sin\theta. \end{aligned} \tag{A3.11}$$

A similar expression is obtained for the stress vector, viz.

$$\{\sigma'\} = [T_\sigma]\{\sigma\}. \tag{A3.12}$$

However, $[T_\varepsilon]$ and $[T_\sigma]$ can be related by equating virtual strain energy densities in both coordinate systems as given below

$$\{\partial\varepsilon'\}^T\{\sigma'\} = \{\partial\varepsilon\}^T\{\sigma\}. \tag{A3.13}$$

From Eq. (A3.9) one can write

$$\{\partial\varepsilon'\}^T = \{\partial\varepsilon\}^T[T_\varepsilon]^T. \tag{A3.14}$$

Substituting from Eqs. (A3.12) and (A3.14) into Eq. (A3.13), we obtain

$$\left[T_{\sigma}\right]=\left[T_{\varepsilon}\right]^{-T}. \tag{A3.15}$$

Now, the orientation transformation of elasticity matrix $\left[C'\right]$ in material coordinate system to $\left[C\right]$ in reference coordinate system can be derived as follows:

$$\left\{\sigma'\right\}=\left[C'\right]\ \left\{\varepsilon'\right\}. \tag{A3.16}$$

Using relations in Eqs. (A3.9), (A3.12) and (A3.15), we get

$$\left[T_{\varepsilon}\right]^{-T}\left[C\right]\ \left\{\varepsilon\right\}=\left[C'\right]\ \left[T_{\varepsilon}\right]\ \left\{\varepsilon\right\} \tag{A3.17}$$

or,
$$\left[C\right]=\left[T_{\varepsilon}\right]^{T}\left[C'\right]\ \left[T_{\varepsilon}\right]. \tag{A3.18}$$

Now, replacing $\left[C'\right]$ by $\left[\bar{C}\right]$ and dropping the subscript ε in $\left[T_{\varepsilon}\right]$, we have

$$\left[C\right]=\left[T\right]^{T}\left[\bar{C}\right]\ \left[T\right]. \tag{A3.19}$$

Like $\left[\bar{C}\right]$, $\left[C\right]$ matrix is also symmetric. After substituting the values from Eq. (A3.10) into Eq. (A3.19), we get

$$C_{11}=\bar{C}_{11}m^4+2(\bar{C}_{12}+2\bar{C}_{66})m^2n^2+\bar{C}_{22}n^4$$

$$C_{12}=(\bar{C}_{11}+\bar{C}_{22}-4\bar{C}_{66})m^2n^2+\bar{C}_{12}(m^4+n^4)$$

$$C_{13}=\bar{C}_{13}m^2+\bar{C}_{23}n^2$$
$$C_{14}=0$$
$$C_{15}=0$$

$$C_{16}=m^3n\bar{C}_{11}-mn^3\bar{C}_{22}-mn(m^2-n^2)(\bar{C}_{12}+2\bar{C}_{66})$$

$$C_{22}=\bar{C}_{11}n^4+2(\bar{C}_{12}+2\bar{C}_{66})m^2n^2+\bar{C}_{22}m^4$$

$$C_{23}=\bar{C}_{13}n^2+\bar{C}_{23}m^2$$
$$C_{24}=0$$
$$C_{25}=0$$

$$C_{26}=mn^3\bar{C}_{11}-m^3n\bar{C}_{22}+mn(m^2-n^2)(\bar{C}_{12}+2\bar{C}_{66})$$

$$
\begin{aligned}
C_{33} &= \bar{C}_{33} \\
C_{34} &= 0 \\
C_{35} &= 0 \\
C_{36} &= (\bar{C}_{13} - \bar{C}_{23})mn \\
C_{44} &= \bar{C}_{44}\, m^2 + \bar{C}_{55}\, n^2 \\
C_{45} &= (\bar{C}_{55} - \bar{C}_{44})mn \\
C_{46} &= 0 \\
C_{55} &= \bar{C}_{44}\, n^2 + \bar{C}_{55}\, m^2 \\
C_{56} &= 0 \\
C_{66} &= (\bar{C}_{11} + \bar{C}_{22} - 2\bar{C}_{12})m^2 n^2 + \bar{C}_{66}(m^2 - n^2)^2 \,.
\end{aligned}
\tag{A3.20}
$$

Appendix-4

Variational Principle

A4.1 INTRODUCTION

The application of calculus of variations has been significantly used in the finite element formulation. Such an approach makes it possible to apply finite element techniques to solve, in general, field problems. Also it helps to study mathematically the convergence and other properties of elements and assemblage in a rigorous manner. The use of variational statements makes it possible to concentrate in a single functional all of the intrinsic features of the problem at hand: the governing differential equations as well as the boundary and initial conditions. This ensures a simultaneous and consistent approximation to all the equations in case an approximate solution is sought. In this Appendix, a brief introduction is given to the calculus of variation and its application in the area of solid mechanics. Details can be found in Dym and Shames (1973) and Washizu (1965).

A4.2 CALCULUS OF VARIATIONS

In the calculus of variations we are concerned with functionals and their extreme values. A functional, in simple terms, is a function of functions. More specifically, a functional is defined as an expression that takes on a particular value which is dependent on the function used in the functional. In many areas of applied mathematics, a functional is defined by an integral form as

$$I(u) = \int_V F\left(u, u_{,1}, u_{,2}, u_{,3}\right) dV + \int_{S_F} f(u)\, dS, \tag{A4.1}$$

in which u is the unknown scalar function defined in a domain V, and F and f are specified operators. S is the domain boundary and u is prescribed on part of the boundary S_U. S_F is the part of the boundary where f operates and is the complementary part of S with respect to S_U. In the above equation, we use the convention $u_{,i} = \dfrac{\partial u}{\partial x_i}$.

In the variational problem, the objective is to find u which makes the above functional $I(u)$ stationary (*i.e.* to be possibly a maximum or a minimum) for a small variation δu. Here δu, the first variation in u is defined as an infinitesimal change in u for a fixed value of the independent variable x. For the functional $I(u)$ to attain a stationary value its first variation must be zero (Forry, 1968). Thus, we have

$$\delta I(u) = 0. \tag{A4.2}$$

More generally, if above equation is satisfied, the functional is stationary for the current value of u (it is not necessarily a minimum or a maximum). The functional is now differentiated in the same way as a function of four independent arguments, *i.e.*

$$\delta I(u) = \int_V \left(\frac{\partial F}{\partial u}\delta u + \sum_{i=1}^{3} \frac{\partial F}{\partial u_{,i}}\delta u_{,i} \right) dV + \int_{S_F} \frac{\partial f}{\partial u}\delta u dS = 0. \tag{A4.3}$$

In Eq. (A4.3), each term of the summation with index i can be transformed using Green's theorem (See Appendix 1). Thus we have

$$\int_V \frac{\partial F}{\partial u_{,i}}\delta u_{,i} dV = \int_S \frac{\partial F}{\partial u_{,i}} n_i \delta u dS - \int_V \frac{\partial}{\partial x_i}\left(\frac{\partial F}{\partial u_{,i}}\right)\delta u dV, \tag{A4.4}$$

where n_i is the unit outward normal to the surface S. Putting Eq. (A4.4) into Eq. (A4.3), we get

$$\delta I(u) = \int_V \left\{ \frac{\partial F}{\partial u} - \sum_{i=1}^{3} \frac{\partial}{\partial x_i}\left(\frac{\partial F}{\partial u_{,i}}\right) \right\}\delta u dV + \int_{S_F} \frac{\partial f}{\partial u}\delta u dS + \int_S \sum_{i=1}^{3} \frac{\partial F}{\partial u_{,i}} n_i \delta u dS = 0$$

(A4.5)

Since the variation δu is arbitrary, the terms must vanish individually. First putting $\delta u = 0$ on S, second and third integrals in the above equation vanish. Thus, for the first integral to be null for any $\delta u \neq 0$ inside *V*, the following equation must hold:

$$\frac{\partial F}{\partial u}-\sum_{i=1}^{3}\frac{\partial}{\partial x_i}\left(\frac{\partial F}{\partial u_{,i}}\right)=0, \tag{A4.6}$$

which is the partial differential equation (PDE) that the function u satisfies in *V*. Now, we choose any δu which is equal to zero on the part of the boundary S_U. Thus, it can be concluded from Eqs. (A4.6) and (A4.5) that the following equation must hold:

$$\frac{\partial f}{\partial u}+\sum_{i=1}^{3}\frac{\partial F}{\partial u_{,i}}n_i=0, \tag{A4.7}$$

which is the natural boundary condition (NBC) that the function u must satisfy on S_F.

Although u is considered as scalar in the above discussion, generalization to vector function is straightforward. Using similar notations as for scalar function, we define the functional I by

$$I(\bar{u})=\int_V F(\bar{u},\bar{\varepsilon})dV+\int_{S_F} f(\bar{u})dS, \tag{A4.8}$$

where $\bar{u}=(u_1,u_2,u_3)$ is the unknown vector function, and $\bar{\varepsilon}$ is the second order tensor and is given as: $\varepsilon_{ij}=u_{i,j}$. Here the convention, $u_{i,j}=\dfrac{\partial u_i}{\partial x_j}$ is used.

The functional is now differentiated to give

$$\delta I(\bar{u})=\int_V\left(\frac{\partial F}{\partial u_i}\delta u_i+\sum_{j=1}^{3}\frac{\partial F}{\partial u_{i,j}}\delta u_{i,j}\right)dV+\int_{S_F}\frac{\partial f}{\partial u_i}\delta u_i dS=0. \tag{A4.9}$$

In the above equation, the summation is over index i.

The equation can also be written as (using summation convention)

$$\delta I(\bar{u})=\int_V\left(\frac{\partial F}{\partial u_i}\delta u_i+\frac{\partial F}{\partial \varepsilon_{ij}}\delta\varepsilon_{ij}\right)dV+\int_{S_F}\frac{\partial f}{\partial u_i}\delta u_i dS=0. \tag{A4.10}$$

After a treatment that is formally equivalent to that for the scalar function and noting that each component of variations δu_i is individually arbitrary, the following equations for PDE and NBC are obtained for each component i :

$$\frac{\partial F(\overline{u},\overline{\varepsilon})}{\partial u_i} - \sum_{j=1}^{3} \frac{\partial}{\partial x_j}\left(\frac{\partial F(\overline{u},\overline{\varepsilon})}{\partial \varepsilon_{ij}}\right) = 0 \tag{A4.11}$$

and

$$\frac{\partial f(\overline{u})}{\partial u_i} + \sum_{j=1}^{3} \frac{\partial F(\overline{u},\overline{\varepsilon})}{\partial \varepsilon_{ij}} n_j = 0. \tag{A4.12}$$

A4.3 VARIATIONAL CALCULATION FOR MECHANICAL PROBLEMS

Let us identify u_i as displacement vector and ε_{ij} as strain tensor in the above derivation (Eqs. (A4.11) and (A4.12)). Further, the terms involving the partial derivatives of the function F with respect to the components of strain tensor can be identified as stress tensor so that

$$\frac{\partial F(\overline{u},\overline{\varepsilon})}{\partial \varepsilon_{ij}} = \sigma_{ij}. \tag{A4.13}$$

Equation (A4.11) is thus the equilibrium equation with body forces given as (see Eq. (A2.1))

$$\frac{\partial F(\overline{u},\overline{\varepsilon})}{\partial u_i} = -b_i. \tag{A4.14}$$

The boundary condition in terms of stress vector is obtained from Eqs. (A4.12) and (A4.13) as

$$t_i = \sum_{j=1}^{3} \sigma_{ij} n_j = -\frac{\partial f(\overline{u})}{\partial u_i}. \tag{A4.15}$$

Now putting these values in Eq. (A4.10), we get

$$\delta I(\overline{u}) = \int_V \sigma_{ij} \delta\varepsilon_{ij} dV - \int_V b_i \delta u_i dV - \int_{S_F} t_i \delta u_i dS = 0. \tag{A4.16}$$

The above equation can be identified as the virtual work principle which states that the virtual work of the external forces is equal to the virtual work of the internal forces. The first integral represents the total internal virtual work done by the stresses σ_{ij} due to virtual strains $\delta\varepsilon_{ij}$ (caused by virtual displacements δu_i), and the second and third integrals represent the external virtual work done by the body forces b_i and traction forces t_i respectively due to virtual displacements δu_i.

Appendix-5

Numerical Integration

A5.1. INTRODUCTION

Numerical integration is an important part of the finite element method. Volume integration as well as surface integration should be carried out in order to represent the element characteristics equations in a simple matrix form. Numerical integration requires, in general, that the integrand be evaluated at a finite number of points, called integration points, within the integration limits. The number of integration points can be reduced, while achieving the same degree of accuracy, by determining the locations of integration points appropriately. In evaluating integration in the characteristics matrices, it is necessary to use an integration formula that requires the least number of integrand evaluations. Gaussian quadrature formula is known to require the minimum number of integration points.

A5.2. GAUSSIAN QUADRATURE APPROACH

This method has proved most useful in finite element work. Let us consider a one-dimensional scalar function $f(\xi)$ defined in $-1 \leq \xi \leq 1$ (Figure A5.1). In the Gaussian quadrature, the integration of $f(\xi)$ can be evaluated by

$$\int_{-1}^{1} f(\xi)\, d\xi = \sum_{I=1}^{n} w_I f(\xi_I), \tag{A5.1}$$

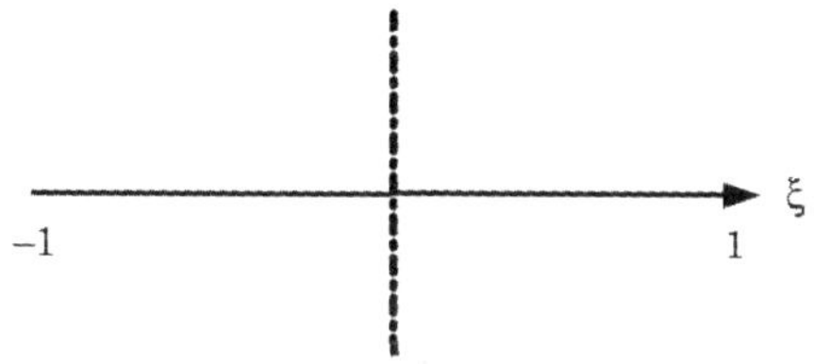

Fig. A5.1 Integration limit in one-dimension.

where ξ_I is the coordinate of an integration point called sampling point or Gauss point, w_I is a weight factor, and the summation is carried out over n integration points. n is termed as order of integration. Assuming $f(\xi)$ as a polynomial function, the formula given by Eq. (A5.1) provides the exact integration of polynomial of degree (2n-1).

The integration of a scalar function $f(\xi,\eta)$ defined by the natural coordinates $(-1 \le \xi \le 1, -1 \le \eta \le 1)$ in the two-dimensional space (Figure A5.2) can be obtained by applying the integration formula successively, *i.e.*

$$\int_{-1}^{1}\int_{-1}^{1} f(\xi,\eta)\,d\xi d\eta = \int_{-1}^{1}\left[\sum_{I=1}^{n} w_I f(\xi_I,\eta)\right]d\eta = \sum_{J=1}^{m}\sum_{I=1}^{n} w_I w_J f(\xi_I,\eta_J), \qquad \text{(A5.2)}$$

where m is the number of integration points in η direction.

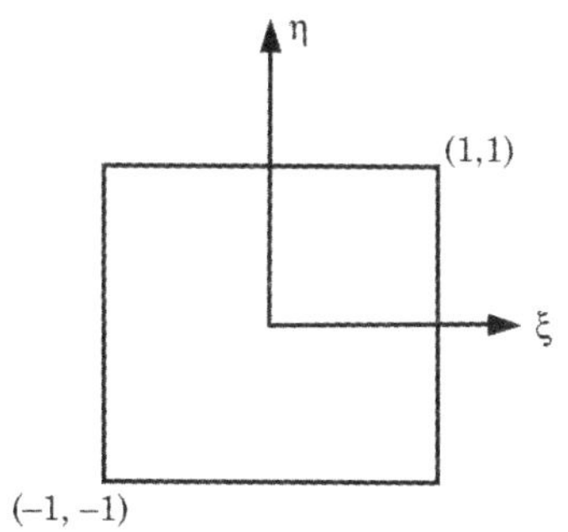

Fig. A5.2 Integration limit in two-dimension.

Similarly, in the three- dimensional space $(-1 \le \xi \le 1, -1 \le \eta \le 1, -1 \le \zeta \le 1)$ we have

$$\int_{-1}^{1}\int_{-1}^{1}\int_{-1}^{1} f(\xi,\eta,\zeta)\,d\xi d\eta d\zeta = \sum_{K=1}^{l}\sum_{J=1}^{m}\sum_{I=1}^{n} w_I w_J w_K f(\xi_I,\eta_J,\zeta_K). \qquad \text{(A5.3)}$$

In Eq. (A5.3), l is the number of integration points in ζ direction.

The optimum Gauss points and corresponding weight factors are given in Table A5.1 for different orders of integration.

Table: A5.1 Gauss points and weights for Gaussian quadrature.

Number of points, n	*Location of gauss points,* ξ_I or η_J or ζ_K	*Weights,* w_I or w_J or w_K
1	0.0	2.0
2	± 1/√3 = ± 0.5773502692	1.0
3	± 0.7745966692	0.5555555556
	0.0	0.8888888889
4	± 0.8611363116	0.3478548451
	± 0.3399810436	0.6521451549
5	± 0.9061798459	0.2369268851
	± 0.5384693101	0.4786286705
	0.0	0.5688888889
6	± 0.9324695142	0.1713244924
	± 0.6612093865	0.3607615730
	± 0.2386191861	0.4679139346

A5.2.1 Two-point formula

As an example, considering the formula in equation (A5.3) for integration order 2 in all directions, we get the integrand as

$$
\begin{aligned}
I \approx & \sum_{I=1}^{2} \sum_{J=1}^{2} \sum_{K=1}^{2} w_I w_J w_K \, f(\xi_I, \eta_J, \zeta_K) \\
= & \; w_1 w_1 w_1 \, f(\xi_1, \eta_1, \zeta_1) + w_1 w_1 w_2 \, f(\xi_1, \eta_1, \zeta_2) + w_1 w_2 w_1 \, f(\xi_1, \eta_2, \zeta_1) + \\
& w_1 w_2 w_2 \, f(\xi_1, \eta_2, \zeta_2) + w_2 w_1 w_1 \, f(\xi_2, \eta_1, \zeta_1) + w_2 w_1 w_2 \, f(\xi_2, \eta_1, \zeta_2) + \\
& w_2 w_2 w_1 \, f(\xi_2, \eta_2, \zeta_1) + w_2 w_2 w_2 \, f(\xi_2, \eta_2, \zeta_2) \, , \qquad \text{(A5.4)}
\end{aligned}
$$

where

$$w_1 = w_2 = 1.0$$

and

$$
\begin{aligned}
\xi_1 &= \eta_1 = \zeta_1 = -0.57735 \\
\xi_2 &= \eta_2 = \zeta_2 = +0.57735 .
\end{aligned}
$$

A5.2.2 Application of integration rule

The order of the numerical integration is important from two points of view. First, the cost of the numerical evaluation of characteristics matrices increases considerably with the increase in the order of integration. Secondly the results can be affected by selecting different orders of integration since it is suggested that a higher order integration generally provides accurate evaluation of integrands. Therefore, a proper choice should be the lowest possible order of integration that doesn't introduce much error into the results.

The integration order required to evaluate the integrands can be determined by studying order of the function to be integrated. The minimum order of integration that ensures convergence has been established for different types of elements and suggested by Bathe and Wilson (1978) for evaluation of element stiffness matrices for two-dimensional elements, as depicted in Table A5.2.

Table A5.2 Suggested order of integration for some two-dimensional elements.

Element	*Suggested order of integration (Bathe and Wilson, 1978)*
Four noded rectangle	2×2
Four noded quadrilateral	2×2
Eight noded rectangle	2×2
Eight noded quadrilateral	3×3

It may be noted that the above suggested order can be used for one dimensional and three-dimensional cases by deducing the appropriate integration orders for each plane of the element. Also the element mass matrices and nodal load vectors can be computed using the above order of integration.

A5.3. GAUSSIAN QUADRATURE OVER TRIANGULAR DOMAIN

For a triangle, the integrals are of the form (Figure A5.3)

$$\iint f(\xi,\eta)\,d\xi d\eta = \int_0^1 \int_0^{1-\xi} f(\xi,\eta)\,d\eta d\xi \,. \tag{A5.5}$$

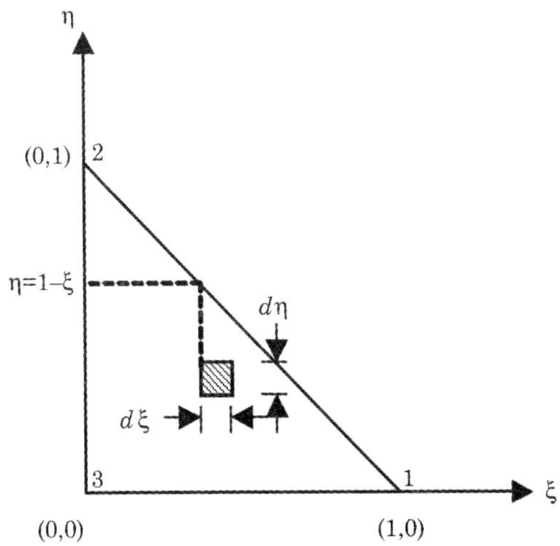

Fig. A5.3 Integration limit in triangular domain.

In many cases, the above integration has to be carried out numerically. However, the procedure described earlier for rectangular domain is not applicable directly since the limits of integration involve the variable itself. A great deal of work has been carried out on the development of suitable integration formula and a series of formulae of different orders has been derived by Cowper (1973). The integration formula is

$$\int_0^1 \int_0^{1-\xi} f(\xi,\eta)\,d\eta\,d\xi = \frac{1}{2}\sum_{I=1}^{n} w_I f(\xi_I,\eta_I)\,. \tag{A5.6}$$

Typical values of sampling points and weights are given in Table A5.3.

Table A5.3 Sampling points and weights for Gaussian quadrature over triangular domain (Cowper, 1973).

Order	*Number of sampling points, n*	*Location of sampling points,* (ξ_I,η_I)	*Weights,* w_I
Linear	1	$\left(\frac{1}{3},\frac{1}{3}\right)$	1
Quadratic	3	$\left(\frac{1}{2},0\right)$	$\frac{1}{3}$
		$\left(\frac{1}{2},\frac{1}{2}\right)$	$\frac{1}{3}$
		$\left(0,\frac{1}{2}\right)$	$\frac{1}{3}$

A5.4. GAUSSIAN QUADRATURE OVER TETRAHEDRON DOMAIN

For a tetrahedron, the integrals are of the form

$$\iiint f(\xi,\eta,\zeta)\,d\xi\,d\eta\,d\zeta = \int_0^1 \int_0^{1-\xi} \int_0^{1-\xi-\eta} f(\xi,\eta,\zeta)\,d\zeta\,d\eta\,d\xi \tag{A5.7}$$

and the integration formula is given as

$$\int_0^1 \int_0^{1-\xi} \int_0^{1-\xi-\eta} f(\xi,\eta,\zeta)\,d\zeta\,d\eta\,d\xi = \frac{1}{6}\sum_{I=1}^{n} w_I f(\xi_I,\eta_I,\zeta_I). \tag{A5.8}$$

Typical values of sampling points and weights are given in Table A5.4 (Hammer *et al.*, 1956).

Table A5.4 Sampling points and weights for Gaussian quadrature over tetrahedral region.

Order	*Number of sampling points, n*	*Location of sampling points,* $\left(\xi_I, \eta_I, \zeta_I\right)$	*Weights,* w_I
Linear	1	$\left(\frac{1}{4}, \frac{1}{4}, \frac{1}{4}\right)$	1
Quadratic	4	(α, β, β)	$\frac{1}{4}$
		(β, α, β)	$\frac{1}{4}$
		(β, β, α)	$\frac{1}{4}$
		(β, β, β)	$\frac{1}{4}$
		Here: $\alpha = 0.58541020$ $\beta = 0.13819660$	

Appendix-6

SB-FEM Elements

A6.1 INTRODUCTION

The different element types implemented in the SB-FEM code are listed in this Appendix. Each element type is identified by a name (alphanumeric character string) that indicates element category, number of nodes in the element and the type of analysis to be used in. Some general features of these elements including input data required and loading associated, are briefly mentioned. The valid key options which are to be specified in the *ET* command (*i.e. ET, IDNo, name, kOptI1, kOptI2, kOptF1, kOptF2, kOptF3*), are also stated in this listing. Detailed descriptions of these elements can be found in an appropriate documentation within the code.

A6.2 SB-FEM ELEMENTS

A6.2.1 link2 (2D two-noded spar element)

Nodal DOF set	UX, UY
Element attributes	1: Cross-sectional width; 2: Cross-sectional height
Material properties	EX
Concentrated forces	FX, FY

A6.2.2 stBeam3 (Three-noded beam element)

Nodal DOF set	UX, UZ, ROTXZ
Material properties	EX, PRXY
Concentrated forces	FX, FZ, MXZ
kOptF1	Cross-sectional width
kOptF2	Cross-sectional height

A6.2.3 stBeam3lay (Three-noded layered beam element)

Nodal DOF set, Material properties and Concentrated forces: Same as stbeam3 element
kOptF1 and kOptF2: Same as stbeam3 element

A6.2.4 stQuad4 (2D 4-noded structural element)

Nodal DOF set	UX, UY
Material properties	EX, EY, EZ, PRXY, GXY
Concentrated forces	FX, FY
kOptI1	1: Incompatible modes are added to the shape functions. 0: Incompatible modes are not added.
kOptI2	0: Plane stress 1: Plane strain
kOptF1	Width (dimension perpendicular to the plane) of the element. If this value is set negative (–1 or –2), the element is an axisymmetric element (–1: y-axis symmetry; –2: x-axis symmetry).

A6.2.5 flQuad4 (2D 4-noded fluid element)

Nodal DOF set	UX, UY
Material properties	VISCMU
kOptF1	Same as stquad4 element

A6.2.6 mfQuad4 (2D 4-noded metal forming element)

Nodal DOF set	UX, UY
Material properties	FLOWSTRESS
Concentrated forces	FX, FY
kOptF1	Same as stquad4 element

A6.2.7 thQuad4 (2D 4-noded thermal element)

Nodal DOF set	TEMP
Material properties	KX, KY; KXT, KYT
kOptF1	Same as stquad4 element

A6.2.8 stQuad8 (2D 8-noded structural element)

Nodal DOF set, Material properties and Concentrated forces: same as stquad4 element
kOptI2 and kOptF1: Same as stquad4 element

A6.2.9 flQuad8 (2D 8-noded fluid element)

Nodal DOF set and Material properties: Same as flquad4 element
kOptF1: Same as stquad4 element

A6.2.10 mfQuad8 (2D 8-noded metal forming element)

Nodal DOF set, Material properties and Concentrated forces: Same as mfquad4 element
kOptF1: Same as stquad4 element

A6.2.11 thQuad8 (2D 8-noded thermal element)

Nodal DOF set and Material properties: Same as thquad4 element
kOptF1: Same as stquad4 element

A6.2.12 stshellQuad8 (8-noded shell element)

Nodal DOF set	UX, UY, UZ, ROTXZ, ROTYZ
Material properties	EX, EY, PRXY, GXY, GYZ, GZX
Concentrated forces	FX, FY, FZ, MXZ, MYZ
kOptF1	Depth (dimension perpendicular to the curvilinear plane) of the element
kOptF2	Radius along x-direction
kOptF3	Radius along y-direction

A6.2.13 stshellQuad8Lay (8-noded layered shell element)

Nodal DOF set, Material properties and Concentrated forces: Same as stshellQuad8 element
kOptF1, kOptF2 and kOptF3: Same as stshellQuad8 element

A6.2.14 stQuad9 (2D 9-noded structural element)

Nodal DOF set, Material properties and Concentrated forces: Same as stquad4 element
kOptI2 and kOptF1: Same as stquad4 element

A6.2.15 flQuad9 (2D 9-noded fluid element)

Nodal DOF set and Material properties: Same as flquad4 element
kOptF1: Same as stquad4 element

A6.2.16 mfQuad9 (2D 9-noded metal forming element)

Nodal DOF set, Material properties and Concentrated forces: Same as mfquad4 element
kOptF1: Same as stquad4 element

A6.2.17 thQuad9 (2D 9-noded thermal element)

Nodal DOF set and Material properties: Same as thquad4 element
kOptF1: Same as stquad4 element

A6.2.18 StshellQuad9 (9-noded shell element)

Nodal DOF set, Material properties and Concentrated forces: Same as stshellQuad8 element
kOptF1, kOptF2 and kOptF3: Same as stshellQuad8 element

A6.2.19 stshellQuad9Lay (9-noded layered shell element)

Nodal DOF set, Material properties and Concentrated forces: Same as stshellQuad9 element
kOptF1, kOptF2 and kOptF3: Same as stshellQuad9 element

A6.2.20 stBrick8 (3D 8-noded structural brick element)

Nodal DOF set	UX, UY, UZ
Material properties	EX, EY, EZ, PRXY, PRYZ, PRXZ, GXY, GYZ, GZX
Concentrated forces	FX, FY, FZ
kOptI1	1: Incompatible modes are added to the shape functions. 0: Incompatible modes are not added.

A6.2.21 stBrick8Lay (3D 8-noded layered structural brick element)

Nodal DOF set, Material properties and Concentrated forces: Same as stbrick8 element
kOptI1: Same as stbrick8 element

A6.2.22 flBrick8 (3D 8-noded hexahedral fluid element)

Nodal DOF set	UX, UY, UZ
Material properties	VISCMU

A6.2.23 stBrick20 (3D 20-noded structural brick element)

Nodal DOF set, Material properties and Concentrated forces: Same as stbrick8 element

---

A6.2.24 stBrick20Lay (3D 20-noded layered structural brick element)

Nodal DOF set, Material properties and Concentrated forces: Same as stbrick20 element

Appendix-7

SB-FEM Commands

A7.1 INTRODUCTION

The SB-FEM commands and their tasks are listed briefly in this Appendix. The list here is not exhaustive and a comprehensive list along with detailed descriptions of the commands can be found in an appropriate documentation within the code. The list of commands below has been arranged to a possible extent by a functional grouping.

A7.2 GENERAL COMMANDS

This group consists of some general commands which define analysis title, input and output filenames and so on. The list of these commands are as follows.

inpEx	Executes a command file. This command is an alternative of the *input* command.
inpFile Name	Defines an input file name for all the input data sub-sequently needed for the analysis.
input	Executes a command file specified.
outFile Name	Defines an output file name for all the output data to be stored during the analysis.
Title	Defines a title of the problem.

A7.3 ENTITIES SELECTION COMMANDS

These commands are used to select subsets of model entities for further operations. These commands can be used inside any of the processors (preprocessor, solution or postprocessor).

asel	Select a subset of areas. The type of selection is identified by the Label Seltype specified in the command (*asel, Seltype, Item, Itemcomp, Item_Vmin, Item_ Vmax, Item_Vinc*).

esel	Select a subset of elements.
ksel	Select a subset of keypoints.
lsel	Select a subset of lines.
nsel	Select a subset of nodes.
vsel	Select a subset of volumes.

A7.4 DATA LISTING COMMANDS

These commands list out the different entities and other data.

alist	Lists all the selected areas in the range specified.
alistSt	Lists areas of all the stuctures (master (lower and upper dies) and slave (the workpiece)) in the range specified.
dnlist	Lists nodal constraints for the nodes in the range specified.
dnlistSt	Lists nodal constraints for the nodes of all the structures in the range specified.
eLayList	Lists layer information (data) of the layered elements in the range specified.
elist	Lists all the selected elements in the range specified.
elistSt	Lists elements of all the structures in the range specified.
etlist	Lists element types in the range specified.
fnlist	Lists nodal forces for the nodes in the range specified.
fnlistSt	Lists nodal forces for the nodes of all the structures (master (lower and upper dies) and slave (the workpiece)) in the range specified.
klist	Lists all the selected keypoints in the range specified.
klistSt	Lists keypoints of all the structures in the range specified.
llist	Lists all the selected lines in the range specified.
llistSt	Lists lines of all the structures in the range specified.
mpList	Lists all the material properties in the material property sets available in the given range.
nlist	Lists all the selected nodes in the range specified.
nlistSt	Lists the selected nodes of all the structures in the range specified.
vlist	Lists all the selected volumes in the range specified.

A7.5 ENTITIES PLOTTING COMMANDS

These commands are used to display different entities.

aplot	Displays areas in the range specified.
aplotSt	Displays areas of all the structures master (lower and upper dies) and slave (the workpiece)) in the range specified.
eplot	Displays elements in the range specified.
eplotSt	Displays elements of all the stuctures in the range specified.

kplot	Displays keypoints in the range specified.
kplotSt	Displays keypoints of all the structures in the range specified.
lplot	Displays lines in the range specified.
lplotSt	Displays lines of all the structures in the range specified.
nplot	Displays all the selected nodes in the range specified.
nplotSt	Displays selected nodes of all the structures in the range specified.
replot	Redraws the current plot (model, mesh, contour, etc.).
vplot	Displays volumes in the range specified.

A7.6 GRAPHICS COMMANDS

These graphics commands are used to control the graphics, and include those used for setting line and fill colors and controlling the view in which model and results are displayed.

grFillCol	Sets the current fill color for the drawing.
grLinCol	Sets the current line drawing color.
grLinSty	Sets the current line drawing style.
grLinTh	Sets the current line drawing thickness.
grView	Sets the current viewing direction for the display.
matCol	Sets the colors of materials (starting from material reference number 1) for the display purpose.

A7.7 PREPROCESSOR COMMANDS

These preprocessor commands are required to build the finite element model.

aadd	Creates an area by adding two areas.
adele	Deletes areas having identity numbers in the range specified.
ak	Defines an area by connecting keypoints.
al	Defines an area bounded by previously defined lines.
amesh	Performs area meshing and generates nodes and area elements within the areas in the given range.
arc	Creates an elliptical arc with given centre, radii, and start point and end point angles.
arc2	Creates a circular arc with given radius and being tangent to two straight lines
arc3	Creates an arc at an offset of another arc. This offset is either in terms of centre translation or radius difference.
asub	Creates an area by subtracting one area from another area.
ATrA	Creates an area at an offset and scaled with respect to another (template) area.
block	Creates a block (hexahedral) volume with given coordinates of first and second corners.

block Hollow	Creates a hollow block (box) volume with given coordinates of first and second corners of outer and inner blocks.
cylinder	Creates a cylindrical volume segment with given centre coordinates, radii, and start point and end point angles of the two opposite elliptical faces.
cylinder2	Creates a cylindrical volume segment. This command is identical to the *cylinder* command except that different kinds of volume segments are created by the two commands.
cylHollow	Creates a hollow cylindrical volume segment with given centre coordinates, inner and outer radii and start point and end point angles of the two opposite hollow elliptical faces.
edele	Deletes elements having numbers in the range specified.
ellipse	Creates an elliptical area segment with given centre, radii, and start point and end point angles.
ellipse2	Creates an elliptical segment. This command is identical to the *ellipse* command except that different kinds of area segments are created by the two commands.
EllRing	Creates an elliptical ring area segment with given centre, inner and outer radii and start point and end point angles.
eModify	This command modifies certain attributes of all the selected elements having numbers in the specified range.
ereadf	Reads element data from a given file.
et	Defines element type with given identification number, type name and certain key options.
etdele	Deletes element types having identification numbers in the given range.
fieldCons	Specifies whether field consistency is to be used in the finite element solution. This feature is available only for few element types.
intOrder	Sets the integration order for the element matrices and vectors (identified as integral numbers), for the current element type.
k	Defines a new keypoint at a given location.
kbk	Defines a new keypoint at a parametric location between the two keypoints.
kl	Defines a new keypoint at a parametric location in a given line.
kmesh	Performs keypoint meshing and generates nodes and point elements within the keypoints in the range specified.
ldele	Deletes lines having numbers in the given range.
lesize	Defines number of elements/element edges in all the selected lines having numbers in the given range.

line	Defines a straight line between two given points (locations) in 2D plane.
line3d	Defines a straight 3D line between two given points (locations).
lineccat	Creates a concatenated line of several lines.
lk	Defines a straight line between the two given keypoints.
lmesh	Performs line meshing and generates nodes and line elements within the lines in the given range.
merge	Merges coincident model entities (nodes and elements) by retaining the lowest numbered coincident entity and deleting all higher numbered coincident entities.
mp	Defines material properties for a material property set.
mpDele	Deletes material property sets having identification numbers in the specified range.
n	Defines a new node located at a given point.
nbn	Defines a new node at a parametric location between the two nodes.
ndele	Deletes the nodes having numbers in the given range.
nl	Defines a new node at a parametric location in a given line.
nmodify	modifies the nodal coordinates of a node with new coordinate data.
nreadf	Reads nodal data from a given file.
rect	Creates a rectangular area with given coordinates of its two corners.
selectInt	Specifies whether selective integration is to be used for the current element type.
SetCurCS	Sets and activates the current coordinate system number.
SetCurEA	Sets and activates the current element attribute reference number.
SetCurET	Sets and activates the current element type number.
SetCurMP	Sets and activates the current material property reference number.
SetCurST	Sets and activates the current structure type (master (lower or upper die) or slave (the workpiece)) identified as an integer number.
triangle	Creates a triangular area with given coordinates of its three corners.
unSym Form	Specifies whether unsymmetric formulation is to be used in the finite element solution. This feature is available only for limited element types.
va	Defines a volume bounded by existing areas (upto a maximum of six).

vdele	Deletes all the volumes having numbers in the range given.
vextrud	Creates a volume by extruding an area along a straight line joining the centroid of the area and a point at a given offset from the centroid.
vTwo Areas	Defines a volume created by two opposite faces defined by two given areas.
Work Plane	Defines a working plane coordinate system which can be used to transform the model geometrical data defined in global coordinate system to working plane coordinate system.

A7.8 SOLUTION COMMANDS

These commands are used to apply the loads and to obtain the solution.

antype	Specifies the analysis type.
cnvTol	Sets convergence values for a non-linear analysis.
da	Specifies surface constraints on selected areas in the given range.
dl	Specifies surface constraints on selected lines in the given range.
dn	Specifies degree of freedom constraints at the selected nodes in the given range.
elResEx	Specifies scheme for extrapolation of integration point results to the nodes.
fn	Specifies point forces at the selected nodes in the given range.
kmatOpt	Specifies whether characteristics (K) matrix is to be calculated at the beginning of each iteration or at the beginning of each step, in a non-linear analysis.
massOpt	Specifies whether consistent or lumped mass formulation is to be used in a transient dynamic analysis.
nlGeom	Specifies whether non-linear geometry (large deformation effects) should be included in a static or a transient structural analysis.
noTime Steps	Specifies the number of time steps to be taken for a multiple loadstep/transient analysis.
sfa	Specifies surface load on selected areas in the given range.
sfe	Specifies surface load on selected elements in the given range.
sfl	Specifies surface load on selected lines in the given range.
sfn	Specifies surface load on selected nodes in the given range.
solve	Initiates the solution based on the current analysis type and options.

A7.9 POSTPROCESSOR COMMANDS

These postprocessor commands are used to review the results.

noPlRng	Sets the number of ranges in a contour plot.
plesol	Plots element solutions for the given label in a particular loadstep.
plnsol	Plots nodal solutions for the given label in a particular loadstep.
PlRng Prec	Sets the precision of range values in the legend of a contour plot.
PlRngVal	Sets the values of ranges in a contour plot.
plvnsol	Draws a vector plot (depicting magnitude and direction) of nodal solutions for the given label in a particular loadstep.
presol	Prints element solutions for the given label in a particular loadstep.
presolt	Prints time-varying solutions for a given label and for a given element number.
prnsol	Prints nodal solutions for the given label in a particular loadstep.
prnsolt	Prints time-varying solutions for a given label and for a given node number.

A7.10 AUXILIARY COMMANDS

These are the auxiliary commands that can be used during an analysis.

getAreaArea	Calculates area of a given area.
getAreaCent	Calculates the centroid of a given area.
getKPAtLoc	Retrieves keypoint number at a given location.
getnodAtLoc	Retrieves node number at a given location.
getNodCoord	Retrieves coordinates of a given node.
getNoNod	Retrieves total number of nodes in the structure.
getNoNodDof	Retrieves number of degrees of freedom associated with a node.
getSelStat	Retrieves selection status of a specified entity number for a given entity (keypoint, line, area, node, etc.).
getTotNodDof	Retrieves in a given array degree of freedom values for all the nodes in the structure.
setTotNodDof	Sets from a given array degree of freedom values for all the nodes in the structure.

A7.11 MATHEMATICAL CALCULATION COMMANDS

As stated earlier, C library functions including mathematical ones are not directly supported by the SB-FEM code. However, there are several interface commands defined within the code for the execution of mathematical calculations. In general, all the mathematical commands begin with a prefix m_ and followed by the actual C math library function names. As an example, the following command is used to calculate the arctangent of PAR_1/PAR_2:

```
double dResult = m_atan2(PAR_1, PAR_2);
```

Here PAR_1 and PAR_2 are either floating point constants or floating point variables.

Appendix-8

Some SB-FEM Input Files

A8.1 INTRODUCTION

In this appendix, SB-FEM input (command) files are listed for a few selected problems solved in different chapters. The listing here is merely typical and is given for the sake of completeness. The command files for the entire example problems along with few others are given in a suitable example subdirectory within the SB-FEM program.

A8.2 LISTING OF SOME SB-FEM INPUT FILES

A8.2.1 Example problem (Section 4.6.1)

```
/*
A tapered aluminum alloy bar of square cross section (width × depth)
and length 'Leng' is suspended from a ceiling. An axial load of
10000 units is applied to the free end of the bar.
The output is the maximum axial deflection in the bar and contour of
the axial deflection.
*/

// DEFINE SOME GLOBAL VARIABLES.
float Leng = 10;
float width = 2;
float depth = 2;
float ScaleFactor = 0.5;
float VertLoadAtTip = 10000;

// ~~~~~ GENERAL COMMANDS ~~~~~~~~
int gen ( ) {
    // SET ANALYSIS TITLE.
    title("Elongation of a solid tapered bar using bar element");
    // SPECIFY INPUT AND OUTPUT FILE NAMES.
    inpfilename(,"BarElong1D");
```

```
    outfilename(,"BarElong1D");
    return 1;
}
// ~~~~~ PREPROCESSOR ~~~~~~~~

int prep ( ) {
    // DEFINE ELEMENT TYPE AND OPTION.
    et(1,"link2");
    elresex("on");
    // DEFINE ELEMENT ATTRIBUTES.
    int noDivs = 16;
    float CSWidth[16],CSDepth[16];
    float width1 = width;
    float width2 = width1*ScaleFactor;
    float SlopeWidth = (width1-width2)/noDivs;
    float depth1 = depth;
    float depth2 = depth1*ScaleFactor;
    float SlopeDepth = (depth1-depth2)/noDivs;
    float WidthAtNode,WidthAtNextNode,DepthAtNode,
    DepthAtNextNode;
    int i;
    for (i=0;i<noDivs;++i) {
        WidthAtNode = width1-SlopeWidth*i;
        WidthAtNextNode = width1-SlopeWidth*(i+1);
        CSWidth[i] = (WidthAtNode+WidthAtNextNode)/2;
        DepthAtNode = width1-SlopeDepth*i;
        DepthAtNextNode = width1-SlopeDepth*(i+1);
        CSDepth[i] = (DepthAtNode+DepthAtNextNode)/2;
        ea(i+1,1,2,,CSWidth[i],CSDepth[i]);
    }

    //
    // SPECIFY MATERIAL PROPERTIES.
    mp(1,,"ex",10.4e6);
    // DEFINE GEOMETRY.
    line(0,0,Leng,0);
    // SPECIFY MESH CONTROL.
    lesize(1,1,,noDivs,1);
    // MESH THE LINE.
    lmesh("all");
    // MODIFY (SET) ELEMENT ATTRIBUTES.
    for (i=0;i<noDivs;++i) {
        emodify(i+1,i+1,,"attr",i+1);
    }
    return 1;
}

// ~~~~~ SOLUTION ~~~~~~~~
int solu ( ) {
    // DEFINE ANALYSIS TYPE AND OPTION.
    antype("static");
    nlgeom("off");
    // APPLY BOUNDARY CONDITIONS AT FIXED END.
```

```
    nsel("s","loc","x",0.0,0.0);
    dn("all",,,"ux",0.0);
    nsel("all");
    dn("all",,,"uy",0.0); // must be there to avoid matrix singularity.
    //
    // APPLY LOAD AT THE OTHER (FREE) END.
    // Select node at distance between Leng and Leng.
    nsel("s","loc","x",Leng,Leng);
    // Apply above force at the node selected.
    fn("all",,,"fx",VertLoadAtTip);
    nsel("all");
    //
    // INITIATE THE SOLUTION.
    solve("front");
    return 1;
}

// ~~~~~ POSTPROCESSOR ~~~~~~~~
int post( ) {
    // PRINT NODAL SOLUTION FOR DISPLACEMENTS.
    prnsol("u");
    // PLOT NODAL SOLUTION FOR DISPLACEMENT COMPONENT UX.
    plnsol("ux");
    return 1;
}

// ~~~~~ MAIN FUNCTION ~~~~~~~~
int main ( ) {
    gen ( );
    prep ( );
    solu ( );
    post ( );
    return 1;
}
```

A8.2.2 Example problem (Section 5.5.1)

```
/*
Rectangular cantilever beam (L=10,W=2,T=2) with different loading
cases:
(1) Concentrated bending moment of -4000 units at the free end.
(2) Vertical tip force of 600 units at the free end.
(3) Uniformly distributed load per unit width (pressure) of 80
    units.
NOTE: The three loading cases can be applied by choosing different
options in the procedure solu().
*/

float Leng = 10;
float width = 2;
float depth = 2;

int gen ( ) {
    title("Cantilever beam using beam element");
    inpfilename(,"BeamCL1D");
```

```
    outfilename(,"BeamCL1D");
    return 1;
}

// ~~~~~Preprocessor ~~~~~~~~

int prep( ) {
    et(1,"stbeam3",,,width,depth);
    selectint("on");
    elresex("on");
    mp(1,,"ex",1500);
    mp(1,,"prxy",0.0);
    line3d(0,0,0,Leng,0,0);
    lesize(1,1,,8,1);
    lmesh("all");
    return 1;
}

// ~~~~~Solution ~~~~~~~~
int solu( ) {
    antype("static");
    // boundry conditions at one end.
    nsel("s","loc","x",0.0,0.0);
    dn("all",,,"ux",0.0);
    dn("all",,,"uz",0.0);
    dn("all",,,"ROTXZ",0.0);
    nsel("all");
    //
    // Apply load.
    int option = 3;
    if (option == 1) {
        float moment = -4000;
        // select node at distance between Leng and Leng.
        nsel("s","loc","x",Leng,Leng);
        // Apply above moment at the node selected.
        fn("all",,,"MXZ",moment);
        nsel("all");
    }
    else if (option == 2) {
        float force = 600;
        // select node at distance between Leng and Leng.
        nsel("s","loc","x",Leng,Leng);
        // Apply above force at the node selected.
        fn("all",,,"fz",force);
        nsel("all");
    }
    else if (option == 3) {
        float surfload = 80;
        // Apply above surfload at all the lines (here only one).
        sfl("all",,,"press",,,surfload,,1);
    }
    //
    solve("front");
    return 1;
```

```
}

// ~~~~~Postprocessor ~~~~~~~~
int post ( ) {
   prnsol("u");
   plnsol("uz");
   return 1;
}

int main ( ) {
   gen ( );
   prep ( );
   solu ( );
   post ( );
   return 1;
}
```

A8.2.3 Example problem (Section 6.4.3)

```
/*
A square plate with a circular hole at the centre and subjected to
uniform tension on both sides.
Length = 300 units,
width = 300 units,
Thickness = 25 units,
Diameter of circular hole = 20 units,
E = 2.0e5 units
and Load = 100 units.

This is a stress concentration problem.
*/

double Length = 300;
double Width = 300;
double HalfLength = Length/2.0;
double HalfWidth = Width/2.0;
double Thickness = 25;
double HoleDia = 20;

int LineNo1,LineNo2,LineNo3,LineNo4,LineCatNo,AreaNo1;

int gen( ) {
   title("Plate with a hole problem using 4,8 and 9 noded quadrilateral
element");
   inpfilename(,"PlateWithHole2D");
   outfilename(,"PlateWithHole2D");
   return 1;
}

// ~~~~~Preprocessor ~~~~~~~~
int prep( ) {
   // define element type.
   int ElemOpt = 0;
   if (ElemOpt == 0) {
      int incompatible_modes = 1;
      et(1,"stquad4",incompatible_modes,,Thickness);
```

```
    }
    else if (ElemOpt == 1) {
        //intorder(1,1,,"red1");
        //selectint("on");
        et(1,"stquad8",,,Thickness);
    }
    else if (ElemOpt == 2) {
        //intorder(1,1,,"red1");
        //selectint("on");
        et(1,"stquad9",,,Thickness);
    }
    elresex("on");
    //
    // define material properties.
    mp(1,,"ex",2.0e5);
    mp(1,,"prxy",0.3);
    //
    int noSubArcs;
    int SubArcs[8];
    arc(0,0,0,HoleDia/2.0,HoleDia/2.0,0,90,,0,&noSubArcs, SubArcs);
    print(noSubArcs);
    print(SubArcs[0]);
    LineNo1 = line(HoleDia/2.0,0,HalfLength,0.0);
    LineNo2 = line(HalfLength,0,HalfLength,HalfLength);
    LineNo3 = line(HalfLength,HalfLength,0,HalfLength);
    LineNo4 = line(0,HoleDia/2.0,0,HalfLength);
    print(LineNo1," ",LineNo2," ",LineNo3," ",LineNo4);

    // Concatenate two lines.
    LineCatNo = lineccat(LineNo3,LineNo2);
    print(LineCatNo);
    AreaNo1 = al(SubArcs[0],LineNo1,LineCatNo,LineNo4);
    //
    int NoDivArc,NoDivLengWid;
    NoDivArc = 16;
    NoDivLengWid = 16;
    //
    lsel("s","ltype",,"arc");
    lesize("all",,,NoDivArc,1);
    lsel("all");
    //
    lsel("s","num",,LineNo2,LineNo2);
    lsel("a","num",,LineNo3,LineNo3);
    lesize("all",,,NoDivArc/2,1);
    lsel("all");
    //
    lsel("s","num",,LineNo1,LineNo1);
    lsel("a","num",,LineNo4,LineNo4);
    lesize("all",,,NoDivLengWid,0.2);
    lsel("all");
    //
    SetCurET(1); // if not given, the last one created. Case is not
important.
```

```
    amesh("all");
    return 1;
}
// ~~~~~Solution ~~~~~~~~
int solu() {
    antype("static");
    //
    // Apply boundary conditions.
    lsel("s","num",,LineNo1,LineNo1);
    nsel("s","line");
    dn("all",,,"uy",0.0);
    nsel("all");
    lsel("all");
    //
    lsel("s","num",,LineNo4,LineNo4);
    nsel("s","line");
    dn("all",,,"ux",0.0);
    nsel("all");
    lsel("all");
    //
    // Apply load.
    float surfload = 100;
    // Apply above surfload at selected lines.
    lsel("s","num",,LineNo2,LineNo2);
    sfl("all",,,"press",,,surfload,,1);
    lsel("all");
    //
    solve("front");
    return 1;
}

// ~~~~~Postprocessor ~~~~~~~~
int post() {
    plnsol("ux");
    plesol("sigx");
    return 1;
}
int main() {
    gen();
    prep();
    solu();
    post();
    return 1;
}
```

A8.2.4 Example problems (Sections 7.6.1 and 7.6.2)

```
/*
A square plate of size (Leng x Leng) is clamped/simply-supported at
all edges and is subjected to a unit point load at the centre or a
unit uniformly distributed load.
Calculate the displacement at centre of the plate.
*/
```

```
float Side = 2175;
float HalfLeng = Side/2;
float HalfWidth = Side/2;
float depth;
int LoadOption = 1;

int gen() {
   title("Bending of a square plate using 8 noded plate element");
   inpfilename(,"PlateBend2DPl8");
   outfilename(,"PlateBend2DPl8");
   return 1;
}

// ~~~~~Preprocessor ~~~~~~~~
int prep() {
   int case = 1;
   if (case == 0) depth = 2;
   else depth = 40;
   et(1,"stplatequad8",,,depth);
   //intorder(1,1,,"red1");
   selectint("on");
   elresex("on");
   //
   mp(1,,"ex",2.0e5);
   mp(1,,"prxy",0.3);
   rect(0.0,0.0,HalfLeng,HalfWidth);
   //
   int NoDivLenWid = 16;
   float scaleFact = 1;
   if (LoadOption == 1) scaleFact = 0.4;
   //
   // Horizontal lines
   lsel("s","loc","y",0,0);
   lesize("all",,,NoDivLenWid,scaleFact);
   lsel("all");
   //
   lsel("s","loc","y",HalfWidth,HalfWidth);
   lesize("all",,,NoDivLenWid,1.0/scaleFact);
   lsel("all");
   //
   // Vertical lines
   lsel("s","loc","x",0,0);
   lesize("all",,,NoDivLenWid,1.0/scaleFact);
   lsel("all");
   //
   lsel("s","loc","x",HalfLeng,HalfLeng);
   lesize("all",,,NoDivLenWid,scaleFact);
   lsel("all");
   //
   SetCurET(1); // if not given, the last one created.
   amesh("all");
   return 1;
}
```

```
// ~~~~~Solution~~~~~~~
int solu() {
    antype("static");
    //
    int BCOption = 0;
    if (BCOption == 0) {
        // clamped conditions on two edges.
        nsel("s","loc","x",HalfLeng,HalfLeng);
        dn("all",,,"uz",0.0);
        dn("all",,,"ux",0.0);
        dn("all",,,"uy",0.0);
        dn("all",,,"ROTYZ",0.0);
        dn("all",,,"ROTXZ",0.0);
        nsel("all");
        nsel("s","loc","y",HalfWidth,HalfWidth);
        dn("all",,,"uz",0.0);
        dn("all",,,"ux",0.0);
        dn("all",,,"uy",0.0);
        dn("all",,,"ROTYZ",0.0);
        dn("all",,,"ROTXZ",0.0);
        nsel("all");
    }
    else {
        // Simply-supported conditions on two edges.
        nsel("s","loc","x",HalfLeng,HalfLeng);
        dn("all",,,"uz",0.0);
        dn("all",,,"ROTXZ",0.0);
        nsel("all");
        nsel("s","loc","y",HalfWidth,HalfWidth);
        dn("all",,,"uz",0.0);
        dn("all",,,"ROTYZ",0.0);
        nsel("all");
    }
    //
    // symmetric b.c. for quarter symmetry.
    nsel("s","loc","x",0.0,0.0);
    dn("all",,,"ux",0.0);
    dn("all",,,"ROTYZ",0.0);
    nsel("all");
    nsel("s","loc","y",0.0,0.0);
    dn("all",,,"uy",0.0);
    dn("all",,,"ROTXZ",0.0);
    nsel("all");
    // Apply load.
    if (LoadOption == 1) {
        float pointLoad = 1;
        // Apply above pointLoad at centre.
        int NodeNo = getnodAtLoc(0,0.0,0.0,0.0); // case is not
important in getnodAtLoc.
        fn(NodeNo,NodeNo,,"fz",pointLoad/4); // Note: /4 (Quarter
symmetry)
    }
```

```
    else if (LoadOption == 2) {
        float surfLoad = 1;
        // Apply above surfload at all areas (in present case only
        1).
        sfa("all",,,"press",,,,surfLoad,1);
    }
    solve("front","all");
    return 1;
}

// ~~~~~Postprocessor ~~~~~~~~
int post() {
    int NodeNo = getnodAtLoc(0,0.0,0.0,0.0); // case is not important
in getnodAtLoc.
    prnsol("u",,NodeNo);
    plnsol("uz");
    return 1;
}
int main() {
    gen();
    prep();
    solu();
    post();
    return 1;
}
```

A8.2.5 Example problem (Section 8.6.1)

```
/*
A tapered aluminum alloy bar of square cross section (width x depth)
and length 'Leng' is suspended from a ceiling. An axial load of
10000 units is applied to the free end of the bar after converting it
to an equivalent surface load.
The output is the maximum axial deflection in the bar and contour of
axial deflection.
*/

float Leng = 10;
float width = 2;
float depth = 2;
float ScaleFactor = 0.5;
int AreaNo1,AreaNo2;
float VertLoadAtTip = 10000;

int gen( ) {
    title("Elongation of a solid tapered bar using 8 noded/20 noded
brick element");
    inpfilename(,"BarElong3D");
    outfilename(,"BarElong3D");
    return 1;
}

// ~~~~~Preprocessor ~~~~~~~~

int prep() {
```

```
    /*
    Here one work plane is defined to transform the model so that
    base (bottom) of the bar lies in -y plane instead of z-plane.
    Although it is not necessary, it has been done here so that the
    model looksvsame as the actual body in its normal view.
    */
    int WPNo = 1;
    WorkPlane(WPNo,0,0,0,100,0,0,100,0,100); // 100 is arbitrary
    number.
    AreaNo1 = rect(0,0,width,depth,,WPNo);
    print(AreaNo1);
    AreaNo2 = AtrA(AreaNo1,0,-Leng,0,ScaleFactor);
    print(AreaNo2);
    VTwoAreas(AreaNo1,0,AreaNo2,0);
    //
    int NoDivLeng,NoDivWid;
    NoDivLeng = 16;
    NoDivWid = 4;
    asel("s","num",,AreaNo1,AreaNo2,AreaNo2-AreaNo1);// AreaNo2-
AreaNo1 is the increment (Here it will be 1).
    lsel("s","area");
    lesize("all",,,NoDivWid,1);
    lsel("inv");
    lesize("all",,,NoDivLeng,1);
    lsel("all");
    asel("all");
    //
    int option = 2;
    if (option == 1) {
        int incopatibleMode = 1;
        et(1,"stbrick8",incopatibleMode);
    }
    else {
        et(1,"stbrick20");
    }
    elresex("on");
    mp(1,,"ex",10.4e6);
    mp(1,,"prxy",0.3);
    vmesh("all");
    print("I am here 1");
    return 1;
}

// ~~~~~Solution ~~~~~~~~
int solu() {
    antype("static");
    // boundry conditions at top face.
    // select nodes at top face.
    nsel("s","areano",,AreaNo1,AreaNo1);
    dn("all",,,"ux",0.0);
    dn("all",,,"uy",0.0);
    dn("all",,,"uz",0.0);
    nsel("all");
```

```
    int option = 2;
    if (option == 1) { // will give wrong result.
        // Apply force at the centre node of bottom face.
        double LoadPt[3];
        getAreaCent(AreaNo2,LoadPt); // case is not important in get
AreaCent.
        int NodeNo = getnodAtLoc(0,LoadPt[0],LoadPt[1], LoadPt[2]);
// case is not important in getnodAtLoc.
        nsel("s","num",,NodeNo,NodeNo);
        fn("all",,,"fy",-VertLoadAtTip);
        nsel("all");
    }
    else {
        double surfArea;
        getAreaArea(AreaNo2,&surfArea);
        float surfload = VertLoadAtTip/surfArea;
        asel("s","num",,AreaNo2,AreaNo2);
        // Apply above surfload at selected area.
        sfa("all",,,"press",,,,surfload,1);
        /*
        // Alternatively, we can select node corresponding to the
area and apply surface pressure on selected
        // nodes.
        nsel("s","area");
        sfn("all",,,"press",,,,surfload,1,2);
        nsel("all");
        */
        asel("all");
    }
    print("I am here 2");
    solve("front");
    return 1;
}

int post() {
    // ~~~~~Postprocessor starts ~~~~~~~~
    double LoadPt[3];
    getAreaCent(AreaNo2,LoadPt);//case is not important in getAreaCent.
    int NodeNo = getnodAtLoc(0,LoadPt[0],LoadPt[1],LoadPt [2]); //
case is not important in getnodAtLoc.
    prnsolt("u",,NodeNo);
    plnsol("uy");
    return 1;
}

int main() {
    gen();
    prep();
    solu();
    post();
    return 1;
}
```

A8.2.6 Example problem (Section 9.5.2)

```
/*
An example of an isotropic rectangular steel plate of length 200 mm,
width 200 mm and thickness 8 mm simply supported on its edges is
considered.
The plate is subjected to an impact induced by a 20 mm diameter steel
ball with initial velocity 1 m/s.
The material properties are taken as:
DENS = 7800 kg/m^3, EX =207 GPa; PRXY = 0.3.
*/

float length,width,halflength,halfwidth;
double PlateThk;

int model = 0;
float imprad = 10e-3;
float impdens = 7800;
float impex = 207e9;
float impnu = 0.3;
float impinitvel = 1;
float tardens = 7800;
float tarex = 207e9;
float tarnu = 0.3;
float deltim;
float nosteps = 20;
float NMalpha1,NMalpha2,NMbeta1,NMbeta2;
float PI = 3.1415926;
int noNode,nonoddof,totnoDof;

int gen() {
    title("Impact on a steel plate");
    outfilename(,"IsoPlate3DImp");
    inpfilename(,"IsoPlate3DImp");
    return 1;
}
// ~~~~~Preprocessor ~~~~~~~~
int prep() {
    PlateThk = 0.008;
    int noDivThick = 4;
    int ElemOpt = 0;
    if (ElemOpt == 0) {
        int incompatible_modes = 1;
        et(1,"stbrick8",incompatible_modes);
    }
    else if (ElemOpt == 1) {
        et(1,"stbrick20");
    }
    elresex("on");
    //
    mp(1,,"dens",7800);
    mp(1,,"ex",207e9);
    mp(1,,"prxy",0.3);
    //
```

```
    // Quarter symmetry.
    length = 0.2;
    halflength = length/2.0;
    width = 0.2;
    halfwidth = width/2.0;
    block(0.0,0.0,0.0,halflength,halfwidth,PlateThk);
    int NoDivLengWid = 8;
    lesize(1,1,,NoDivLengWid,0.2);
    lesize(2,2,,NoDivLengWid,0.2);
    lesize(3,3,,NoDivLengWid,5);
    lesize(4,4,,NoDivLengWid,5);
    lesize(5,5,,NoDivLengWid,0.2);
    lesize(6,6,,NoDivLengWid,0.2);
    lesize(7,7,,NoDivLengWid,5);
    lesize(8,8,,NoDivLengWid,5);
    //
    lsel("s","leng","z",PlateThk,PlateThk);
    lesize("all",,,noDivThick,1);
    lsel("all");
    //
    SetCurET(1); // if not given, the last one created. Case is not
important.
    vmesh("all");
    // ~~~~~Preprocessor finishes ~~~~~~~~
    return 1;
}

// ~~~~~Solution ~~~~~~~~
int solu() {
    float kk1 = (1-m_pow(impnu,2.0))/(PI*impex);
    float kk2 = (1-m_pow(tarnu,2.0))/(PI*tarex);
    float MatConst_k = 4.0*m_pow(imprad,0.5)/(3.0*PI*(kk1+kk2));
    float impmass = (4.0/3.0)*PI*m_pow(imprad, 3.0)*impdens;
    float tarmass = length*width*PlateThk*tardens;
    float invmass = 1/impmass + 1/tarmass;
    float   alphamax   =   m_pow(5*m_pow(impinitvel,2.0)/
(4*invmass*MatConst_k),(2.0/5.0));
    float timem = 2.94*alphamax/impinitvel;
    deltim = timem/24;
    print("deltim = ",deltim); //deltim = 2.87172e-6.
    //
    antype("transient");
    //massopt("lump");
    notimesteps(nosteps);
    time(nosteps*deltim);
    //
    // Simply-supported conditions on two edges.
    nsel("s","loc","x",halflength,halflength);
    nsel("r","loc","z",PlateThk/2,PlateThk/2);
    dn("all",,,"ux",0.0);
    dn("all",,,"uz",0.0);
    nsel("all");
```

```
nsel("s","loc","y",halfwidth,halfwidth);
nsel("r","loc","z",PlateThk/2,PlateThk/2);
dn("all",,,"uy",0.0);
dn("all",,,"uz",0.0);
nsel("all");
// symmetric b.c. for quarter symmetry.
nsel("s","loc","x",0.0,0.0);
dn("all",,,"ux",0.0);
nsel("all");
nsel("s","loc","y",0.0,0.0);
dn("all",,,"uy",0.0);
nsel("all");
//
NMalpha1 = 0.5;  NMalpha2 = 0.5;
NMbeta1 = 0.25;  NMbeta2 = 0.25;
float force = 1;

int startStepNo = 1;
int endStepNo = nosteps;
//
float exponent1 = 1.5;
float exponent2 = 2.5;
float exponent3 = 1.5;
float PermIndExponent = 2.0/5.0;

float term = m_pow(deltim,2.0)/impmass;
float* F,*V,*W;
F  = new float[nosteps+1];
V  = new float[nosteps+1];
W = new float[nosteps+1];

int i,j,ii,jj;
for (i=0;i<=nosteps;++i) {
    F[i]  = 0;
    V[i]  = 0;
    W[i] = 0;
}
V[0] = impinitvel;

double* Disp1,*Disp2;
int noNode = getNoNod();
int nonoddof = getNoNodDof();
int totnoDof = noNode*nonoddof;
Disp1 = new double[totnoDof+1];
Disp2 = new double[totnoDof+1];
//
int midNodeAtImpPt = getnodAtLoc(0,0.0,0.0,PlateThk/2.0);
int variab = 3;
int dofnum = (midNodeAtImpPt-1)*nonoddof+variab; // 1-based.
print("dofnum = ", dofnum);
float
unit,dofdisp,alpha,oldf,G,DG,Alph,Alph1,diff,fmax,alpha0,
alpha02,alphacr;
alpha0 = 0;
```

```
    solstart();
    //
    for (i=startStepNo;i<=endStepNo;++i) {
        stepstart(i);

        if (i == startStepNo) {
            nsel("s","loc","x",0.0,0.0);
            nsel("r","loc","y",0.0,0.0);
            nsel("r","loc","z",0.0,0.0);
            // Apply force at the node selected.
            fn("all",,,"fz",force);
            nsel("all");
            //
            effLoadCal("off");

            stepsolve("front",i);
            getTotNodDof(Disp1);
            unit = Disp1[dofnum];
            if (model == 0) unit = unit/4.0;
            //print("unit = ",unit);
            //
            effLoadCal("on");
            nsel("s","loc","x",0.0,0.0);
            nsel("r","loc","y",0.0,0.0);
            nsel("r","loc","z",0.0,0.0);
            // Delete force at the node selected.
            fn("all",,,"fz",0);
            nsel("all");
        }
        //
        stepsolve("front",i);
        getTotNodDof(Disp2);
        //
        dofdisp = Disp2[dofnum];
        //print("dofdisp = ",dofdisp);
        W[i]  =  W[i-1]+deltim*V[i-1]-NMbeta1*term*F[i-1]-
NMbeta2*term*F[i-1];
        alpha = W[i]-dofdisp-F[i-1]*unit;
        if (alpha <= 0 && i > 1) {
            alpha = W[i]-dofdisp-(2*F[i-1]-F[i-2])*unit;
        }
        //print("alpha = ",alpha);
        //
        if (alpha > 0) {
            F[i] = MatConst_k*m_pow(alpha,exponent1);
            for (;;) {
                oldf = F[i];
                Alph   =   W[i]-NMbeta2*term*(F[i]-F[i-1])-
dofdisp-F[i]*unit;
                if (Alph <= 0) {
                    F[i] = 0;
                    break;
                }
```

```
        G = F[i] - MatConst_k*m_pow(Alph,exponent1);
            DG = 1 -
  exponent1*MatConst_k*m_pow(Alph,(exponent1-1))*(-NMbeta2*term-
unit);
            F[i] = F[i] - G/DG;
            if (F[i] > 0) {
                diff = m_fabs(F[i]-oldf)/F[i];
            }
            else {
                F[i] = 0; break;
            }
            if (diff <= 1e-5) break;
        }
        //
      }
      else {
        print("F[i] = ", F[i]);
        F[i] = 0;
      }
      //print("F[i] = ",F[i]);
      W[i] = W[i] - NMbeta2*term*(F[i]-F[i-1]);
      //print("W[i] = ",W[i]);
      V[i] = V[i-1] - (deltim/impmass)*(NMalpha1*F[i-
1]+NMalpha2*F[i]);
      //
      for (ii=0; ii<=totnoDof; ++ii) {
        if (model == 0) Disp2[ii] = Disp2[ii] + (F[i]/
4.0)*Disp1[ii];
        else if (model == 1) Disp2[ii] = Disp2[ii]+
F[i]*Disp1[ii];
      }
      setnodDof(Disp2);
      //
      stepfinish(i);
    }
    solfinish( );
    //
    delete[ ] Disp1;
    delete[ ] Disp2;
    delete[ ] F;
    delete[ ] V;
    delete[ ] W;
    return 1;
}

// ~~~~~Postprocessor ~~~~~~~~
int post() {
    int NodeNo = getnodAtLoc(0,0.0,0.0,PlateThk/2.0);
// case is not important in getnodAtLoc.
    prnsolt("u",,NodeNo);
    plnsol("uz"); // current load step.
    return 1;
}
```

```
int main( ) {
    gen( );
    prep( );
    solu( );
    post( );
    return 1;
}
```

A8.2.7 Example problem (Section 10.4.2)

```
/*
A 16-ply graphite-epoxy square laminate of size b = 254 mm and
thickness h = 2.114 mm is considered as investigated by Noor (1985).
The plate is clamped on all edges and subject to uniform load q. The
material properties of graphite-epoxy unidirectional ply are taken
as:
EX =131 GPa;EY=EZ=13.03 GPa; GXY=GYZ=GXZ = 6.41 GPa; PRXY = PRYZ =
PRXZ = 0.38.
*/

double length,width,halflength,halfwidth;
double Laminatethk;
double EY;

int gen( ) {
    title("graphite/epoxy plate subjected to a uniform load");
    inpfilename(,"GrEpPlateNoor3D");
    outfilename(,"GrEpPlateNoor3D");
    return 1;
}

// ~~~~~Preprocessor ~~~~~~~~
int prep() {
    Laminatethk = 0.002114;
    int TotNoLays = 16;
    int noDivThick = 4;
    int NoLayPerElem;
    int ElemOpt = 1;
    if (ElemOpt == 0) {
        int incompatible_modes = 1;
        et(1,"stbrick8lay",incompatible_modes);
        NoLayPerElem = TotNoLays/noDivThick;
    }
    else if (ElemOpt == 1) {
        et(1,"stbrick20lay");
        NoLayPerElem = TotNoLays/(noDivThick/2);
    }
    elresex("on");
    //
    int NoLayData = TotNoLays/NoLayPerElem;
    float eachlayth = Laminatethk/TotNoLays;
    //
    float Orient[16];
    Orient[0] = 45;
```

```
Orient[1] = -45;
Orient[2] = 0;
Orient[3] = 0;
Orient[4] = -45;
Orient[5] = 45;
Orient[6] = 90;
Orient[7] = 90;
Orient[8] = 90;
Orient[9] = 90;
Orient[10] = 45;
Orient[11] = -45;
Orient[12] = 0;
Orient[13] = 0;
Orient[14] = -45;
Orient[15] = 45;
//
int MatRef = 1;
int n,ii = -1,i;
for (n=1;n<=NoLayData;++n) {
    lay(n,NoLayPerElem);
    for (i=1;i<=NoLayPerElem;++i) {
        laymore(i,eachlayth,Orient[++ii],MatRef);
    }
}
//
EY = 13.03e9; // This is needed in solution phase.
mp(1,,"dens",1536);
mp(1,,"ex",131e9);
mp(1,,"ey",EY);
mp(1,,"ez",EY);
mp(1,,"prxy",0.38);
mp(1,,"pryz",0.38);
mp(1,,"prxz",0.38);
mp(1,,"gxy",6.41e9);
mp(1,,"gyz",6.41e9);
mp(1,,"gzx",6.41e9);
mp(1,,"yt",45e6);
mp(1,,"yc",252e6);
mp(1,,"si",105e6);
//
// Quarter symmetry.
length = 0.254;
halflength = length/2.0;
width = 0.254;
halfwidth = width/2.0;
block(0.0,0.0,0.0,halflength,halfwidth, Laminatethk);
int NoDivLengWid = 8;
lsel("s","leng","x",halflength,halflength);
lesize("all",,,NoDivLengWid,1);
lsel("all");
//
lsel("s","leng","y",halfwidth,halfwidth);
```

```
    lesize("all",,,NoDivLengWid,1);
    lsel("all");
    //
    lsel("s","leng","z",Laminatethk,Laminatethk);
    lesize("all",,,noDivThick,1);
    lsel("all");
    //
    SetCurET(1); // if not given, the last one created. Case is not
important.
    vmesh("all");
    return 1;
}

// ~~~~~Solution ~~~~~~~~
int solu() {
    antype("multistep");
    nlgeom("on");
    int noSteps = 50;
    notimesteps(noSteps);
    //kmatopt("step");
    cnvtol("d-r","dof",0.02,0.05);  // 2% and 5%
    cnvtol("d-r","force",0.02,0.05);// 2% and 5%
    cnvtol("n-r","dof",0.01,0.03);  // 1% and 3%
    cnvtol("n-r","force",0.01,0.03);// 1% and 3%
    //
    // clamped conditions on two edges.
    nsel("s","loc","x",halflength,halflength);
    dn("all",,,"ux",0.0);
    dn("all",,,"uz",0.0);
    nsel("all");
    nsel("s","loc","y",halfwidth,halfwidth);
    dn("all",,,"uy",0.0);
    dn("all",,,"uz",0.0);
    nsel("all");
    // symmetric b.c. for quarter symmetry.
    nsel("s","loc","x",0.0,0.0);
    dn("all",,,"ux",0.0);
    nsel("all");
    nsel("s","loc","y",0.0,0.0);
    dn("all",,,"uy",0.0);
    nsel("all");
    //
    //print(Laminatethk);
    //
    // Apply load.
    double term = m_pow(length,4.0)/(EY*m_pow (Laminatethk, 4.0));
    print(term);
    double surfLoadAtEndOfAllSteps = 2000/term;
    print(surfLoadAtEndOfAllSteps);
    // Apply above surfload at selected area.
    ksel("s","loc","z",0.0,0.0);
    asel("s","kp");
```

```
    //alist("all"); // If uncommented, lists the area selected which
can be verified.
    // Apply above surfload at selected area.
    sfa("all",,,"press",,,,surfLoadAtEndOfAllSteps,1);
    asel("all");
    solve("front");
    return 1;
}

// ~~~~~Postprocessor ~~~~~~~~
int post() {
    int NodeNo = getnodAtLoc(0,0.0,0.0,0.0); // case is not important
in getnodAtLoc.
    prnsolt("u",,NodeNo);
    plnsol("uz"); // current load step.
    return 1;
}

int main ( ) {
    gen ( );
    prep ( );
    solu ( );
    post ( );
    return 1;
}
```

A8.2.8 Example problem (Section 11.4.1)

```
/*
Large deformation of hinged-hinged and pinned-pinned beams. A uniform
beam of length L=100, cross-section dimensions of 1x1, made of a
material with E = 30E6 is considered. The beam is simply supported
at both ends and subjected to a uniformly distributed load of intensity
q per unit length. The units are consistently chosen, so that the
exact deflection at the middle of the beam in linear bending theory
is 0.5208, when q = 1. In linear bending theory, the beam is assumed
to undergo pure bending (i.e. there is no axial deformation). Hinged-
hinged (HH) condition is designated for the case where there is no
axial restraint at both ends and pinned-pinned (PP) condition for
the one where there is full restraint. In the former case, inextensional
bending occurs which is largely of a linear nature, and in the
latter case, bending with extension exists.
NOTE: The two conditions (HH and PP) can be set by choosing option in
solu( ).
*/

float HalfLeng = 50;
float width = 1;
float depth = 1;

int gen() {
    title("simply-supported beam using 8 noded shell element");
    inpfilename(,"BeamSS2DPl8NL");
    outfilename(,"BeamSS2DPl8NL");
```

```
    return 1;
}

// ~~~~~Preprocessor ~~~~~~~~
int prep() {
    et(1,"stshellquad8",,,depth);
    //intorder(1,1,,"red1");
    selectint("on");
    elresex("on");
    //
    mp(1,,"ex",3.0e7);
    rect(0.0,0.0,HalfLeng,width);
    //
    lsel("s","leng","x",HalfLeng,HalfLeng);
    lesize("all",,,8,1);
    lsel("all");
    //
    lsel("s","leng","y",width,width);
    lesize("all",,,2,1);
    lsel("all");
    //
    SetCurET(1); // if not given, the last one created.
    amesh("all");
    return 1;
}

// ~~~~~Solution ~~~~~~~~
int solu() {
    antype("multistep");
    nlgeom("on");
    int noSteps = 10;
    notimesteps(noSteps);
    //kmatopt("step");
    cnvtol("d-r","dof",0.02,0.05);  // 2% and 5%
    cnvtol("d-r","force",0.02,0.05); // 2% and 5%
    cnvtol("n-r","dof",0.01,0.03);  // 1% and 3%
    cnvtol("n-r","force",0.01,0.03); // 1% and 3%
    //
    int HHOrPP = 1;
    if (HHOrPP == 1) {
        // HH conditions
        // boundry conditions at one end.
        nsel("s","loc","x",0.0,0.0);
        dn("all",,,"uz",0.0);
        nsel("all");
        // boundry conditions at symmetrical half length.
        nsel("s","loc","x",HalfLeng,HalfLeng);
        dn("all",,,"ux",0.0);
        //dn("all",,,"uy",0.0); // not useful.
        dn("all",,,"ROTYZ",0.0); // ROTYZ, case is not important.
        nsel("all");
    }
    else if (HHOrPP == 2) {
```

```
        // PP conditions
        // boundry conditions at one end.
        nsel("s","loc","x",0.0,0.0);
        dn("all",,,"uz",0.0);
        dn("all",,,"ux",0.0);
        nsel("all");
        // boundry conditions at symmetrical half length.
        nsel("s","loc","x",HalfLeng,HalfLeng);
        dn("all",,,"ux",0.0);
        //dn("all",,,"uy",0.0); // not useful.
        dn("all",,,"ROTYZ",0.0);
        nsel("all");
    }
    // Apply load.
    float surfLoadAtEndOfAllSteps = 10;
    // Apply above surfload at all area (in present case only 1).
    sfa("all",,,"press",,,,surfLoadAtEndOfAllSteps,1);
    solve("front","all");
    return 1;
}
// ~~~~~Postprocessor ~~~~~~~~

int post() {
    int NodeNo = getnodAtLoc(0,HalfLeng,width/2,0.0);
// case is not important in getnodAtLoc.
    prnsolt("u",,NodeNo);
    plnsol("uz"); // current load step (10).
    return 1;
}

int main() {
    gen();
    prep();
    solu();
    post();
    return 1;
}
```

A8.2.9 Example problem (Section 12.5.2)

```
/*
Heat transfer analysis of a blast furnace hearth.
*/

int FurnaceId = 1;
int CalculationType = 0;
// 0: Coke bed geometry not included, 1: geometry included with
floating, 2: geometry included with sitting
int CokeBedGeomOption = 0;
int WallBCOption = 1;
int UHBCOption = 1;
double r[17],z[15];
float tol;
float temphm1,temphm2,tempsm;
```

```
int CBPartZ;
int *Arcoke,*Ariron;
double HeighthmUnderCoke;
int MatNumIronPool,MatNumCokeBed,MatNumCorner,MatNum Deposit;

int gen() {
    title("Heat transfer analysis in a blast furnace hearth");
    inpfilename(,"BFAHearth1");
    outfilename(,"BFAHearth1");
    return 1;
}

// ~~~~~Preprocessor ~~~~~~~~

int prep() {
    //
    float cmass = 0.050;
    r[1] = 0.0;
    r[2] = 3.3;   // coke bed size
    r[3] = (11.570/2.0)-2.000;
    r[4] = r[3]+0.200-cmass;
    r[5] = r[4]+cmass;
    r[6] = r[5]+0.200-cmass;
    r[7] = r[6]+cmass;
    r[8] = r[7]+0.200-cmass;
    r[9] = r[8]+cmass;
    r[10] = r[9]+0.200-cmass;
    r[11] = r[10]+cmass;
    r[12] = r[11]+0.200-cmass;
    r[13] = r[12]+cmass;
    r[14] = 9.750/2;
    r[15] = r[14]+0.910;
    double cmasside = 0.150;
    r[16] = r[15]+cmasside;
    z[1] = 0.0;
    z[2] = cmass;
    z[3] = z[2]+1.600;
    double halbt = 0.551;
    double hcbt = 0.551;
    z[4] = z[3]+halbt;
    z[5] = z[4]+halbt;
    z[6] = z[5]+halbt;
    z[7] = z[6]+halbt;
    z[8] = z[7]+halbt;
    z[9] = z[8]+hcbt;
    z[10] = z[9]+hcbt;
    z[11] = z[10]+hcbt;
    z[12] = z[8]+1.761; // upto the taphole
    //
    //z[13] = z[11]+hcbt;
    //z[14] = z[13]+hcbt;
    //
    HeighthmUnderCoke = 0.5; // Assumed for height of hot metal
beneath coke-bed
```

```
int i,j;
for (i=1; i<=15; ++i) {
    for (j=1; j<=7; ++j) {
        rect(r[i], z[j], r[i+1], z[j+1]);
    }
}
for (i=3; i<=15; ++i) {
    rect(r[i],z[8],r[i+1],z[9]);
}
int strrec;
if (CokeBedGeomOption == 0)  strrec = 5;
else  strrec = 3;
for (i=strrec; i<=15; ++i) {
    rect(r[i],z[9],r[i+1],z[10]);
}
if (CokeBedGeomOption== 0)  strrec = 14;
else  strrec = 3;
for (i=strrec; i<=15; ++i) {
    for (j=10; j<=11; ++j) {
        rect(r[i],z[j],r[i+1],z[j+1]);
    }
}
CBPartZ = 4; // Coke-bed partition in Z-axis
//
float *xccok,*yccok,*xcoke,*ycoke,*xside,*yside;
int *kpccok,*kpcoke,*kpside;
xccok = new float[CBPartZ+1];
yccok = new float[CBPartZ+1];
xcoke = new float[CBPartZ+1];
ycoke = new float[CBPartZ+1];
xside = new float[CBPartZ+1];
yside = new float[CBPartZ+1];
//
kpccok = new int[CBPartZ+1];
kpcoke = new int[CBPartZ+1];
kpside = new int[CBPartZ+1];
//
Arcoke = new int[CBPartZ];
Ariron = new int[CBPartZ];
//
int Csys = 0;
if (CokeBedGeomOption > 0) {
    for (i=0;i<=CBPartZ;++i) {
        xccok[i] = r[1];
        yccok[i] = z[8+i];
        ycoke[i] = z[8+i];
    }
    for (i=0;i<=CBPartZ;++i) {
        xside[i] = r[3];
        yside[i] = z[8+i];
    }
    xcoke[0] = r[2];
```

```
        xcoke[1] = 3.2;
        xcoke[2] = 3.1;
        xcoke[3] = 3.0;
        xcoke[4] = 2.9;
        int kpno;
        for (i=0;i<=CBPartZ;++i) {
            kpno = 0;
            kpccok[i] = k(kpno,xccok[i],yccok[i],0);
            kpno = 0;
            kpcoke[i] = k(kpno,xcoke[i],ycoke[i],0);
            kpno = 0;
            kpside[i] = k(kpno,xside[i],yside[i],0);
        }
bspline("keypoint",Csys,kpcoke[0],kpcoke[1],kpcoke[2],
kpcoke[3],kpcoke[4]);
    for (j=0; j<CBPartZ; ++j) {
        Arcoke[j] =
ak(kpccok[j],kpcoke[j],kpcoke[j+1],kpccok[j+1]);
        Ariron[j] =
ak(kpcoke[j],kpside[j],kpside[j+1],kpcoke[j+1]);
        }
    }
    //
    print("I am here 1");
    int mul = 1;
    tol = 1e-8;

    int elszy[12];
    elszy[0] = 0;   elszy[1] = 2;
    elszy[2] = 4;   elszy[3] = 2;
    elszy[4] = 2;   elszy[5] = 2;
    elszy[6] = 2;   elszy[7] = 2;
    elszy[8] = 2;   elszy[9] = 2;
    elszy[10] = 2;  elszy[11] = 2;
    float scraty[12];
    scraty[0] = 1;   scraty[1] = 1;
    scraty[2] = 1;   scraty[3] = 1;
    scraty[4] = 1;   scraty[5] = 1;
    scraty[6] = 1;   scraty[7] = 1;
    scraty[8] = 1;   scraty[9] = 1;
    scraty[10] = 1;  scraty[11] = 1;
    //
    for (i=1; i<=16; ++i) {
        for (j=1; j<=11; ++j) {
            lsel("s","loc","x",r[i]-tol, r[i]+tol);
            lsel("r","loc","y",z[j],z[j+1]);
            lesize("all","","",elszy[j]*mul,scraty[j]);
            lsel("all");
        }
    }
    if (CokeBedGeomOption > 0) {
        for (i=0; i<CBPartZ-1; ++i) {
```

```
        lsel("s","loc","x",xcoke[i], xcoke[i+1]);
        lsel("r","loc","y",z[8+i],z[9+i]);
        lesize("all","","",elszy[8+i]*mul, scraty[8+i]);
        lsel("all");
    }
    lsel("s","loc","x",xcoke[3], xcoke[4]);
    lsel("r","loc","y",z[11],z[12]);
    lesize("all","","",elszy[11]*mul,0.5);
    lsel("all");
}
int elszx[16];
elszx[0] = 0;    elszx[1] = 8;
elszx[2] = 4;    elszx[3] = 2;
elszx[4] = 2;    elszx[5] = 2;
elszx[6] = 2;    elszx[7] = 2;
elszx[8] = 2;    elszx[9] = 2;
elszx[10] = 2;   elszx[11] = 2;
elszx[12] = 2;   elszx[13] = 2;
elszx[14] = 3;   elszx[15] = 2;
float scratx[16];
scratx[0] = 1;    scratx[1] = 1;
scratx[2] = 1;    scratx[3] = 1;
scratx[4] = 1;    scratx[5] = 1;
scratx[6] = 1;    scratx[7] = 1;
scratx[8] = 1;    scratx[9] = 1;
scratx[10] = 1;   scratx[11] = 1;
scratx[12] = 1;   scratx[13] = 1;
scratx[14] = 1;   scratx[15] = 1;
for (i=1; i<=12; ++i) {
    for (j=1; j<=15; ++j) {
        lsel("s","loc","y",z[i]-tol,z[i]+tol);
        lsel("r","loc","x",r[j],r[j+1]);
        lesize("all","","",elszx[j]*mul,scratx[j]);
        lsel("all");
    }
}
if (CokeBedGeomOption > 0) {
    for (i=0; i<=CBPartZ; ++i) {
        lsel("s","loc","y",z[8+i]-tol,z[8+i]+tol);
        lsel("r","loc","x",xcoke[i], xside[i]);
        lesize("all","","",elszy[2]*mul,scraty[2]);
        lsel("all");
    }
}
print("I am here 2");
et(1,"thquad4",1,"",-1.0);
float thC[7][3];
thC[0][0]=0.8; thC[0][1]=0.8; thC[0][2]=0.8;
thC[1][0]=0.9*101; thC[1][1]=101; thC[1][2] = 0.9*101;
thC[2][0]=3.55; thC[2][1]=3.55; thC[2][2]=3.55;
thC[3][0]=3.15; thC[3][1]=3.15; thC[3][2]=3.15;
thC[4][0]=2.0; thC[4][1]=2.0; thC[4][2]=2.0; // coke
```

```
    thC[5][0] = 28.0;  thC[5][1] = 28.0; thC[5][2] = 28.0; // molten
iron at 1150 C
    thC[6][0] = 40.0; thC[6][1] = 40.0; thC[6][2] = 40.0;
// molten iron at 1450 C
    //
    float alphC[5][3];
    alphC[0][0] = 0.00056; alphC[0][1] = 0.00056; alphC[0][2] =
0.00056;
    alphC[1][0] = -0.9*0.0577; alphC[1][1] = -0.0577; alphC[1][2] = -
0.9*0.0577;
    alphC[2][0] = 0.00179; alphC[2][1] = 0.00179; alphC[2][2] =
0.00179;
    alphC[3][0] = 0.00167; alphC[3][1] = 0.00167; alphC[3][2] =
0.00167;
    alphC[4][0] = 0.0;   alphC[4][1] = 0.0; alphC[4][2] = 0.0;
    //
    float betaC = 0;
    float RefTemp[5];
    for (i=0;i<5;++i) RefTemp[i] = 273;
    //
    string kk[3];
    kk[0] = "kx"; kk[1] = "ky"; kk[2] = "kz";
    for (i=0;i<5;++i) {
        for (j=0; j<3;++j) {
            mp(i+1,"",kk[j],thC[i][j],alphC[i][j],betaC, RefTemp[i]);
        }
    }
    kk[0] = "kxt"; kk[1] = "kyt"; kk[2] = "kzt";
    temphm1 = 1723;
    temphm2 = 1773;
    tempsm = 1423;
    for (j=0; j<3;++j) {
        mp(6,"",kk[j],thC[5][j],tempsm,thC[6][j], temphm2);
    }
    MatNumIronPool = 6;
    // Coke-Bed Properties
    float volperchm;
    volperchm = volperchm = 0.4; // porosity
    float volperccoke = 1-volperchm;
    // First continuous, then discrete
    mp(7,"porousth", 2,6,volperchm,5,volperccoke);
    MatNumCokeBed = 7;
    //
    // Solidified Skull Properties
    volperchm = 0.4;
    if (volperchm < 0 || volperchm > 1) volperchm = 1;
    if (volperchm == 0) MatNumDeposit = 5;
    else if (volperchm == 1) MatNumDeposit = 6;
    else {
        volperccoke = 1-volperchm;
        // First continuous, then discrete
        mp(8,"porousth", 2,6,volperchm,5,volperccoke);
```

```
        MatNumDeposit = 8;
    }
    //
    delete[] xccok;
    delete[] yccok;
    delete[] xcoke;
    delete[] ycoke;
    delete[] xside;
    delete[] yside;
    //
    delete[] kpccok;
    delete[] kpcoke;
    delete[] kpside;
    //
    amesh("all");
    return 1;
}
// ~~~~~Solution ~~~~~~~~

int solu() {
    antype("static");
    //*********** boundary conditions at the hot metal/refractory
interface *************
    // Hot metal temperature
    if (CokeBedGeomOption == 0) {
        nsel("s","loc","x",r[1]-tol,r[3]+tol);
        nsel("r","loc","y",z[8]-tol,z[8]+tol);
        dn("all","","","temp",temphm1);
        nsel("all");
        //
        nsel("s","loc","x",r[3]-tol,r[3]+tol);
        nsel("r","loc","y",z[8]-tol,z[9]+tol);
        dn("all","","","temp",temphm1);
        nsel("all");
        //
        nsel("s","loc","x",r[3]-tol,r[5]+tol);
        nsel("r","loc","y",z[9]-tol,z[9]+tol);
        dn("all","","","temp",temphm1);
        nsel("all");
        //
        nsel("s","loc","x",r[5]-tol,r[5]+tol);
        nsel("r","loc","y",z[9]-tol,z[10]+tol);
        dn("all","","","temp",temphm1);
        nsel("all");
        //
        nsel("s","loc","x",r[5]-tol,r[14]+tol);
        nsel("r","loc","y",z[10]-tol,z[10]+tol);
        dn("all","","","temp",temphm1);
        nsel("all");
        //
        nsel("s","loc","x",r[14]-tol,r[14]+tol);
        nsel("r","loc","y",z[10]-tol,z[12]+tol);
        dn("all","","","temp",temphm1);
```

```
        nsel("all");
    }
    else if (CokeBedGeomOption > 0) {
        nsel("s","loc","x",r[1]-tol,r[14]+tol);
        nsel("r","loc","y",z[12]-tol,z[12]+tol);
        dn("all","","","temp",temphm2);
        nsel("all");
    }
    // ****Surface b.c.****
    // UnderHearth
    double TempAir = 273+35;
    double HTCair;// heat transfer coefficient for air
    HTCair = 10;
    if (UHBCOption == 1) {   //     Convective
        lsel("s","loc","x",r[1]-tol,r[16]+tol);
        lsel("r","loc","y",z[1]-tol,z[1]+tol);
        //sfl("all","","","conv",TempAir,HTCair);
        nsel("s","line");
        sfn("all","","","conv",TempAir,HTCair,,1);
        lsel("all");
        nsel("all");
    }
    else if (UHBCOption == 2) {
        // **** Constant Temperature ****
        nsel("s","loc","x",r[1]-tol,r[16]+tol);
        nsel("r","loc","y",z[1]-tol,z[1]+tol);
        dn("all","","","temp",TempAir);
        nsel("all");
    }
    //
    // Side-wall
    double TempWater = 273+30;
    double HTCwater; //heat transfer coefficient for water
    HTCwater = 300;
    //
    if (WallBCOption == 1) { //     Convective
        lsel("s","loc","y",z[1]-tol,z[12]+tol);
        lsel("r","loc","x",r[16]-tol,r[16]+tol);
        //sfl("all","","","conv",TempWater,HTCwater);
        nsel("s","line");
        sfn("all","","","conv",TempWater,HTCwater,,1);
        lsel("all");
        nsel("all");
    }
    //
    else if (WallBCOption == 2) {
        // **** Constant Temperature ****
        nsel("s","loc","y",z[1]-tol,z[12]+tol);
        nsel("r","loc","x",r[16]-tol,r[16]+tol);
        dn("all","","","temp",TempWater);
        nsel("all");
    }
```

```
int i,j,matno[8];
for (i=0; i<8; ++i) matno[i] = i;
if (CokeBedGeomOption==1) {
   MatNumCorner = MatNumIronPool;
}
else {
   MatNumCorner = MatNumCokeBed;
}
//
double toler = 1e-2,differ,maxdiffer;
int count = 0,flag = 0,from;
int MaxCount = 30;
for (;;) {
    ++count;
    emodify("all","","","mat",matno[1]);
    //
    asel("s","loc","x",r[1]-tol,r[15]+tol);
    asel("r","loc","y",z[2],z[3]);
    esel("s","area");
    emodify("all","","","mat",matno[2]);
    asel("all");
    esel("all");
    //
    for (i=3; i<=7; ++i) {
        asel("s","loc","y",z[i]-tol,z[i+1]+tol);
        from = 15-2*(i-2);
        asel("r","loc","x",r[from]-tol,r[15]+tol);
        esel("s","area");
        emodify("all","","","mat",matno[3]);
        asel("all");
        esel("all");
    }
    //
    asel("s","loc","y",z[8]-tol,z[9]+tol);
    asel("r","loc","x",r[3]-tol,r[15]+tol);
    esel("s","area");
    emodify("all","","","mat",matno[3]);
    asel("all");
    esel("all");
    //
    asel("s","loc","y",z[9]-tol,z[10]+tol);
    asel("r","loc","x",r[5]-tol,r[15]+tol);
    esel("s","area");
    emodify("all","","","mat",matno[3]);
    asel("all");
    esel("all");
    //
    asel("s","loc","y",z[10]-tol,z[12]+tol);
    asel("r","loc","x",r[14]-tol,r[15]+tol);
    esel("s","area");
    emodify("all","","","mat",matno[3]);
    asel("all");
```

```
esel("all");
//
asel("s","loc","x",r[1]-tol,r[15]+tol);
asel("r","loc","y",z[1],z[2]);
esel("s","area");
emodify("all","","","mat",matno[4]);
asel("all");
esel("all");
//
asel("s","loc","x",r[15]-tol,r[16]+tol);
asel("r","loc","y",z[1],z[12]);
esel("s","area");
emodify("all","","","mat",matno[4]);
asel("all");
esel("all");
//
for (i=3; i<=7; ++i) {
    asel("s","loc","y",z[i]-tol,z[i+1]+tol);
    from = 15-2*(i-2);
    asel("r","loc","x",r[from-1]-tol,r[from]+tol);
    esel("s","area");
    emodify("all","","","mat",matno[4]);
    asel("all");
    esel("all");
}
if (CokeBedGeomOption > 0) {
    for (j=0; j<CBPartZ; ++j) {
        asel("s","num","",Arcoke[j],Arcoke[j]);
        esel("s","area");
        emodify("all","","","mat",MatNum
        CokeBed);
        asel("all");
        esel("all");
    }
    if (CokeBedGeomOption == 1) {
        esel("s","mat","",MatNumCokeBed,MatNumCokeBed);
        nsel("s","loc","y",z[8]-tol,z[8]+
        HeighthmUnderCoke+tol);
        esel("r","node");
        emodify("all","","","mat",MatNumIron Pool);
        esel("all");
    }
    //
    for (j=0; j<CBPartZ; ++j) {
        asel("s","num","",Ariron[j],Ariron[j]);
        esel("s","area");
        emodify("all","","","mat",Mat Num Corner);
        asel("all");
        esel("all");
    }
    asel("s","loc","x",r[3]-tol,r[14]+tol);
    asel("r","loc","y",z[8]-tol,z[12]+tol);
```

```
            esel("s","area");
            esel("r","mat","",matno[1],matno[1]);
            emodify("all","","","mat",MatNumCorner);
            asel("all");
            esel("all");
        }
        if (count > 1) {
            nsel("s","val","temp",tempsm,temphm2);
            esel("s","node");
            emodify("all","","","mat",MatNumDeposit);
            esel("all");
            nsel("all");
        }
        solve("front","all");
        if (CalculationType == 0) break; // Initial design, no moving
interface
        //if (CalculationType == 3 && flag == 1) break;
        // With elements modified
        //if (CalculationType > 0 && count > MaxCount) break;
    }
    return 1;
}

// ~~~~~Postprocessor ~~~~~~~~
int post() {
    plnsol("temp");
    return 1;
}

int main() {
    gen();
    prep();
    solu();
    post();
    //
    delete[] Arcoke;
    delete[] Ariron;
    return 1;
}
```

A8.2.10 Example problem (Section 13.7.1)

```
/* Upsetting between flat/inclined dies. */

int NoTimeSteps;
int startStepNo = 1,endStepNo = 20;
float RamVelocity;
float TotalRamDisp,RamDispInOneStep,TotalTime;
int LowerDiekeypt[4],LowerDiehline[2],LowerDievline[2];
int UpperDiekeypt[4],UpperDiehline[2],UpperDievline[2];
int keypt[4],hline[2],vline[2];
float DegtoRad(float deg) {
    float rad = deg*3.141592653589/180.0;
```

```
    return rad;
}

int gen() {
    title("Upsetting between flat/inclined dies.");
    inpfilename(,"Upset1");
    outfilename(,"Upset1");
    return 1;
}

// ~~~~~Preprocessor ~~~~~~~~
int prep() {
    float radius = 1.0;
    float height = 1.0;
    float PercentReductionInHeight = 0;
    if (PercentReductionInHeight == 0)
PercentReductionInHeight = 40;
    TotalRamDisp = height*PercentReductionInHeight/100.0;
    float PercentReductionInOneStep = 2.0;
    NoTimeSteps=
m_ceil(PercentReductionInHeight/PercentReduction InOneStep);
    print("NoTimeSteps = ",NoTimeSteps);
    RamDispInOneStep = TotalRamDisp/NoTimeSteps;
    RamVelocity = 0;
    if (RamVelocity <= 0) RamVelocity = 1.0;
    TotalTime = TotalRamDisp/RamVelocity;
    float UpperDieHeight = 4*RamDispInOneStep;
    float LowerDieHeight = UpperDieHeight;
    float UpperDieSpan = 2*radius;
    float LowerDieSpan = UpperDieSpan;
    float Offset = 4*RamDispInOneStep;
    float lowerangle,upperangle;
    lowerangle = 0;
    upperangle = 0;
    lowerangle = DegtoRad(lowerangle);
    upperangle = DegtoRad(upperangle);
    //
    //***********Lower Die******************
    SetCurSt("lower");
    float LowerDieX[4],LowerDieY[4],LowerDieZ[4];
    LowerDieX[0]=0; LowerDieY[0]=0; LowerDieZ[0]=0;
    LowerDieX[1]=LowerDieX[0]+LowerDieSpan;LowerDieY[1]=
LowerDieSpan*m_tan(lowerangle); LowerDieZ[1] = 0;
    LowerDieX[2]=LowerDieX[0]+LowerDieSpan;LowerDieY[2]=
LowerDieY[1]+LowerDieHeight; LowerDieZ[2] = 0;
    LowerDieX[3]=LowerDieX[0];LowerDieY[3]=LowerDieY[0]+
LowerDieHeight; LowerDieZ[3] = 0;
    //
    int i;
    for (i=0;i<4; ++i) {
        LowerDiekeypt[i]= k(,LowerDieX[i],LowerDieY[i], LowerDieZ[i]);
    }
    LowerDiehline[0] = lk(LowerDiekeypt[0], LowerDiekeypt[1]);
```

```
    LowerDiehline[1] = lk(LowerDiekeypt[2], LowerDiekeypt[3]);
    LowerDievline[0] = lk(LowerDiekeypt[3], LowerDiekeypt[0]);
    LowerDievline[1] = lk(LowerDiekeypt[1], LowerDiekeypt[2]);
    //
    // Element Sizes
    int LowerDieelszx[1],LowerDieelszy[1];
    float LowerDiescratx[1],LowerDiescraty[1];
    LowerDieelszx[0] = 20;
    LowerDiescratx[0] = 1;
    LowerDieelszy[0] = 1;
    LowerDiescraty[0] = 1;
    //
    for (i=0;i<2;++i) {
        lesize(LowerDiehline[i],LowerDiehline[i],,
LowerDieelszx[0],LowerDiescratx[0]);
        lesize(LowerDievline[i],LowerDievline[i],,
LowerDieelszy[0],LowerDiescraty[0]);
    }
    int LowerDieArea[1];
    LowerDieArea[0]=al(LowerDiehline[0], LowerDievline[1],Lower
Diehline[1],LowerDievline[0]);
    int incompatible_modesMasterLower = 0;
    float ThicknessMasterLower = -1; // -1: axisymmetric and y-axis sym
    et(1,"mfquad4",incompatible_modesMasterLower,"",
ThicknessMasterLower);
    //
    amesh("all");
    //
    // ***********End LowerDie***********
    //
    //***********Upper Die***********
    SetCurSt("upper");
    float UpperDieX[4],UpperDieY[4],UpperDieZ[4];
    UpperDieX[0] = 0; UpperDieY[0] = height+LowerDieHeight;
UpperDieZ[0] = 0;
    UpperDieX[1] = UpperDieX[0]+UpperDieSpan; UpperDieY[1]=
UpperDieY[0]-UpperDieSpan*m_tan(upperangle); UpperDieZ[1] = 0;
    UpperDieX[2] = UpperDieX[0]+UpperDieSpan; UpperDieY[2]=
UpperDieY[1]+UpperDieHeight; UpperDieZ[2] = 0;
    UpperDieX[3] = UpperDieX[0]; UpperDieY[3] =
UpperDieY[0]+UpperDieHeight; UpperDieZ[3] = 0;
    //
    for (i=0;i<4; ++i) {
        UpperDiekeypt[i] = k(,UpperDieX[i],UpperDieY[i],
UpperDieZ[i]);
    }
    UpperDiehline[0] = lk(UpperDiekeypt[0], UpperDiekeypt[1]);
    UpperDiehline[1] = lk(UpperDiekeypt[2], UpperDiekeypt[3]);
    UpperDievline[0] = lk(UpperDiekeypt[3], UpperDiekeypt[0]);
    UpperDievline[1] = lk(UpperDiekeypt[1], UpperDiekeypt[2]);
    //
    // Element Sizes
```

```
    int UpperDieelszx[1],UpperDieelszy[1];
    UpperDieelszx[0] = 20;
    UpperDieelszy[0] = 1;
    float UpperDiescratx[1],UpperDiescraty[1];
    UpperDiescratx[0] = 1;
    UpperDiescraty[0] = 1;
    //
    for (i=0;i<2;++i) {
        lesize(UpperDiehline[i],UpperDiehline[i],"", Upper
Dieelszx[0],UpperDiescratx[0]);
        lesize(UpperDievline[i],UpperDievline[i],"",
UpperDieelszy[0],UpperDiescraty[0]);
    }
    int UpperDieArea[1];
    UpperDieArea[0]          =          al(UpperDiehline[0],
UpperDievline[1],UpperDiehline[1],UpperDievline[0]);
    int incompatible_modesMasterUpper = 0;
    float ThicknessMasterUpper = -1; // -1: axisymmetric and y-axis sym
    et(1,"mfquad4",incompatible_modesMasterUpper,"", Thickness
MasterUpper);
    //
    amesh("all");
    //
    //******Workpiece******
    SetCurSt("slave");
    float X[4],Y[4],Z[4];
    X[0]=Offset; Y[0] =
LowerDieHeight+X[0]*m_tan(lowerangle); Z[0] = 0;
    X[1]=X[0]+radius; Y[1] =
Y[0]+radius*m_tan(lowerangle); Z[1]= 0;
    X[2]  =  X[0]+radius;  Y[2]  =  LowerDieHeight+height-
(X[0]+radius)*m_tan(upperangle); Z[2] = 0;
    X[3] = X[0]; Y[3] = LowerDieHeight+height-X[0]*m_tan(upperangle);
Z[3] = 0;
    int kpno;
    //
    for (i=0;i<4; ++i) {
        keypt[i] = 0;
        keypt[i] = k(keypt[i],X[i],Y[i],Z[i]);
    }
    hline[0] = lk(keypt[0],keypt[1]);
    hline[1] = lk(keypt[2],keypt[3]);
    vline[0] = lk(keypt[3],keypt[0]);
    vline[1] = lk(keypt[1],keypt[2]);
    //
    // Element Sizes
    int elszx[2];
    elszx[0] = 4;
    elszx[1] = 4;
    float scratx[2];
    scratx[0] = 1; //5;
    scratx[1] = 1; //0.2;
```

```
    //
    int elszy[2];
    elszy[0] = 4;
    elszy[1] = 4;
    float scraty[2];
    scraty[0] = 1; //0.2;
    scraty[1] = 1; //5;
    //
    for (i=0;i<2;++i) {
        lesize(hline[i],hline[i],"", elszx[i],scratx[i]);
        lesize(vline[i],vline[i],"", elszy[i],scraty[i]);
    }
    int Area[1];
    Area[0] = al(hline[0],vline[1],hline[1],vline[0]);
    //
    int incompatible_modes = 0;
    float Thickness = -1; // -1: axisymmetric and y-axis sym
    et(1,"mfquad4",incompatible_modes,"",Thickness);
    //
    float StngCoeff = 10;
    float StrRatSens = 0.1;
    float dummy = 0.0;
    float LimitingStrainRateValue = 0.01;
    mp(1,"","FLOWSTRESS",StngCoeff,dummy,dummy, StrRatSens,
LimitingStrainRateValue);
    //
    amesh("all");
    print("I am here 1");
    //
    return 1;
}

//
// ~~~~~Solution ~~~~~~~~
int solu() {
    antype("multistep");
    int NoNode,NoNodDOF,TotNoDof;
    int SolnCount;
    NoNode = getNoNod();
    NoNodDOF = getNoNodDof();
    TotNoDof = NoNode*NoNodDOF;
    cnvtol("d-r","dof",0.0003,0.003);
    cnvtol("d-r","force",0.0003,0.003);
    cnvtol("n-r","dof",0.0001,0.001);
    cnvtol("n-r","force",0.0001,0.001);
    noiter("d-r",100);
    noiter("n-r",50);
    penalcoeff(10000);
    time(TotalTime);
    Notimesteps(endStepNo);
    //
    print("I am here 2");
```

```
// Boundary Conditions
//~~~~~~~~~~~~~~~ Begin B.C.s. ~~~~~~~~~~~~~~~~~~~~~~ ~~~~~~~~~
float frictionfact = 0.5;
float bcux1 = 0,bcuy1 = 0;
float bcuy2 = -RamVelocity;
//
// ******Die constraints*******
int ConstType = 2; // normal
int HalfSymmetry = 1;
if (HalfSymmetry == 0) {
   SetCurSt("lower");
   lsel("s","num","",LowerDiehline[1], LowerDiehline[1]);
   dl("all","","",ConstType,0.0,-bcuy2,0); // Global CSYS.
   lsel("all");
   SetCurSt("slave");
}
else {
   nsel("s","lineno","",hline[0],hline[0]);  // Bottom face.
   dn("all","","","uy",bcuy1);
   nsel("all");
}
//******************************
/*
nsel("s","lineno","",hline[1],hline[1]);
dn("all","","","uy",bcuy2);
nsel("all");
*/
//
/*
ConstType = 2; // normal
lsel("s","num","",hline[1],hline[1]);
dl("all","","",ConstType,0.0,bcuy2, 0); //Global CSYS
lsel("all");
*/
// ******Die constraints*******
SetCurSt("upper");
ConstType = 2;
lsel("s","num","",UpperDiehline[0],UpperDiehline[0]);
dl("all","","",ConstType,0.0,bcuy2,0); //Global CSYS
lsel("all");
SetCurSt("slave");
//******************************
//
/*
lsel("s","num","",hline[1],hline[1]);
sfl("all","","","frict1",frictionfact,0.0,bcuy2,  0); //  t_s,
tangential stress
lsel("all");
*/
//
SetCurSt("upper");
lsel("s","num","",UpperDiehline[0], UpperDiehline[0]);
```

```
    sfl("all","","","frictl",frictionfact,0.0, bcuy2, 0.0, 0); //
t_s, tangential stress, Global CSYS
    lsel("all");
    SetCurSt("slave");
    //
    if (HalfSymmetry == 0) {
        SetCurSt("lower");
        lsel("s","num","",LowerDiehline[1], LowerDiehline[1]);
        sfl("all","","","frictl",frictionfact,0.0,-bcuy2, 0.0, 0); /
/ t_s, tangential stress
        lsel("all");
        SetCurSt("slave");
    }
    //
    nsel("s","lineno","",vline[0],vline[0]);
    dn("all","","","ux",bcux1);
    nsel("all");
    //
    float UpperDieVelocity[2],LowerDieVelocity[2];
    UpperDieVelocity[0] = 0; UpperDieVelocity[1] = -RamVelocity;
    LowerDieVelocity[0] = 0; LowerDieVelocity[1] = RamVelocity;
    SetCurSt("upper");
    dieVel(UpperDieVelocity[0],UpperDieVelocity[1],0);
    if (HalfSymmetry == 0) {
        SetCurSt("lower");
        dieVel(LowerDieVelocity[0],LowerDieVelocity [1],0);
    }
    SetCurSt("slave");
    solve("front","all");
    //~~~~~~~~~~~~~~~ End B.C.s. ~~~~~~~~~~~~~~~~~~~~~~~~
    print("***** Solution Complete *****");
    return 1;

}
// ~~~~~Postprocessor ~~~~~~~~
int post() {
    plnsol("uy"); // current load step.
    return 1;
}

int main() {
    gen();
    prep();
    solu();
    post();
    return 1;
}
```

Subject Index

F

G

H

I

L

M

N

T

U

V

W